T0282712

LONDON MATHEMATICAL SOCIETY LECTURE NOTE SERIES

Managing Editor: Professor I.M. James,
Mathematical Institute, 24-29 St Giles, Oxford

1. General cohomology theory and K-theory, P.HILTON
4. Algebraic topology, J.F.ADAMS
5. Commutative algebra, J.T.KNIGHT
8. Integration and harmonic analysis on compact groups, R.E.EDWARDS
9. Elliptic functions and elliptic curves, P.DU VAL
10. Numerical ranges II, F.F.BONSALL & J.DUNCAN
11. New developments in topology, G.SEGAL (ed.)
12. Symposium on complex analysis, Canterbury, 1973, J.CLUNIE & W.K.HAYMAN (eds.)
13. Combinatorics: Proceedings of the British Combinatorial Conference 1973, T.P.McDONOUGH & V.C.MAVRON (eds.)
15. An introduction to topological groups, P.J.HIGGINS
16. Topics in finite groups, T.M.GAGEN
17. Differential germs and catastrophes, Th.BROCKER & L.LANDER
18. A geometric approach to homology theory, S.BUONCRISTIANO, C.P. BOURKE & B.J.SANDERSON
20. Sheaf theory, B.R.TENNISON
21. Automatic continuity of linear operators, A.M.SINCLAIR
23. Parallelisms of complete designs, P.J.CAMERON
24. The topology of Stiefel manifolds, I.M.JAMES
25. Lie groups and compact groups, J.F.PRICE
26. Transformation groups: Proceedings of the conference in the University of Newcastle-upon-Tyne, August 1976, C.KOSNIOWSKI
27. Skew field constructions, P.M.COHN
28. Brownian motion, Hardy spaces and bounded mean oscillations, K.E.PETERSEN
29. Pontryagin duality and the structure of locally compact Abelian groups, S.A.MORRIS
30. Interaction models, N.L.BIGGS
31. Continuous crossed products and type III von Neumann algebras, A.VAN DAELE
32. Uniform algebras and Jensen measures, T.W.GAMELIN
33. Permutation groups and combinatorial structures, N.L.BIGGS & A.T.WHITE
34. Representation theory of Lie groups, M.F. ATIYAH et al.
35. Trace ideals and their applications, B.SIMON
36. Homological group theory, C.T.C.WALL (ed.)
37. Partially ordered rings and semi-algebraic geometry, G.W.BRUMFIEL
38. Surveys in combinatorics, B.BOLLOBAS (ed.)
39. Affine sets and affine groups, D.G.NORTHCOTT
40. Introduction to Hp spaces, P.J.KOOSIS
41. Theory and applications of Hopf bifurcation, B.D.HASSARD, N.D.KAZARINOFF & Y-H.WAN
42. Topics in the theory of group presentations, D.L.JOHNSON
43. Graphs, codes and designs, P.J.CAMERON & J.H.VAN LINT
44. Z/2-homotopy theory, M.C.CRABB
45. Recursion theory: its generalisations and applications, F.R.DRAKE & S.S.WAINER (eds.)
46. p-adic analysis: a short course on recent work, N.KOBLITZ
47. Coding the Universe, A.BELLER, R.JENSEN & P.WELCH
48. Low-dimensional topology, R.BROWN & T.L.THICKSTUN (eds.)

49. Finite geometries and designs, P.CAMERON, J.W.P.HIRSCHFELD & D.R.HUGHES (eds.)
50. Commutator calculus and groups of homotopy classes, H.J.BAUES
51. Synthetic differential geometry, A.KOCK
52. Combinatorics, H.N.V.TEMPERLEY (ed.)
53. Singularity theory, V.I.ARNOLD
54. Markov processes and related problems of analysis, E.B.DYNKIN
55. Ordered permutation groups, A.M.W.GLASS
56. Journées arithmétiques 1980, J.V.ARMITAGE (ed.)
57. Techniques of geometric topology, R.A.FENN
58. Singularities of smooth functions and maps, J.MARTINET
59. Applicable differential geometry, F.A.E.PIRANI & M.CRAMPIN
60. Integrable systems, S.P.NOVIKOV et al.
61. The core model, A.DODD
62. Economics for mathematicians, J.W.S.CASSELS
63. Continuous semigroups in Banach algebras, A.M.SINCLAIR
64. Basic concepts of enriched category theory, G.M.KELLY
65. Several complex variables and complex manifolds I, M.J.FIELD
66. Several complex variables and complex manifolds II, M.J.FIELD
67. Classification problems in ergodic theory, W.PARRY & S.TUNCEL
68. Complex algebraic surfaces, A.BEAUVILLE
69. Representation theory, I.M.GELFAND et. al.
70. Stochastic differential equations on manifolds, K.D.ELWORTHY
71. Groups - St Andrews 1981, C.M.CAMPBELL & E.F.ROBERTSON (eds.)
72. Commutative algebra: Durham 1981, R.Y.SHARP (ed.)
73. Riemann surfaces: a view toward several complex variables, A.T.HUCKLEBERRY
74. Symmetric designs: an algebraic approach, E.S.LANDER
75. New geometric splittings of classical knots (algebraic knots), L.SIEBENMANN & F.BONAHON
76. Linear differential operators, H.O.CORDES
77. Isolated singular points on complete intersections, E.J.N.LOOIJENGA
78. A primer on Riemann surfaces, A.F.BEARDON
79. Probability, statistics and analysis, J.F.C.KINGMAN & G.E.H.REUTER (eds.)
80. Introduction to the representation theory of compact and locally compact groups, A.ROBERT
81. Skew fields, P.K.DRAXL

London Mathematical Society Lecture Note Series. 79

Probability, Statistics and Analysis

Edited by

J.F.C. KINGMAN

Professor of Mathematics, University of Oxford

and

G.E.H. REUTER

Professor of Mathematics, Imperial College of Science and Technology

CAMBRIDGE UNIVERSITY PRESS

Cambridge

London New York New Rochelle

Melbourne Sydney

CAMBRIDGE UNIVERSITY PRESS
Cambridge, New York, Melbourne, Madrid, Cape Town, Singapore, São Paulo

Cambridge University Press
The Edinburgh Building, Cambridge CB2 8RU, UK

Published in the United States of America by Cambridge University Press, New York

www.cambridge.org
Information on this title: www.cambridge.org/9780521285902

First published 1983
Re-issued in this digitally printed version 2008

A catalogue record for this publication is available from the British Library

Library of Congress Catalogue Card Number: 82–19731

ISBN 978-0-521-28590-2 paperback

These papers are dedicated to David Kendall (Professor D.G. Kendall, F.R.S., Professor of Mathematical Statistics in the University of Cambridge) for his sixty-fifth birthday. We hope that he will enjoy reading them, and will accept them as a sign of our admiration and friendship.

J.F.C. Kingman, G.E.H. Reuter
and the authors of these papers

CONTENTS

THE ASYMPTOTIC SPEED AND SHAPE OF A PARTICLE SYSTEM*

David Aldous and Jim Pitman

1 Introduction

We study in this paper the asymptotics as $k \to \infty$ of the motion of a system of k particles located at sites labelled by the integers. This section gives an informal description of the particle system and our results, and the original motivation for the study.

The particles will be referred to as *balls*, and the sites as *boxes*. The motion may be described as follows. Initially the k balls are distributed amongst boxes in such a way that the set of occupied boxes is *connected*. (A box may contain many balls, but there is no empty box between two occupied boxes.) At each move, a ball is taken from the left-most occupied box and placed one box to the right of a ball chosen uniformly at random from among the k balls, the successive choices being mutually independent. It is clear that the set of occupied boxes remains connected, and that the collection of balls drifts off to infinity. It is easy to see that for each k the k-ball motion drifts off to infinity at an almost certain average speed s_k, defined formally by (2.3) below. Our main result is that $s_k \sim e/k$ as $k \to \infty$. To be more precise:

THEOREM 1.1 *As* k *increases to infinity,* ks_k *increases to* e.

This result was conjectured by Tovey (private communication), and informal arguments supporting the conjecture have been given by Keller (1980) and Weiner (1980). Our method of proof (Sections 2-4) is to use coupling to compare the k-ball process with a certain, more easily analysed, pure growth process (defined at (3.3)).

Secondly, for fixed k we can define (Section 5) a random vector $(\pi_0, \pi_1, \pi_2, \ldots)$ describing the equilibrium proportions of balls in the (0th,

---

* Research supported in part by National Science Foundation grants MCS80-02698 and MCS78-25301.

1st, 2nd,...) box from the leftmost occupied box at times when the leftmost box has just been cleared. We conjecture (5.9) that as $k \to \infty$, $(\pi_0, \pi_1, \pi_2, \ldots)$ converges in distribution to a certain sequence $(\hat{p}_0, \hat{p}_1, \hat{p}_2, \ldots)$ of constants with $\hat{p}_0 = e^{-1}$. This would imply that for large k the process of proportions in the k-ball process evolves almost deterministically. In Section 6 we show how this conjecture is related to problems concerning a certain transformation of probability measures on the positive integers.

The origin of the k-ball process came in work of Tovey (1980) on abstractions of local improvement algorithms. Consider functions f defined on the vertices of the d-dimensional cube, with distinct real values, and with the *local-global* property:

f has no local maximum except the global maximum.

(Here a *local* maximum is a vertex i such that $f(i) > f(j)$ for each *neighbor* j of i, and vertices are *neighbors* if they are connected by an edge.) There is an obvious algorithm to locate the maximum of a local-global function: move from a vertex v to the neighbor v' for which $f(v')$ is largest; unless v is a local maximum, in which case it must be the global maximum. How good is this algorithm "on average"? In other words, what is the expected number of steps required to locate the maximum of a function f picked at random according to some distribution μ on local-global functions? Now any function f induces an ordering $v_1, v_2, v_3, \ldots$ of vertices such that $f(v_1) > f(v_2) > f(v_3) > \ldots$, and f is local-global iff

$$v_n \text{ is a neighbor of at least one of } \{v_1, \ldots, v_{n-1}\},\ n \geq 2. \tag{1.2}$$

Thus a distribution μ on local-global functions induces a random ordering $V_1, V_2, V_3, \ldots$ of vertices satisfying (1.2). Let N_i denote the number of steps required by the algorithm started at vertex V_i, so that $\max_i N_i$ is the number of steps required from the "worst" initial vertex. Plainly (N_i) satisfies the recurrence

$$\begin{aligned} &N_1 = 1 \\ &N_{i+1} = 1 + N_j, \text{ where } j \leq i \text{ is the least integer for which} \\ &\qquad V_j \text{ is a neighbor of } V_i \end{aligned} \tag{1.3}$$

But we can think of (N_i) as a process of balls in boxes, where ball i corresponds to vertex V_i and the box containing ball i corresponds to the number N_i of steps in the algorithm. Then (1.3) says the process evolves by the $(i+1)^{st}$ ball being placed in the box to the right of the box containing ball j, where j is chosen in some random way from the existing balls. And $\max_i N_i$, the number of steps required from the worst initial vertex, is the position of the rightmost occupied box after 2^d balls have been used. To estimate this, we need an upper bound on the speed of the rightmost occupied box of this new "balls in boxes" process. This process differs from the k-ball process we study-- because j is chosen in some complicated random way involving the distribution μ on local-global functions, and the number of balls increases. But coupling arguments in the spirit of those of Section 3 can be used to show that for certain distributions μ the speed of the rightmost occupied box of the new process is less than the speed of the k-ball process: see Tovey (1980).

2. Preliminaries

The configuration of a finite number of balls in boxes numbered 0,1,2,... will be described by a k-tuple of nonnegative integers

$$\underset{\sim}{x} = (x_1, x_2, \ldots, x_k)$$

where $k = \#\underset{\sim}{x}$ is the total number of balls; the balls are assumed to be labelled $1,\ldots,k$, and x_j is the box number of ball number j. The distribution of balls among boxes without regard to labelling is recorded by the *counting measure* $\underset{\sim}{N}\underset{\sim}{x} = (N_i\underset{\sim}{x},\ i=0,1,\ldots)$ defined by

$$N_i\underset{\sim}{x} = \#\{j:\ x_j = i\}, \quad i=0,1,\ldots\ .$$

So $N_i\underset{\sim}{x}$ is the number of balls in box i for configuration $\underset{\sim}{x}$. The *left end* $L\underset{\sim}{x}$ and *right end* $R\underset{\sim}{x}$ of configuration $\underset{\sim}{x}$ are defined by

$$L\underset{\sim}{x} = \min_j x_j = \min\{i:\ N_i\underset{\sim}{x} > 0\},$$

$$R\underset{\sim}{x} = \max_j x_j = \max\{i:\ N_i\underset{\sim}{x} > 0\},$$

and $\underset{\sim}{x}$ is *connected* if

$$N_i\underset{\sim}{x} > 0 \text{ for } L\underset{\sim}{x} < i < R\underset{\sim}{x} \ .$$

The set of connected configurations of k balls will be denoted C_k, and we put $C_* = \cup_k C_k$. For $\underset{\sim}{x} \in C_k$, $i = 1,\ldots,k$, define a new configuration $\underset{\sim}{x}^i$ by

$$\begin{aligned} x_j^i &= x_j && \text{except if } j = \hat{j} \\ &= x_i + 1 && \text{if } j = \hat{j} \end{aligned} \tag{2.1}$$

where $\hat{j} = \hat{j}(\underset{\sim}{x})$ is the number of the lowest numbered ball in the left end box $L\underset{\sim}{x}$. That is to say, $\underset{\sim}{x}^i$ is obtained from $\underset{\sim}{x}$ by removing ball $\hat{j}$ from box $L\underset{\sim}{x}$ and replacing it in the box to the right of ball i. Clearly $\underset{\sim}{x}^i \in C_k$ for all $i = 1,\ldots,k$. The *discrete* k-*ball process* is the discrete time Markov chain with countable state space C_k and one step transition matrix $p_k(\underset{\sim}{x},\underset{\sim}{y})$ defined by

$$p_k(\underset{\sim}{x},\underset{\sim}{x}^i) = 1/k \ , \quad i = 1,\ldots,k \ . \tag{2.2}$$

The *speed* s_k of the discrete k-ball process is the constant

$$s_k = \lim_{m\to\infty} m^{-1} L\underset{\sim}{X}(m) = \lim_{m\to\infty} m^{-1} R\underset{\sim}{X}(m) \ , \tag{2.3}$$

where $(\underset{\sim}{X}(m),\ m = 0,1,\ldots)$ is a discrete k-ball process. We assert that the limits exist almost surely and do not depend on the initial configuration $\underset{\sim}{X}(0)$. To see why, consider the *left counting process*

$$(\underset{\sim}{N}^L \underset{\sim}{X}(m),\ m = 0,1,\ldots)$$

where for a configuration $\underset{\sim}{x}$ the *left count of* $\underset{\sim}{x}$ is the vector $\underset{\sim}{N}^L\underset{\sim}{x} = (N_i^L\underset{\sim}{x},\ i = 0,1,\ldots)$ defined by

$$N_i^L\underset{\sim}{x} = N_{L\underset{\sim}{x}+i}\underset{\sim}{x} \ , \quad i \geq 0 \ . \tag{2.4}$$

Indeed, for given k the left counting process is a Markov chain whose *finite* state space is the set of counting vectors $\underset{\sim}{n} = (n_i, i \geq 0)$ such $\sum_i n_i = k$ and there exists $d \geq 1$ with

$$n_i > 0 \text{ for } i \leq d, \; n_i = 0 \text{ for } i > d .$$

This motion is easily seen to be irreducible and aperiodic, so there is a unique equilibrium distribution, λ_k say. Now from the definition of the k-ball process, for $m \geq 1$ the random variable $L\underset{\sim}{X}(m) - L\underset{\sim}{X}(m-1)$ is identical to the indicator of the event $(N_0^L\underset{\sim}{X}(m-1)=1)$, which we shall refer to as a *clearance at move* m. Thus

$$\lim_{m\to\infty} m^{-1} L\underset{\sim}{X}(m) = \lim_{m\to\infty} m^{-1} \#\{j \leq m : N_0^L\underset{\sim}{X}(j)=1\} \tag{2.5}$$

$$= \lambda_k(\underset{\sim}{n}: n_0=1) \text{ a.s.}$$

This justifies (2.3), since obviously

$$0 \leq R\underset{\sim}{X}(m) - L\underset{\sim}{X}(m) \leq k .$$

There are several other expressions for the speed s_k: Section 7 describes some we do not use, but let us here give only one, which will be the basis of developments in Section 5. Recall that for an irreducible Markov chain $Y(0), Y(1), \ldots$ with equilibrium distribution λ, if $Y(0)$ is given the distribution $\lambda|A$ obtained by conditioning λ on a set of states A, then

(i) the return time

$$T_A = \inf\{m: Y_m \in A\}$$

has expectation $1/\lambda(A)$, and

(ii) the distribution of $Y(T_A)$ is $\lambda|A$.

See, for example, Freedman (1971), Section 2.5. Applying this fact to $Y(m) = \underset{\sim}{N}(m)$ and $A = (\underset{\sim}{n}: n_0=1)$, the definition of the k-ball process implies that T_A is identical to $N_0(1)$. Let ν_k be the distribution of $\underset{\sim}{N}(1)$ when $\underset{\sim}{N}(0)$ has distribution $\lambda_k|(n: n_0=1)$, and call ν_k the *clearance equilibrium* of the left counting process. After a change of variables, (i) above in conjunction with (2.5) yields the formula

$$1/s_k = E_{\nu_k} N_0 , \tag{2.6}$$

where the right side denotes the expectation of N_0 when N has distribution ν_k. We record also for later use a consequence of (ii) above. For the left counting process $(\underset{\sim}{N}(0), \underset{\sim}{N}(1), \ldots)$,

$$\text{if } \underset{\sim}{N}(0) \text{ has distribution } \nu_k, \text{ then so does } \underset{\sim}{N}(N_0(0)) . \tag{2.7}$$

For an arbitrary initial distribution, the distribution ν_k can still be interpreted as the limiting distribution of $\underset{\sim}{N}(M_n)$ as $n \to \infty$, where M_n is the time of the nth clearance:

$$M_{n+1} = M_n + N_0(M_n), \quad n \geq 0, \quad M_0 = 0.$$

Further, ν_k is the almost sure limit as $n \to \infty$ of the empirical distribution of the sequence $\underset{\sim}{N}(M_1), \underset{\sim}{N}(M_2), \ldots, \underset{\sim}{N}(M_n)$.

3. Speed comparisons

To facilitate comparison of the speeds s_k for different values of k, we introduce now the *continuous k-ball process*. This is the Markov process with countable state space C_k and continuous time parameter $t \geq 0$ which is specified by the transition rates.

$$\underset{\sim}{x} \to \underset{\sim}{x}^i, \text{ rate } 1, \ i=1,\ldots,k . \tag{3.1}$$

Here, and later in similar descriptions of transition rate matrices, off-diagonal rates not explicitly mentioned are assumed to be zero, and the diagonal entries are taken to make the row sums zero. Put another way, $(\underset{\sim}{X}_t, t\geq 0)$ is a continuous k-ball process iff

$$\underset{\sim}{X}_t = \underset{\sim}{Y}_{M(t)} , \quad t \geq 0$$

where $\underset{\sim}{Y}_0, \underset{\sim}{Y}_1, \ldots$ is a discrete k-ball process and $(M(t), t\geq 0)$ is an independent Poisson process with rate k.

Since $t^{-1}M(t) \to k$ a.s., a comparison with (2.3) above shows that the continuous k-ball process has speed

$$\lim_{t\to\infty} L\underset{\sim}{X}_t/t = \lim_{t\to\infty} R\underset{\sim}{X}_t/t = ks_k . \tag{3.2}$$

Notice that both $L\underset{\sim}{X}_t$ and $R\underset{\sim}{X}_t$ are determined by the counting measure $N\underset{\sim}{X}_t$, so the speed is determined by the counting measure process $(N\underset{\sim}{X}_t, t\geq 0)$. Now so far as these counts are concerned, one can view the transition $\underset{\sim}{x} \to \underset{\sim}{x}^j$ of a k-ball process as the creation of a new ball in box x_j+1, together with simultaneous annihilation of a ball in the leftmost occupied box $L\underset{\sim}{x}$. From this point of view the motion proceeds as if each ball were splitting at rate 1 into two balls, independently of other balls. These two balls are a "mother" ball remaining in the original box and a "daughter" ball appearing one box to the right, with annihilation of one ball in the leftmost occupied box simultaneous with each split. This suggests comparing the continuous k-ball process with the Markov process whose state space is $C_* = \cup_k C_k$ which evolves according to the splitting rules described above but with no annihilations. We call this the *lateral birth process*. Its transition rates are

$$\underset{\sim}{x} \to \underset{\sim}{x}^{i+} \text{ at rate } 1, \quad i=1,\ldots,\#\underset{\sim}{x} , \tag{3.3}$$

where $x^{i+} \in C_{\#x+1}$ is defined by

$$\begin{aligned} x_j^{i+} &= x_j , \quad j=1,\ldots,\#\underset{\sim}{x} \\ &= x_i + 1, \quad j=\#\underset{\sim}{x}+1 . \end{aligned}$$

In the terminology of Mollison (1978), the lateral birth process is a particularly simple contact birth process. The next result is a special case of more general results for branching processes and Markovian contact processes due to Kingman (1975) and Mollison (1978), but for the sake of completeness we shall provide a proof in Section 4.

PROPOSITION 3.4. *For a lateral birth process* $(\underset{\sim}{B}_t, t\geq 0)$,

$$\lim_{t\to\infty} R\underset{\sim}{B}_t/t = e \text{ a.s.}$$

To compare the progress of different processes of balls in boxes we introduce a partial ordering of configurations. Say $\underset{\sim}{x}$ is behind $\underset{\sim}{y}$, or $\underset{\sim}{y}$ is ahead of $\underset{\sim}{x}$, and write $\underset{\sim}{x} \le \underset{\sim}{y}$ iff

$$\sum_{j\ge h} x_j \le \sum_{j\ge h} y_j \ , \quad h = 0,1,\ldots \ .$$

So $\underset{\sim}{x} \le \underset{\sim}{y}$ iff $\underset{\sim}{y}$ has more balls than $\underset{\sim}{x}$ to the right of h for each box h. (To avoid ambiguity we say box i is to the right of box h iff $i \ge h$, strictly to the right of box h iff $i > h$, and just to the right of h iff $i = h+1$, with a similar convention on the left.) Given a configuration $\underset{\sim}{x}$, let the balls of $\underset{\sim}{x}$ be ranked primarily according to their position, and secondarily according to their label. That is to say, the rank of ball j in configuration $\underset{\sim}{x}$ is one plus the number of balls in boxes strictly to the right of ball j plus the number of balls in the same box as ball j whose labels exceed j. For each configuration $\underset{\sim}{x}$ this gives a total ordering of the balls comprising $\underset{\sim}{x}$. It is easy to see that $\underset{\sim}{x} \le \underset{\sim}{y}$ iff $\#\underset{\sim}{x} \le \#\underset{\sim}{y}$ and for each $r = 1,2,\ldots,\#\underset{\sim}{x}$, the ball of rank r in $\underset{\sim}{x}$ is to the left of the ball with rank r in $\underset{\sim}{y}$. In particular, taking $r = 1$ shows that $\underset{\sim}{x} \le \underset{\sim}{y}$ implies $R\underset{\sim}{x} \le R\underset{\sim}{y}$. But, $\underset{\sim}{x} \le \underset{\sim}{y}$ does not imply $L\underset{\sim}{x} \le L\underset{\sim}{y}$, except when $\#\underset{\sim}{x} = \#\underset{\sim}{y}$.

Consider now two continuous time Markov chains M and $\hat{M}$ with state spaces C and $\hat{C}$ which are subsets of C_*, and bounded transition rate matrices Q and $\hat{Q}$ indexed by C and $\hat{C}$ respectively. Say M stays behind $\hat{M}$ (or $\hat{M}$ stays ahead of M) if for every pair of initial configurations $\underset{\sim}{x}$ and $\hat{\underset{\sim}{x}}$ with $\underset{\sim}{x} \le \hat{\underset{\sim}{x}}$ there exists an M-chain $(\underset{\sim}{X}_t, t\ge 0)$ and an $\hat{M}$-chain $(\hat{\underset{\sim}{X}}_t, t\ge 0)$ defined on the same probability space such that

$$\underset{\sim}{X}_0 = \underset{\sim}{x}, \ \hat{\underset{\sim}{X}}_0 = \hat{\underset{\sim}{x}}, \ \text{and}$$
$$\underset{\sim}{X}_t \le \hat{\underset{\sim}{X}}_t \quad \text{for all} \quad t \ge 0.$$

We shall make use of the following Lemma, which is valid for Markov chains on an arbitrary partially ordered countable set C_*. (The idea here is folklore amongst coupling theorists; Liggett (1977) applies the same idea with a different partial ordering in investigating infinite particle systems.)

LEMMA 3.6. *Suppose that* $\hat{x} \le \hat{y}$ *whenever* $\hat{Q}(\hat{x},\hat{y}) > 0$. *Suppose also that for every pair of states* $(x,\hat{x}) \in C\times\hat{C}$ *with* $x < \hat{x}$ *and every state* y *with* $Q(x,y) > 0$ *there is a state* $\hat{y} = f(x,\hat{x},y)$ *such that*

(i) $y \le \hat{y}$

(ii) $Q(x,y) \le \hat{Q}(\hat{x},\hat{y})$

(iii) *For fixed* x *and* $\hat{x}$ *the map* $y \to \hat{y}$ *is one-to-one.*

Then $\hat{M}$ *stays ahead of* M.

PROOF. Construct $(X,\hat{X})$ as the Markov chain on $C\times\hat{C}$ with transition rates

$$\begin{aligned}(x,\hat{x}) &\to (y,\hat{y}), \text{ rate } Q(x,y)\\ &\to (x,\hat{y}), \text{ rate } \hat{Q}(\hat{x},\hat{y}) - Q(x,y)\\ &\to (x,y'), \text{rate } \hat{Q}(\hat{x},y'),\end{aligned}$$

where $\hat{y} = f(x,\hat{x},y)$ and y' is an arbitrary state not in the range of $f(x,\hat{x},\cdot)$. These transitions stay within the set $\{x\le\hat{x}\}$.

REMARK 3.7. In the applications below, (ii) holds with equality. The process $(X,\hat{X})$ above can then be described more simply by saying that $\hat{X}$ is an $\hat{M}$-chain and that X is derived from $\hat{X}$ by letting X make a transition from x to y iff $\hat{X}$ makes a transition from $\hat{x}$ to $\hat{y} = f(x,\hat{x},y)$.

PROPOSITION 3.8. *For each* k *the continuous* k*-ball process stays*

(i) *behind the continuous* $\hat{k}$*-ball process if* $k < \hat{k}$,

(ii) *behind the lateral birth process,*

(iii) *ahead of the lateral birth process, stopped at the time* T_k *that it first attains size* k.

PROOF. These are simple applications of the Lemma. In each case the mapping $\hat{y} = f(x,\hat{x},y)$ is defined like this: if y is obtained from x by putting a ball just to the right of the ball ranked r in x, where $1 \le r \le \#x$, then $\hat{y}$ is obtained from $\hat{x}$ by putting a ball just to the right of the ball ranked r in $\hat{x}$.

Let σ_k now denote the speed of the continuous k-ball process, so from (3.2) we have $\sigma_k = ks_k$ where s_k is the speed of the discrete k-ball process.

COROLLARY 3.9.

(i) $\sigma_k \le \sigma_{\hat{k}}$ if $k \le \hat{k}$.

(ii) $\sigma_k \le e$

(iii) $\sigma_k \ge ER\underset{\sim}{B}(T_k)/ET_k$,

where $(\underset{\sim}{B}(t),\ t\ge 0)$ *is a lateral birth process starting with a single ball in box* 0, *and* $T_k = \inf\{t: \#B(t)=k\}$.

PROOF. In view of (3.4), (i) and (ii) follow at once from the corresponding parts of the proposition. For (iii), consider a continuous k-ball process $(\underset{\sim}{X}(t),\ t\ge 0)$. By (3.8)(ii) we can construct a process $(\underset{\sim}{B}_1(t): 0\le t\le T_{1k})$ such that

$\underset{\sim}{B}_1(t) \le \underset{\sim}{X}(t), \quad 0 \le t \le T_{1k}$,

$\underset{\sim}{B}_1(0)$ is the configuration with one ball in box $R\underset{\sim}{X}(0)$ and no other balls,

$\underset{\sim}{B}_1$ evolves as a lateral birth process run until the time T_{1k} it first reaches size k.

Now repeat the construction to obtain a second lateral birth process $(\underset{\sim}{B}_2(t): 0 \le t \le T_{2k})$ such that

$\underset{\sim}{B}_2(t) \le X(T_{1k}+t), \ 0 \le t \le T_{2k}$

$\underset{\sim}{B}_2(0)$ is the configuration with one ball in box $R\underset{\sim}{X}(T_{1k})$ and no other balls,

$\underset{\sim}{B}_2$ evolves as a lateral birth process run until the time T_{2k} it first reaches size k.

Repeating indefinitely, one obtains a sequence of stopped lateral birth processes $\underset{\sim}{B}_1, \underset{\sim}{B}_2, \ldots$ running behind portions of the k-ball process. To be precise

$$\underset{\sim}{B}_n(t) \le \underset{\sim}{X}(S_{n-1}+t), \quad 0 \le t \le T_{nk}, \tag{3.9}$$

where $S_n = T_{1k} + \ldots + T_{nk}$, and $\underset{\sim}{B}_n(0)$ is a single ball in box

$$R\underset{\sim}{B}_n(0) = R\underset{\sim}{X}(S_{n-1}). \tag{3.10}$$

Note in particular from (3.9) that

$$R\underset{\sim}{B}_n(T_{nk}) \le R\underset{\sim}{X}(S_n) \ . \tag{3.11}$$

But from the Remark (3.7) on the construction of these processes and the obvious invariance of the lateral birth process under shifts, it is plain that for $n = 1,2,\ldots$ the processes

$$(\underset{\sim}{B}_n(t)-R\underset{\sim}{B}_n(0),\ 0 < t < T_{nk})$$

are independent copies of the lateral birth process $(\underset{\sim}{B}(t):\ 0 \le t \le T_k)$ of (3.9)(iii). Put $R_n = R\underset{\sim}{B}_n(T_{nk}) - R\underset{\sim}{B}_n(0)$. It follows that $T_{1k}, T_{2k}, \ldots$ are i.i.d. with the same distribution as T_k, and $R_1, R_2, \ldots$ are i.i.d. with the same distribution as $R\underset{\sim}{B}(T_k)$, whence

(a) $S_n/n \to ET_k$ a.s.

(b) $\sum_{j=1}^{n} R_j/n \to ER\underset{\sim}{B}(T_k)$ a.s.

Now
$$\begin{aligned}
\sigma_k &= \lim_{t\to\infty} t^{-1}R\underset{\sim}{X}(t) && \text{by (3.2)}\\
&= \frac{1}{ET_k}\lim_{n\to\infty} n^{-1}R\underset{\sim}{X}(S_n) && \text{by (a)}\\
&\ge \frac{1}{ET_k}\lim_{n\to\infty} n^{-1}\sum_{j=1}^{n} R_j && \text{by (3.10) and (3.11).}\\
&= \frac{ER\underset{\sim}{B}(T_k)}{ET_k} && \text{by (b) .}
\end{aligned}$$

This establishes the Corollary.

PROOF OF THE THEOREM. In view of the Corollary above, it only remains to show that

$$\liminf_{k\to\infty} ER\underset{\sim}{B}(T_k)/ET_k \ge e \ . \tag{3.12}$$

But $(\#B(t),\ t \ge 0)$ is the linear birth process on $\{1,2,3,\ldots\}$, so

$$ET_k = \sum_{j=1}^{k} \frac{1}{j}$$
$$\mathrm{var}(T_k) = \sum_{j=1}^{k} \frac{1}{j^2} . \qquad (3.13)$$

Set $t_k = \log k - \log \log k$. From well-known facts about the sums in (3.13), and Chebyshev's inequality

$$ET_k/t_k \to 1 \qquad (3.14)$$
$$P(T_k \geq t_k) \to 1 \qquad (3.15)$$

Fix $\delta > 0$. By (3.4),

$$P(R\underset{\sim}{B}(t_k) \geq t_k(e-\delta)) \to 1 ,$$

so $\quad P(R\underset{\sim}{B}(T_k) \geq t_k(e-\delta)) \to 1$ by (3.15).

Therefore $\quad \liminf_{k\to\infty} ER\underset{\sim}{B}(T_k)/t_k \geq e - \delta,$

and (3.14) implies (3.12).

REMARK. We do not know any explicit lower bound for s_k. In principle one could be obtained from (3.9)(iii), if one could obtain an explicit lower bound for $ER\underset{\sim}{B}_t$, but this seems difficult.

Keller (1980) has an informal argument (different from ours) suggesting that s_k should be approximately equal to the solution of

$$s(1-s)^{s^{-1}-1} = k^{-1} .$$

4. The speed of the lateral birth process

This section offers a quick proof of Proposition 3.4, that the limiting average speed of the right end of a lateral birth process $(\underset{\sim}{B}_t, t\geq 0)$ is almost surely e. If balls in the ith box are regarded as balls of the ith generation in a branching process, this result is a special case of a theorem for more general branching processes due to Kingman (1975). Our argument follows the same lines as Kingman's, but with some shortcuts due to the simple structure of the lateral birth process. We shall make repeated use of the *reproductive property* of the lateral birth process.

By this we mean the fact that if $\#\underset{\sim}{B}_0 = k$ then the counting measure $\underset{\sim\sim}{NB}_t$ may be written as

$$\underset{\sim\sim}{NB}_t = \underset{\sim\sim}{NB}_t^1 + \underset{\sim\sim}{NB}_t^2 + \ldots + \underset{\sim\sim}{NB}_t^k , \qquad (4.1)$$

where for $j = 1,\ldots,k$ the processes $(\underset{\sim}{B}_t^j,\ t \geq 0)$ are independent lateral birth processes starting with $\underset{\sim}{B}_0^j$ a single ball b_j in the box of the jth ball of $\underset{\sim}{B}_0$. Call $(\underset{\sim}{B}_t^j)$ the process of descendants of the ancestor ball b_j. In particular (4.1) implies

$$R\underset{\sim}{B}_t = \max_j R\underset{\sim}{B}_t^j . \qquad (4.2)$$

LEMMA 4.3. $\limsup_{t\to\infty} R\underset{\sim}{B}_t/t \leq e$ a.s.

PROOF. Using (4.1), it suffices to show this in the case where $\underset{\sim}{B}(0)$ has a single ball in box 0 and no other balls. In this case, put $m_i(t) = E\, N_i\underset{\sim}{B}_t$, the expected number of balls in box i at time t. Then $m_0(t) \equiv 1$, and it is easy to see that

$$\frac{d}{dt}\, m_i(t) = m_{i-1}(t) , \quad i \geq 1 ,$$

whence

$$m_i(t) = t^i/i! \quad . \qquad (4.4)$$

Now

$$\begin{aligned} P(R\underset{\sim}{B}_t \geq i) &= P(N_i\underset{\sim}{B}_t \geq 1) \\ &\leq E\, N_i B_t = t^i/i! \ . \end{aligned}$$

Taking $i = \lceil ft \rceil$ for $f > e$ and using Stirling's formula yields

$$\sum_{t=1}^{\infty} P(R\underset{\sim}{B}_t \geq \lceil ft \rceil) < \infty ,$$

whence the conclusion of the Lemma with f instead of e. Finally, let f decrease to e.

LEMMA 4.5 *Let*

$$p(d) = P(\liminf_{t\to\infty} R\underset{\sim}{B}_t/t \geq d) ,$$

where $\underset{\sim}{B}_0$ *is a single ball in box* 0. *Then* $p(d) > 0$ *for* $d < e$.

PROOF. Fix $d < e$. By (4.4) and Stirling's formula, $m_i(i/d) \to \infty$ as $i \to \infty$. We now choose and fix h such that $m_h(h/d) > 1$. Starting from one special ball at time 0 in box 0, for $n = 1,2,\ldots$ declare that a ball is special at time $(n+1)h/d$ if it is a ball in box $(n+1)h$ which is descended from one of the special balls in box nh at time nh/d. Let Z_n be the number of special balls at time nh/d, and note that

$$R\underset{\sim}{B}_{nh/d} \geq nh \quad \text{if} \quad Z_n > 0 .$$

But by repeated application of the reproductive property, (Z_n) is a simple branching process whose offspring distribution is the distribution of $N_h\underset{\sim}{B}_{h/d}$ with mean $m_h(h/d) > 1$. Thus (Z_n) has a positive chance of non-extinction, whence

$$P(R\underset{\sim}{B}_{nh/d} \geq nh \text{ for all } n = 1,2,\ldots) > 0 ,$$

establishing the Lemma.

PROOF OF PROPOSITION 3.4. If $(\underset{\sim}{B}_t)$ is a lateral birth process starting with k balls, Lemma 4.5 and the reproductive property yield

$$P(\liminf_{t\to\infty} R\underset{\sim}{B}_t/t < d) \leq (1 - p(d))^k , \tag{4.6}$$

with $p(d) > 0$ for $d < e$. Applying this fact to $(\underset{\sim}{B}_{T(k)+t})$ instead of $(\underset{\sim}{B}_t)$, where $T(k) = \inf\{t: \#\underset{\sim}{B}_t = k\}$, it is found that (4.6) also holds for $(\underset{\sim}{B}_t)$ starting with less than k balls. But, since k is arbitrary, for $d < e$ the right side of (4.6) can then be replaced by zero, whence

$$\liminf_{t\to\infty} R\underset{\sim}{B}_t/t \geq e \text{ a.s.}$$

Together with Lemma 4.3 this completes the proof.

5. The limiting process of proportions

We discuss in this section the limiting behavior as $k \to \infty$ of the left counting process $(\underset{\sim}{N}(m),\ m = 0,1,\ldots)$ associated with a discrete k-ball process $(\underset{\sim}{X}(m),\ m = 0,1,\ldots)$. That is, as in Section 1, let $\underset{\sim}{N}(m) = (N_i(m),\ i = 0,1,\ldots)$ be the vector of counts of $\underset{\sim}{X}(m)$, with the leftmost occupied box of $\underset{\sim}{X}(m)$ corresponding to $i = 0$.

Consider informally the evolution of a k-ball process over time δk, where δ is small but δk is large, starting with left count vector $\underset{\sim}{n}$. In this time about δk balls are removed from box 0 (counting from the leftmost occupied box), and about $\delta k \cdot n_i/k = \delta n_i$ of them are placed in box $(i+1)$. So if $\underset{\sim}{p} = \underset{\sim}{n}/k$ is the initial vector of *proportions* of balls in each box, then after time δk the proportions are about $\hat{\underset{\sim}{p}}$, where

$$\hat{p}_0 = p_0 - \delta$$

$$\hat{p}_{i+1} = p_{i+1} + \delta p_i$$

So if we consider the process $\underset{\sim}{N}(kt)/k$ describing proportions in the k-ball process speeded up by a factor of k, it is intuitively plausible that as $k \to \infty$ this process is approximated by the *deterministically* evolving process $\underset{\sim}{p}(t)$ given by

$$\begin{aligned} \frac{d}{dt} p_0(t) &= -1 \\ \frac{d}{dt} p_{i+1}(t) &= p_i(t)\ ,\ i \geq 0 \end{aligned} \qquad (5.1)$$

when $p_0(t) > 0$ and such that whenever $p_0(t)$ hits zero, $(p_i(t))$ is replaced by $(p_{i-1}(t))$, because we count from the leftmost occupied box.

To make this approximation precise we introduce some further notation. Let

$$P = \{\underset{\sim}{p} = (p_0,p_1,p_2,\ldots):\ p_i \geq 0,\ \textstyle\sum p_i = 1\}$$

$$SP = \{\underset{\sim}{p} = (p_0,p_1,p_2,\ldots):\ p_i \geq 0,\ \textstyle\sum p_i \leq 1\}\ ,$$

equipped with the topology of pointwise convergence (so SP is compact). From an initial state $\underset{\sim}{p}(0) = \underset{\sim}{p} \in SP$, the deterministic process evolves during $0 \leq t \leq p_0$ as $\underset{\sim}{p}(t) = T_t\underset{\sim}{p}$, where (integrating (5.1)),

$$(T_t p)_0 = p_0 - t$$

$$(T_t p)_{i+1} = p_{i+1} + \int_0^t (T_s \underset{\sim}{p})_i \, ds \tag{5.2}$$

After the shift at time $t = p_0$, the state, $T\underset{\sim}{p}$ say, is given by

$$(T\underset{\sim}{p})_i = (T_{p_0}\underset{\sim}{p})_{i+1} \tag{5.3}$$

The transformation T will be described more explicitly in the next section: see in particular (6.1) and (6.2), from which it is plain that T is a continuous mapping from $\mathcal{SP}$ to $\mathcal{SP}$ which maps $\mathcal{P}$ to $\mathcal{P}$.

Now let $\underset{\sim}{H}$ be the *histogram of proportions* in $N(m)$:

$$\underset{\sim}{H}(m) = \underset{\sim}{N}(m)/k \ .$$

For each m, $\underset{\sim}{H}(m)$ is a random element of $\mathcal{P}$. Let $\underset{\sim}{N} = \underset{\sim}{N}(0)$, $\underset{\sim}{H} = \underset{\sim}{H}(0)$. We assert that the random process $\underset{\sim}{H}(kt)$ and the deterministically evolving process $T_t\underset{\sim}{H}$ are approximately the same over the time interval required to empty box 0.

PROPOSITION 5.4. *For* $j = 0,1,\ldots$

$$E \max_{1 \le m \le N_0} |H_j(m) - (T_{m/k}\underset{\sim}{H})_j|^2 \le 3^{2+j}/k \ ,$$

where for $m = N_0$, $H_{j+1}(m)$ *should read as* $H_j(m)$ *due to the shift at time* N_0.

PROOF. We may suppose the original configuration $\underset{\sim}{N}(0)$ has $N_0(0) = n$, a constant. For $1 \le m \le n$ let

$$\begin{aligned} Y_j(m) &= 1 \text{ if the } m^{\text{th}} \text{ ball is moved into box } j \\ &= 0 \text{ else} \end{aligned}$$

By definition of the motion, for $0 \le m \le n-1$

$$H_0(m) = H_0(0) - m/k \ ,$$
$$H_{j+1}(m) = H_{j+1}(0) + \frac{1}{k}\sum_{r=1}^{m} Y_{j+1}(r) \ , \ j \geq 0 \ . \tag{5.4}$$

Now if F_m is the σ-field generated by $\underset{\sim}{N}(0),\ldots,\underset{\sim}{N}(m)$,

$$E(Y_{j+1}(m+1)|F_m) = H_j(m) \ , \ j \geq 0 \ .$$

Thus for each $j \geq 0$ the process

$$D_j(m) = \frac{1}{k}\sum_{r=1}^{m}[Y_{j+1}(r) - H_j(r-1)] \ , \ 1 \leq m \leq n-1$$

is an (F_m)-martingale. Applying Doob's L^2-maximal inequality to this martingale, and approximating the integrals (5.2) by Riemann sums obtained from (5.3), it is straightforward to deduce (5.4) by induction on j.

Let $\hat{\underset{\sim}{H}} = \underset{\sim}{H}(N_0(0))$ be the proportions after box 0 has been cleared. Proposition 5.4 says, in particular,

$$E|\hat{H}_j - (T\underset{\sim}{H})_j|^2 \leq 3^{2+j}/k \ , \tag{5.5}$$

for $T: SP \to SP$ defined at (5.3). Now let $\underset{\sim}{H}$ have its *clearance equilibrium distribution* π_k, that is the distribution of $\underset{\sim}{H} = \underset{\sim}{N}/k$ when $\underset{\sim}{N}$ has the clearance equilibrium distribution ν_k described above (2.6). Then $\hat{\underset{\sim}{H}}$ has distribution π_k too, by (2.7). By compactness of SP, we can find a subsequential limit π for (π_k). And for any such subsequential limit probability π on SP it is plain from (5.9) and the continuity of T that

$$\pi \text{ is T-invariant.} \tag{5.6}$$

We know also from (2.6) that

$$E_{\pi_k} H_0 = 1/ks_k$$
$$\to 1/e \ \text{ as } \ k \to \infty$$

by Theorem 1.1, whence

$$E_\pi H_0 = 1/e \ . \tag{5.7}$$

Here $E_\pi H_0$ stands for the expectation of H_0 when H has distribution π. Simulations of the k-ball process show that the sequence of histograms at clearance times is subject to less and less variation as k increases, suggesting that π_k converges to a limit distribution $\hat{\pi}$ which is degenerate at a single point $\hat{\underset{\sim}{p}} \in \mathcal{P}$. If so, then by (5.6) and (5.7) this $\hat{\underset{\sim}{p}}$ must be a *fixed point* of T with $\hat{p}_0 = 1/e$. According to Proposition 6.4 below there is a unique such fixed point $\hat{\underset{\sim}{p}}$, namely the probability measure $\underset{\sim}{p}$ defined by the recursion formula (6.3) with $p_0 = 1/e$. The first few probabilities are given approximately by

$\hat{p}_0 = .368$	$\hat{p}_5 = .018$
$\hat{p}_1 = .300$	$\hat{p}_6 = .008$
$\hat{p}_2 = .173$	$\hat{p}_7 = .003$
$\hat{p}_3 = .087$	$\hat{p}_8 = .001$
$\hat{p}_4 = .040$	

Thus our simulations support the following conjecture:

CONJECTURE 5.8. *As* $k \to \infty$, π_k *converges to the distribution degenerate at* $\hat{p}$.

From the argument leading to (5.6) and (5.7), it is plain that the truth of this conjecture would be implied by the truth of

CONJECTURE 5.9. *The probability degenerate at* $\hat{p}$ *is the only* T-*invariant probability measure* π *on* $\mathcal{SP}$ *such that* $E_\pi H_0 = 1/e$.

Some further conjectures which bear on (5.9) are given at the end of the next section.

6. The transformation T

We record in this section some basic properties of the transformation T on $\mathcal{SP}$ defined by (5.3). In particular we describe the fixed points of T, but we are unable to provide the kind of information about invariant measures or iterates of T which is required to settle the conjectures of the previous section. The transformation T is expressed more explicitly by

PROPOSITION 6.1. *For* $i = 0,1,\ldots$

$$(T\underset{\sim}{p})_i = \sum_{j=0}^{i+1} p_{i+1-j} p_0^j / j! - p_0^{i+2}/(i+2)!$$

PROOF. Given $\underset{\sim}{p} = (p_0, p_1, \ldots) \in SP$, consider the associated generating function

$$G(\underset{\sim}{p}, z) = \sum_i p_i z^i \ , \quad |z| \le 1 \ .$$

From (5.1) and (5.2) we obtain

$$\frac{d}{dt} G(T_t\underset{\sim}{p}, z) = -1 + zG(T_t\underset{\sim}{p}, z) \ ,$$

which is readily solved to yield

$$G(T_t\underset{\sim}{p}, z) = (G(\underset{\sim}{p}, z) - z^{-1})e^{tz} + z^{-1} \ .$$

Now from definition (5.3)

$$G(T\underset{\sim}{p}, z) = G(T_{p_0}\underset{\sim}{p}, z)/z \ ,$$

whence $$G(T\underset{\sim}{p}, z) = z^{-1}(G(\underset{\sim}{p}, z) - z^{-1})e^{p_0 z} + z^{-2} \ , \tag{6.2}$$

and collecting coefficients of z^i yields (6.1).

We can now describe the fixed points of T in SP. From (6.1), $\underset{\sim}{p}$ is a fixed point iff for all $i \ge 0$ the following *balance equations* are satisfied:

$$p_{i+1} = p_i - \sum_{j=0}^{i} p_{i-j} p_0^{j+1}/(j+1)! + p_0^{i+2}/(i+2)! \tag{6.3}$$

Clearly these determine $\underset{\sim}{p}$ in terms of p_0. Intuitively, one would expect the extra condition $\sum p_i = 1$ to force only one fixed point $\underset{\sim}{p} \in \mathcal{P}$. But, surprisingly, this is not so:

PROPOSITION 6.4. *For each* $0 < p \le e^{-1}$ *there is a unique fixed point* $\underset{\sim}{p}$ *of* T *with* $p_0 = p$, *and* $\underset{\sim}{p}$ *is a probability. These are the only fixed points in* $S\mathcal{P}$ *apart from the trivial fixed point* $\underset{\sim}{p} = \underset{\sim}{0}$.

PROOF. The following argument, suggested to us by E.J.G. Pitman, is a simplification of our earlier proof. Given $0 \le p \le 1$, let $\underset{\sim}{p} = (p_0, p_1, \ldots)$ be the sequence of real numbers defined by the recursion (6.3) with $p_0 = p$.

(i) If $p = 0$, then $\underset{\sim}{p} = \underset{\sim}{0}$, trivially.

(ii) If $0 < p \le e^{-1}$, then an easy induction on i using (6.3) shows that

$$p_i/p_{i-1} > p \quad \text{for} \quad i = 1,2,\ldots \; . \tag{6.5}$$

It follows easily that $0 < p_i \le p$ for all i. Now the generating function of the bounded sequence $\underset{\sim}{p}$ is readily obtained from (6.3):

$$G(\underset{\sim}{p},z) = (e^{pz}-1)/z(e^{pz}-z) \;, \quad |z| < r \;, \tag{6.6}$$

where $r = r(p) \ge 1$ is the radius of convergence of the series. But $G(\underset{\sim}{p},1) = 1$, so $\underset{\sim}{p}$ is a probability.

(iii) Let $p > e^{-1}$. Suppose, to obtain a contradiction, that $\underset{\sim}{p} \in S\mathcal{P}$. Then the generating function of $\underset{\sim}{p}$ is given by (6.6), where the radius of convergence $r \in [1,\infty]$. If $r < \infty$, a classical theorem states that $z = r$ is a singularity of $G(\underset{\sim}{p},z)$ (see Titchmarsh (1939), p.214). But for $p > e^{-1}$ the equation $e^{zp} - z = 0$ has no real solution. Hence r must equal ∞. But this too is impossible since $G(\underset{\sim}{p},z) \to 0$ as $z \to \infty$ by (6.6), whereas the non-negativity of p_i implies $G(\underset{\sim}{p},z) \ge p$ for $0 \le z < r$.

REMARKS 6.6. Weiner (1980) also observed that the generating function of a fixed point $\underset{\sim}{p}$ is of the form (6.6).

A more detailed analysis shows that for $0 < p \le e^{-1}$, the corresponding fixed probability $\underset{\sim}{p}$ has

$$p_i/p_{i-1} \downarrow r^{-1} \;,$$

where $r = r(p)$, the radius of convergence of the generating function, is the least value $z > 1$ with $e^{zp} = z$. So r increases from 1 to e as p increases from 0 to e^{-1}. It seems that only the fixed point corresponding to $p = e^{-1}$ has any significance for the k-ball process.

We attempted to settle the conjectures of the previous section by studying the behavior of the iterates $T^n \tilde{p}$ as $n \to \infty$ for a given initial probability p, but our results in this direction are rather meagre. Given $\tilde{p}$, write simply t_n for $T^n \tilde{p}$, and $\bar{t}_n$ for $(t_0 + \cdots + t_{n-1})/n$. Using the ergodic theorem, it is easy to show that truth of conjectures (5.8) and (5.9) would be implied by the truth of

CONJECTURE 6.7. *For all* $p \in P$,

$$\limsup_{n\to\infty} \bar{t}_n \le 1/e .$$

Computer calculations suggest that even more may be true:

CONJECTURE 6.8. *For all* $p \in P$,

$$\limsup_{n\to\infty} t_n \le 1/e \simeq .37 .$$

The best we can prove analytically is that for all $p \in P$

$$\limsup_{n\to\infty} t_n \le \frac{1}{2} \log 2 + \frac{1}{8} \approx .47 .$$

7. Miscellany

We should mention two more formulas for the speed s_k of the k-ball process, whose proofs we leave to the reader.

Recall from Section 2 that λ_k is the equilibrium distribution of the left counting k-ball process $\tilde{N}^L \tilde{X}$. For a configuration $\tilde{x}$ let $a\tilde{N}^L \tilde{x} = k^{-1} \sum_{i\ge 0} i N_i^L \tilde{x}$, so that $a\tilde{N}^L \tilde{x}$ is the distance of the "average" of configuration $\tilde{x}$ from the left end.

LEMMA 7.1.

$$s_k = k^{-1} \{1 + E_{\lambda_k} a\tilde{N}^L \tilde{X}\}$$

Informally, just consider the motion of a particular ball. If it is moved when the configuration is $\underset{\sim}{x}$ then the expected distance moved is $1 + aN^L\underset{\sim}{x}$, so the long term average distance moved should be $1 + E_{\lambda_k} aN^L X$; and the ball is moved every k time units, on average. So the expression in Lemma 7.1 represents the average speed of the particular ball.

Second, let us consider the right end of the k-ball process. Analogously to (2.4), define the *right counting process* $\underset{\sim}{N}^R\underset{\sim}{X}(m)$, where for configuration $\underset{\sim}{x}$

$$N^R_i\underset{\sim}{x} = N_{R\underset{\sim}{x}-i}\underset{\sim}{x}\ , \quad i \geq 0 .$$

Thus $N^R_0\underset{\sim}{X}$ is the number of balls in the rightmost occupied box of $\underset{\sim}{X}$, and $N^R_i\underset{\sim}{X}$ the number of balls in the box i to its left. This motion also has a unique equilibrium distribution, ρ_k say.

LEMMA 7.2

$$s_k = k^{-1} E_{\rho_k} N^R_0\underset{\sim}{X} .$$

Informally, consider the right end of the *continuous* k-ball process. In configuration $\underset{\sim}{x}$ there are $N^R_0\underset{\sim}{x}$ balls in the rightmost box, and so the "rate" at which a ball is put into the empty box $R\underset{\sim}{x} + 1$ is $N^R_0\underset{\sim}{x}$. Thus $R\underset{\sim}{X}$ should increase at average rate $E_{\rho_k} N^R_0\underset{\sim}{X}$, which therefore represents the speed $\sigma_k = ks_k$ of the continuous k-ball process.

Finally, let us make a conjecture about the *shape* of the right end. As above, let ρ_k be the equilibrium distribution of the right counting k-ball process. Let $\underset{\sim}{B}_t$ be the lateral birth process, and $\underset{\sim}{N}^R\underset{\sim}{B}_t$ the right counting description of that process.

CONJECTURE 7.3. *There exists a distribution* ρ *such that*

(i) $\rho_k \to \rho$ *weakly as* $k \to \infty$

(ii) $\underset{\sim}{N}^R\underset{\sim}{B}_t \to \rho$ *in distribution as* $t \to \infty$

(iii) $E_\rho N^R_0\underset{\sim}{X} = e$ *(c.f. Lemma 7.2)*

(To be precise here, we must work on the set of configurations $\underset{\sim}{n} = (n_0, n_1, n_2, \ldots)$ where $0 \leq n_i < \infty$ but $\sum n_i$ may be infinite.) In words, we conjecture that the asymptotic (as $k \to \infty$) shape of the right end of the equilibrium k-ball process is the same as the asymptotic (as $t \to \infty$) shape of the right end of the lateral birth process.

Acknowledgements

We thank Craig Tovey and Persi Diaconis for introducing us to the k-ball process. We also thank E.J.G. Pitman for helpful criticisms and suggestions for improvements of the first draft of this paper.

References

Freedman, D. (1971). *Markov Chains*. Holden-Day, San Francisco.

Keller, J.B. (1980). Random drift on a lattice. Unpublished manuscript, Department of Mechanical Engineering, Stanford.

Kingman, J.F.C. (1975). The first birth problem for an age-dependent branching process. *Ann. Prob. 3*, 790-801.

Liggett, T.M. (1977). The stochastic evolution of infinite systems of interacting particles. In: *Ecole d'Eté de Saint-Flour VI*. Springer Lecture Notes in Mathematics 598.

Mollison, D. (1978). Markovian contact processes. *Adv. Appl. Prob. 10*, 85-108.

Titchmarsh, E.C. (1939). *The Theory of Functions*, 2nd Ed. Oxford University Press.

Tovey, C.A. (1980). OATS, BATS and local improvement algorithms. Ph.D. thesis, Department of Operations Research, Stanford.

Weiner, H.J. (1980). On a random point motion problem. Unpublished manuscript, Department of Mathematics, U.C. Davis.

ON DOUBLY STOCHASTIC POPULATION PROCESSES

M.S. Bartlett

ABSTRACT

In a discussion of double stochastic population processes in continuous time, attention is concentrated on transition matrices, or equivalent operators, which are linear in the variable parameters. Difficulties with the extreme case of 'white noise' variability for the parameters are recalled by reference to 'stochasticized' deterministic models, but discussed here also in relation to a general 'switching' model.

The use of the 'backward' equations for determining extinction probabilities is illustrated by deriving various formulae for the (infinite) birth-and-death process with 'white-noise' variability for the birth-and-death coefficients.

1. INTRODUCTION

It seems very apt to discuss doubly stochastic (d.s.) population processes in this volume, as it recalls the mutual interest of David Kendall and myself in population processes many years ago (e.g. Bartlett, 1947, 1949; Kendall, 1948, 1949; Bartlett and Kendall, 1951). In more recent years d.s. population processes have been considered *in discrete time* in connection with extinction probabilities for random environments (see, for example, references in Bartlett, 1978, §2.31) and with genetic problems (e.g. Gillespie, 1974); but the concept of d.s processes *in continuous time* has also become of obvious interest to mathematicians as well as to biologists (e.g. Kaplan, 1973; Keiding, 1975). I might note that my own interest in d.s. processes in continuous time first arose during a visit to Australia in early 1980, when I began to notice in the literature the use of such processes as approximations to genetic or other biological problems for *discrete* generations. Such use is relevant to the question of validity of limiting procedures, especially where apparent ambiguities arise in the extreme case of 'white-noise' variability (see also May, 1973; Turelli, 1977; Ricciardi, 1980); so I should emphasize that my discussion here refers to processes specified directly in continuous time, apart from some remarks at the end of §3. Even in this context the case of 'white-noise' must be handled with care; and I have used the more pedestrian device of a 'switching' model to throw further light on the degree of validity of this extreme case, and to compensate for the heuristic character at times of my discussion. This alternative approach indicates incidentally that the terminology of d.s. processes is in a theoretical sense a convention, but a useful one nevertheless.

To recapitulate my own approach (e.g. in seminars in 1980, and in the Appendix to Bartlett, 1981), I shall denote a (singly) stochastic Markov process, univariate for simplicity, by $\{X(t)\}$ and its characteristic function at time t by $C_t(\theta) \equiv E\{\exp[i\theta X(t)]\}$, where E denotes mathematical expectation. Then the evolution of $X(t)$ in time may be described by the 'forward' equation (Bartlett, 1978, §3.5)

$$\partial C_t(\theta)/\partial t = H_t C_t(\theta) \quad , \tag{1}$$

where H_t is an operator with respect to θ, possibly dependent also on t. For population processes where $X(t) \equiv N(t)$, a positive integer, it is often convenient to specify the distribution of $N(t)$ at time t by the probability vector $\underset{\sim}{p}_t$, or by its probability-generating function (p.g.f.) $\Pi_t(z) \equiv E\{z^{N(t)}\}$; and the forward equation (1) is equivalent to the equation

$$d\underset{\sim}{p}_t/dt = \underset{\sim}{R}_t \underset{\sim}{p}_t \tag{2}$$

for $\underset{\sim}{p}_t$, where $\underset{\sim}{R}_t$ is the transition matrix, or to the equation obtained from (1) by the substitution $i\theta \equiv \log z$. However, for the moment we shall keep to the general equation (1), with formal solution represented by

$$C_t(\theta) = \exp\left\{\int_0^t H_u \, du\right\} C_0(\theta) \quad . \tag{3}$$

The corresponding 'backward' equation for extinction probabilities $\underset{\sim}{P}_t$ at $N(t) = 0$, beginning from states $N(0) = 0,1,2,\ldots$, is

$$dP_t'/dt = P_t' R_t , \tag{4}$$

where P_t' is the transposed (row) vector. (The convention with the transition matrix R_t is that the *column* items sum to zero, as in Bartlett, 1978).

We now suppose that the matrix R_t, or equivalently the operator H_t, is linear in parameters α_i, so that

$$R_t = R + \Sigma_i \alpha_i(t) R_i , \tag{5}$$

$$H_t = H + \Sigma_i \alpha_i(t) H_i . \tag{6}$$

Because of the definition of R_t or H_t in relation to the *infinitesimal* interval $(t, t+dt)$, this restriction is less severe than might appear at first sight, and a wide range of population processes is included. When we regard the $\alpha_i(t)$ as themselves stochastic processes (cf. Kaplan, 1973) we obtain a large class of d.s. processes for which the above formal solutions represent the solutions for *particular* realizations of $\alpha_i(t)$. The *expected* solution (which may become the unique solution under appropriate ergodic conditions) is therefore, in terms of $C_t(\theta)$,

$$C_t(\theta) = E \{\exp[\int_0^t \{H + \Sigma_i \alpha_i(u) H_i\} du]\} C_0(\theta) . \tag{7}$$

If we denote the characteristic functional

$$E \{\exp [\int_0^t \Sigma_i \alpha_i(u) \theta_i(u) du]\}$$

by $\exp K(\theta)$, then

$$C_t(\theta) = \exp[tH + K(H)] C_0(\theta) , \tag{8}$$

where H is the vector of the H_i replacing the $\theta_i(u)$ in the functional $\exp K(\theta)$. A convenient particular class of $\alpha_i(t)$ is to suppose the $\alpha_i(t)$ stationary multivariate Gaussian processes, with mean α_i and autocovariance matrix $\{w_{ij}(u)\}$. Then, as

$$K(\theta) = \Sigma_i \alpha_i \int_0^t \theta_i(u) du + \frac{1}{2} \Sigma_{ij} \int_0^t \int_0^t w_{ij}(u-v) \theta_i(u) \theta_j(v) du\, dv, \tag{9}$$

we obtain

$$C_t(\theta) = \exp\{t[H+\Sigma_i \alpha_i H_i] + \frac{1}{2} \Sigma_{ij}\int_0^t\int_0^t H_i H_j w(u-v)dudv\}C_0(\theta). \quad (10)$$

For population processes, there is clearly some need for caution with this Gaussian assumption for $\alpha_i(t)$, as, in contrast with Gaussian variables, many population parameters, such as birth-and-death rates, are essentially positive. More realistic classes of $\alpha_i(t)$ might be assumed, but the Gaussian assumption is often more convenient, especially in view of the Central Limit Theorem. This even applies, though with even more circumspection, to the extreme 'white-noise' (multivariate) Gaussian form, for which equation (10) is replaced by

$$C_t(\theta) = \exp\{t[H+\Sigma_i \alpha_i H_i + \frac{1}{2}\Sigma_{ij} w_{ij} H_i H_j]\}C_0(\theta) , \quad (11)$$

$\{w_{ij}\}$ replacing the more general $\{w_{ij}(u)\}$. In ergodic cases where an *equilibrium* distribution exists, it must satisfy

$C(\theta) = \exp\{\Delta t[H + \Sigma_i \alpha_i H_i] + \frac{1}{2} \Sigma_{ij}\int_t^{t+\Delta t}\int_t^{t+\Delta t} H_1 H_j w(u-v)dudv\}C(\theta)$ for any Δt. In the 'white-noise' case, this gives the equation

$$H'C(\theta) = 0 \quad (12)$$

where

$$H' \equiv H + \Sigma_i \alpha_i H_i + \frac{1}{2} \Sigma_{ij} w_{ij} H_i H_j . \quad (13)$$

In equations like (12) it is important to notice that operators H_i and H_j (like the corresponding matrices $\underset{\sim}{R}_i$ and $\underset{\sim}{R}_j$ in (5)) do not in general commute.

For the corresponding equations for extinction probabilities, note that, if limiting extinction probabilities exist independently of the realization (see also further remarks in §2), they must satisfy

$$\underset{\sim}{P}' = \underset{\sim}{P}' E\{\exp[\int_t^{t+\Delta t} \underset{\sim}{R}_u du]\}$$

for any Δt. In the particular case of white noise for the $\alpha_i(t)$, this yields

$$P'S = O , \tag{14}$$

where, analogously to (13),

$$S \equiv R + \Sigma_i \alpha_i R_i + \frac{1}{2} \Sigma_{ij} w_{ij} R_i R_j . \tag{15}$$

2. A GENERAL 'SWITCHING' MODEL

I have discussed particular cases of a switching model before (e.g. Bartlett, 1982), but a rather more general formulation is given below.

Consider the temporal evolution of a process depending on a parameter α, and postulate a switching from value α_1 to value α_2 at rate ϵ, and vice versa. More complex switching mechanisms for more than one parameter (or more than two values) could obviously be introduced if necessary, but the case of a single parameter with two possible values is considered for definiteness. Suppose the process were previously characterized by an operator $H(\alpha) = H + \alpha G$. Define $\alpha = \frac{1}{2}(\alpha_1 + \alpha_2)$. The characteristic function for the modified process will be given by $C_t(\theta) = F_1(\theta) + F_2(\theta)$, where $F_1(\theta)$ is the (unstandardized) component of $C_t(\theta)$ when the process at time t is in the 'state' $\alpha = \alpha_1$, and $F_2(\theta)$ similarly when $\alpha = \alpha_2$. The joint forward equations for F_1 and F_2 are

$$\left.\begin{aligned} \partial F_1/\partial t &= H(\alpha_1)F_1 + \epsilon(F_2 - F_1), \\ \partial F_2/\partial t &= H(\alpha_2)F_2 + \epsilon(F_1 - F_2), \end{aligned}\right\} \tag{16}$$

or equivalently, if $D_t(\theta) = F_1 - F_2$,

$$\left.\begin{aligned} \partial C_t(\theta)/\partial t &= H(\alpha)C_t(\theta) + \eta G D_t(\theta) , \\ \partial D_t(\theta)/\partial t &= \eta G C_t(\theta) + [H(\alpha) - 2\epsilon]D_t(\theta) , \end{aligned}\right\} \tag{17}$$

or further, if

$$\underset{\sim}{A}_t = \begin{bmatrix} C_t \\ D_t \end{bmatrix}, \quad \underset{\sim}{B} = \begin{bmatrix} H(\alpha) & \eta G \\ \eta G & H(\alpha)-2\varepsilon \end{bmatrix},$$

$$\partial \underset{\sim}{A}_t/\partial t = \underset{\sim}{B}\underset{\sim}{A}_t \tag{18}$$

with solution

$$\underset{\sim}{A}_t = \exp(\underset{\sim}{B}t)\underset{\sim}{A}_o \ . \tag{19}$$

This exact formal solution is valid for all t, given $\underset{\sim}{A}_o$, so that any approximations, as ε increases, say, may be examined in detail in any particular case.

As t increases, the effective 'eigenvalue' of $\underset{\sim}{B}$ will be

$$\Lambda_1 = H(\alpha) - \varepsilon + \varepsilon[1 + \eta^2G^2/\varepsilon^2]^{1/2} \tag{20}$$

The 'white-noise' solution is equivalent to assuming that

$$\Lambda_1 \sim H(\alpha) + \frac{1}{2}\sigma^2G^2 , \tag{21}$$

where $\sigma^2 = \eta^2/\varepsilon$, and ε large. Note that η^2/ε is the average rate of increase of variance for the parameter α.

A simple case of this model, with the basic process Poisson, was discussed by Bartlett (1982). In this case, if we use the p.g.f. in place of $C_t(\theta)$, we have $H(\alpha) = \alpha(z-1)$, and the white-noise limit of the Gauss-Poisson process is obtained as an approximate solution under the conditions on t, ε and η imposed above. The exact solution is also, however, readily available.

As another example involving the Poisson, consider first the relations holding if an equilibrium distribution exists. The equations (17) now become

$$\left.\begin{aligned} &H(\alpha)C(\theta) + \eta GD(\theta) = 0 , \\ &\eta GC(\theta) + [H(\alpha) - 2\varepsilon]D(\theta) = 0 . \end{aligned}\right\} \tag{22}$$

In particular, when under immigration (rate α) and deaths (death-rate μ

per individual, for convenience standardized to unity by choice of the time-scale), the equilibrium distribution $\Pi(z)$ for α constant is $\exp\{\alpha(z-1)\}$. For α variable, the 'white-noise' solution (Bartlett, 1981) is $\exp\{\alpha(z-1) + \frac{1}{4}\sigma^2(z-1)^2\}$. Equations (22) become

$$\left.\begin{aligned} &[\alpha(z-1) + (1-z)\partial/\partial z]\Pi(z) + \eta(z-1)\Delta(z) = 0, \\ &\eta(z-1)\Pi(z) + [(z-1)(\alpha-\partial/\partial z) - 2\varepsilon]\Delta(z) = 0, \end{aligned}\right\} \tag{23}$$

where $\Delta(z) \equiv D(\theta)$. Write

$$\Pi(z) = \exp\{\alpha(z-1)\}\Pi_o(z), \quad \Delta(z) = \exp\{\alpha(z-1)\}\Delta_o(z).$$

Then we obtain

$$\left.\begin{aligned} &-\partial\pi_o/\partial z + \eta\Delta_o(z) = 0, \\ &\eta(z-1)\Pi_o(z) + (z-1)(-\partial\Delta_o/\partial z) - 2\varepsilon\Delta_o(z) = 0. \end{aligned}\right\} \tag{24}$$

Notice from the first equation of (24) that $\Delta_o(z)$ is not zero unless $\Pi_o(z) = 1$. Eliminating $\Delta_o(z)$, we obtain

$$\gamma(z-1)\frac{\partial^2\Pi_o}{\partial z^2} + \frac{\partial\Pi_o}{\partial z} - \beta(z-1)\Pi_o(z) = 0 \tag{25}$$

where $\beta = \frac{1}{2}\sigma^2, \gamma = \frac{1}{2}/\varepsilon$. If we neglect γ, we obtain the 'white-noise' solution. If we do not, we can solve more precisely from (25), for example, by series expansion in $z-1$. We find

$$\Pi_o(z) = 1 + \frac{\beta(z-1)^2}{2(1+\gamma)} + \frac{\beta^2(z-1)^4}{2.4(1+\gamma)(1+3\gamma)} + \cdots\cdots \tag{26}$$

It should be noted that, as $\Delta_o(1) = [\partial\Pi_o/\partial z]_{z=1} = 0$, the *total* probability content of each of the two components is the same. This ultimate equality of total probability content of the two components is a general result not dependent on equilibrium when the system is 'conservative', for in equations (17) we then have $\partial C_t(0)/\partial t = 0$ for all t, whatever α, whence both H and G vanish when $\theta = 0$; hence $\partial D_t(0)/\partial t = -2\varepsilon D_t(0)$, and $D_t(0) \to 0$.

However, this is not true for the 'moments', even in equilibrium. Thus in the example denote the first two 'factorial moments' of $\Pi(z)$ and $\Delta(z)$ by

$$m_1 = [\partial\Pi(z)/\partial z]_{z=1}, \quad f_1 = [\partial^2\Pi(z)/\partial z^2]_{z=1} ,$$

$$m_2 = [\partial\Delta(z)/\partial z]_{z=1}, \quad f_2 = [\partial^2\Delta(z)/\partial z^2]_{z=1} ;$$

then we find, by differentiating the equations for $\Pi(z)$ and $\Delta(z)$,

$$\begin{aligned} m_1 &= \alpha, \; m_2 = \frac{\eta}{1+2\varepsilon}, \; f_1 = \alpha m_1 + \eta m_2 = \alpha^2 + \frac{\eta^2}{1+2\varepsilon} \\ f_2 &= \frac{\eta m_1 + \alpha m_2}{1+2\varepsilon} = \frac{\eta\alpha}{1+2\varepsilon} + \frac{\eta\alpha}{(1+2\varepsilon)^2} . \end{aligned} \tag{27}$$

The value of f_1 may be compared with its value from the 'white-noise' solution, viz. $\alpha^2 + \beta$, where $\beta = \frac{1}{2}\eta^2/\varepsilon$.

Another example involving an equilibrium distribution is considered in the next section. With regard to the relation of the backward equation (14) to the switching model (for a single variable parameter α), let the white-noise solution be determined by

$$\underset{\sim}{P}'[\underset{\sim}{R} + \alpha \underset{\sim}{R}_\alpha + \frac{1}{2}\sigma^2 R_\alpha^2] = 0 . \tag{28}$$

In place of 'white-noise', we allow α as above to switch from values α_1 to α_2, or vice versa, at rate ε. The exact backward equations for $\underset{\sim}{P}_1, \underset{\sim}{P}_2$, where $\underset{\sim}{P}_1, \underset{\sim}{P}_2$ are extinction probabilities starting from 'states' α_1, α_2 are:

$$\left.\begin{aligned} \underset{\sim}{P}_1'[\underset{\sim}{R} + \alpha_1 \underset{\sim}{R}_\alpha] + \varepsilon[\underset{\sim}{P}_2' - \underset{\sim}{P}_1'] = 0 , \\ \underset{\sim}{P}_2'[\underset{\sim}{R} + \alpha_2 \underset{\sim}{R}_\alpha] - \varepsilon[\underset{\sim}{P}_2' - \underset{\sim}{P}_1'] = 0 , \end{aligned}\right\} \tag{29}$$

or equivalently

$$\left.\begin{aligned} \underset{\sim}{P}'[\underset{\sim}{R} + \alpha \underset{\sim}{R}_\alpha] + \eta \underset{\sim}{Q}' \underset{\sim}{R}_\alpha = 0 , \\ \eta \underset{\sim}{P}' \underset{\sim}{R}_\alpha + \underset{\sim}{Q}'[\underset{\sim}{R} + \alpha \underset{\sim}{R}_\alpha - 2\varepsilon] = 0 \end{aligned}\right\} \tag{30}$$

where, as before, $\eta = \frac{1}{2}(\alpha_1 - \alpha_2)$, and $\underset{\sim}{P} = \frac{1}{2}(\underset{\sim}{P}_1 + \underset{\sim}{P}_2)$, $\underset{\sim}{Q} = \frac{1}{2}(\underset{\sim}{P}_1 - \underset{\sim}{P}_2)$. These equations are in principle soluble exactly for $\underset{\sim}{P}, \underset{\sim}{Q}$. However , note that

$$\underset{\sim}{P}'[\underset{\sim}{R} + \alpha\underset{\sim}{R}_\alpha + \frac{1}{2}\eta^2\underset{\sim}{R}_\alpha^2/\varepsilon] + \frac{1}{2}\eta\underset{\sim}{Q}'[\underset{\sim}{R} + \alpha\underset{\sim}{R}]]\underset{\sim}{R}_\alpha/\varepsilon = 0 \ . \tag{31}$$

Hence, if ε increases, but $\eta^2/\varepsilon = \sigma^2$, the equation (28) may be reached as an approximation when the second part of equation (31) can be neglected. It is hardly feasible to discuss the accuracy of this approximation in general, but it may be examined in particular cases. Equation (31) suggests that the 'white-noise' solution should be rather nearer to the exact solution (for which $Q \neq 0$) than our simply putting α constant. Thus, in the rather trivial extinction probability example mentioned in Bartlett (1982), of a finite birth-and-death process with an absorption boundary at $n = 2$ as well as at $n = 0$, with rates λ, μ from $n = 1$ to $n = 2$, 0 respectively, the extinction probability from $n = 1$, viz. $\mu/(\mu+\lambda)$, is modified to $\mu/(\mu+\lambda-a)$ when λ is 'white-noise' variable with $a = \frac{1}{2}\sigma_\lambda^2$. The exact results on the switching model from λ_1 to λ_2 (or vice-versa) at rate ε are, however, also easily calculated. For example, with $\lambda = 10$, $\mu = 1$, $a = 2$, but also $\lambda_1 = 20$, $\lambda_2 = 0$, $\varepsilon = 25$, the numerical values for comparison are:

'white-noise'	$(a = 2)$: $P = 0.1111$	Switching model	$P = 0.1068$
λ constant	$(a = 0)$: $P = 0.0909$		$Q = 0.0350$

3. STOCHASTICIZED DETERMINISTIC MODELS

The general formalism of §1 naturally covers special cases such as a basic deterministic process, but in the 'white-noise' limit in particular (which in practice is an approximation) we may anticipate ambiguities if environmental stochasticity is imposed on basic models which are already limiting approximations such as deterministic or diffusion processes. This is illustrated below on one of the examples in the literature.

May (1973) considered the non-linear logistic model

$$dn_t/dt = n_t(\kappa - n_t) \ , \tag{32}$$

where n_t is treated as a deterministic continuous variable. He imposed 'white-noise' variability on κ by writing $\kappa(t) = \kappa + \gamma(t)$, where $\gamma(t)$ is zero-mean 'white-noise' with variance σ^2 in unit interval. As $\kappa(t)$ is now Gaussian, May rather naturally handled the ensuing process as a diffusion process, with the Sewall Wright equilibrium density $f(n)$ for n_t given by the standard formula

$$f(n) = C(\exp\{\int[2m(n)dn/\sigma^2(n)]\})/\sigma^2(n) , \tag{33}$$

where $m(n) = E\ \{dn_t|n_t = n\}/dt$, $\sigma^2(n) = E\{(dn_t)^2|n_t = n\}/dt$. *If*, with May, we assume $m(n) = \kappa n - n^2$, $\sigma^2(n) = n^2\sigma^2$, we obtain

$$\begin{aligned} f(n) &= C'(\exp\{\int[2(\kappa-n)dn/(n\sigma^2)]\})/n^2 \\ &= C' n^{2\kappa/\sigma^2 - 2}\, e^{-2n/\sigma^2} , \qquad (2\kappa > \sigma^2) . \end{aligned} \tag{34}$$

This derivation has, of course, its counterpart with the corresponding characteristic function equation. If $M_t(\phi) \equiv C_t(\theta)$, where $\phi \equiv i\theta$, the equation corresponding to (32) is

$$\partial M_t(\phi)/\partial t = \phi(\kappa - \partial/\partial\phi)\ \partial M_t(\phi)/\partial\phi ; \tag{35}$$

and, if we still suppose the conditional moments above to be valid, the operator in (35) is replaced by a normal diffusion operator

$$\phi(\kappa - \partial/\partial\phi)\partial/\partial\phi + \frac{1}{2}\sigma^2\phi^2 \frac{\partial^2}{\partial\phi^2}$$

so that in equilibrium

$$[\kappa + (\tfrac{1}{2}\sigma^2\phi - 1)\quad \partial/\partial\phi]\partial M/\partial\phi = 0, \tag{36}$$

leading to the solution

$$\partial M/\partial\phi = A(1 - \frac{1}{2}\sigma^2\phi)^{-2\kappa/\sigma^2}$$

or

$$M = (1 - \frac{1}{2}\sigma^2\phi)^{-2\kappa/\sigma^2+1} \tag{37}$$

in agreement with (34).

However, the formalism of earlier sections would *square* the operator $\phi\frac{\partial}{\partial\phi}$ multiplying $\kappa(t)$ to yield a complete operator which includes the additional term $\frac{1}{2}\sigma^2\phi\frac{\partial}{\partial\phi}$, leading to an index $-2\kappa/\sigma^2$ in (37) in place of $-2\kappa/\sigma^2+1$. This is an example of the well-known ambiguity between the so-called Ito and Stratonovich calculi, and arises from the ambiguity in the conditional moments. Thus, suppose we had written the original deterministic model in the form

$$du_t/dt = \kappa - n_t ,$$

where $u_t = \log n_t$, the stochasticized version is naturally

$$du_t = (\kappa - n_t)dt + dZ_t , \tag{38}$$

where $E\{dZ_t\} = 0$, $E\{(dZ_t)^2\} = \sigma^2 dt$. On averaging this equation, we find

$$(\kappa - n_t)dt = E\{du_t\} = E\{\log[(n_t+dn_t)/n_t]\}$$

$$= E\{dn_t/n_t - \frac{1}{2}(dn_t)^2/n_t^2 \ldots \},$$

so that the natural approximation for $m(n)$ from this relation would be $(\kappa - n)n + \frac{1}{2}\sigma^2 n$, a formula consistent with the *second* solution above.

We may demonstrate the consistency of our general approach for this example by making use again of the switching model (16), or, in its equilibrium form, (22). For the logistic example, write $\alpha = \kappa = \frac{1}{2}(\kappa_1 + \kappa_2)$, $\eta = \frac{1}{2}(\kappa_1 - \kappa_2)$, $H(\alpha) = \phi[\kappa\partial/\partial\phi - \partial^2/\partial\phi^2]$, $G = \phi\partial\{\partial\phi$ in terms of $\phi \equiv i\theta$, so that, with $D(\theta) \equiv F(\phi)$, (22) becomes

$$\left.\begin{aligned} &J(\kappa)\, M(\phi) + \eta\partial F/\partial\phi = 0 , \\ &\eta\phi\partial M/\partial\phi + [\phi J(\kappa) - 2\varepsilon]F(\phi) = 0 \end{aligned}\right\} \tag{39}$$

where $J(\kappa) \equiv \kappa\partial/\partial\phi - \partial^2/\partial\phi^2$. Multiplying the first of these equations by $\phi J(\kappa) - 2\varepsilon$ and the second by $\eta\partial/\partial\phi$ does not immediately enable us to eliminate F, as these two operators do not commute. However, we have

$$[\phi J(\kappa)-2\varepsilon]\eta \frac{\partial}{\partial\phi} - \eta \frac{\partial}{\partial\phi}[\phi J(\kappa)-2\varepsilon] = -\eta J(\kappa) ,$$

and the first equation of (39) yields

$$(\kappa - \partial/\partial\phi)J(\kappa)M(\phi) + \eta J(\kappa)F(\phi) = 0 ,$$

whence we find

$$[\phi J(\kappa)-2\varepsilon]J(\kappa)M(\phi)-\eta^2[1+\phi\partial/\partial\phi]\partial M/\partial\phi + (\kappa - \frac{\partial}{\partial\phi})M(\phi) = 0 . \qquad (40)$$

This equation is exact, but if we now make our usual approximating assumption that ε becomes large, but $\eta^2/\varepsilon = \sigma^2$, we obtain by dividing (40) through by 2ε and assuming that we may neglect other terms of $O(1/\varepsilon)$,

$$J(\kappa)M(\phi) + \frac{1}{2}\sigma^2[1 + \phi\partial/\partial\phi]\partial M/\partial\phi = 0, \qquad (41)$$

in agreement with the second solution above.

We shall continue to adhere to the limiting solutions reached from the superimposed stochastic process for the parameter(s), though recognizing that the extreme 'white-noise' case must be expected to give rise to difficulties and must always be treated with caution. The discussion in this section 3 is in fact in agreement with the general conclusion by previous writers that the Stratonovich calculus is appropriate for processes specified intrinsically in continuous time. A further suggestion that the Ito calculus is more relevant for approximating equations derived from discrete generation models seems, however, more misleading; for it is necessary to consider precisely the approximations involved in passing to the deterministic and diffusion equations, for example, in population genetics problems.

Thus, consider the situation discussed by Gillespie (1974) of two alleles A_1 and A_2 at a locus, with diploid genotypes A_1A_1, A_1A_2, A_2A_2. Let the 'fitnesses' be additive, and be denoted by $1 + s_1$, $1 + \frac{1}{2}(s_1 + s_2)$, $1 + s_2$, where s_1 and s_2 are *small*. By standard formulae (see for example, Ewens, 1979), it may be shown that the change Δx in the relative frequency x of A_1 is, per generation,

$$\Delta x = \frac{\frac{1}{2}x(1-x)(s_1-s_2)}{1+s_1x+s_2(1-x)} .$$

For constant s_1 and s_2, $x \to 0$ or 1 depending on the sign of $s_1 - s_2$. However, if s_1 and s_2 vary such that $E\{s_1\} = \mu_1$, $E\{s_2\} = \mu_2$, with variances and covariances σ_1^2, σ_2^2 and $\rho\sigma_1\sigma_2$, then, under the conditions assumed for s_1 and s_2, with μ_1^2 small compared with σ_1^2, etc. (remember that s_1 and s_2 may be positive and/or negative), we find

$$\left.\begin{aligned} E\{\Delta x|x\} &\sim \frac{1}{2}x(1-x)[\mu_1-\mu_2-x\sigma_1(\sigma_1-\rho\sigma_2)-(1-x)\sigma_2(\rho\sigma_1-\sigma_2)], \\ \mathrm{var}\{\Delta x|x\} &\sim \frac{1}{4}x^2(1-x)^2(\sigma_1^2-2\rho\sigma_1\sigma_2+\sigma_2^2); \end{aligned}\right\} \quad (42)$$

and these are (unambiguously) the appropriate conditional moments to insert in the approximating diffusion equation for x in an infinite population. (In a finite population there will, of course, be an additional contribution to the variance depending on the population size N.) Notice that the approximations (42) have no automatic relation with the deterministic approximating differential equation when s_1 and s_2 are fixed, when the denominator in the expression above for Δx could be neglected.

It may be verified from formula (33) that there is now, for s_1 and s_2 variable, an equilibrium distribution for x if $|\alpha| < 1$, where

$$\alpha = \frac{4[\mu_1-\mu_2 + \frac{1}{2}(\sigma_2^2 - \sigma_1^2)]}{\sigma_1^2 + \sigma_2^2 - 2\rho\sigma_1\sigma_2}$$

of the form

$$F(x) = Cx^{\alpha}(1-x)^{-\alpha} , \quad (43)$$

in conformity with Gillespie's findings.

Clearly, what is most important is to specify the process or model precisely and unambiguously in the first place. If approximating procedures are subsequently introduced, their accuracy may then always be checked.

4. EXTINCTION PROBABILITIES FOR THE BIRTH-AND-DEATH PROCESS

We shall now return to the derivation of extinction probabilities. Consider the (infinite) birth-and-death process, for which in the equation $\underset{\sim}{R} = \lambda \underset{\sim}{R}_\lambda + \mu \underset{\sim}{R}_\mu$ we have

$$\underset{\sim}{R}_\lambda = \begin{bmatrix} 0 & 0 & 0 & 0 & .. \\ 0 & -1 & 0 & 0 & .. \\ 0 & 1 & -2 & 0 & .. \\ 0 & 0 & 2 & -3 & .. \\ . & . & . & . & \\ . & . & . & . & \end{bmatrix} \qquad \underset{\sim}{R}_\mu = \begin{bmatrix} 0 & 1 & 0 & 0 & .. \\ 0 & -1 & 2 & 0 & .. \\ 0 & 0 & -2 & 3 & .. \\ 0 & 0 & 0 & -3 & .. \\ . & . & . & . & \\ .. & . & . & . & \end{bmatrix} \tag{44}$$

In particular, let $a = b = 0$, so that only μ is (white-noise) variable. From (44),

$$\underset{\sim}{R}_\mu^2 = \begin{bmatrix} 0 & -1 & 2 & 0 & .. \\ 0 & 1 & -6 & 6 & .. \\ 0 & 0 & 4 & -15 & .. \\ 0 & 0 & 0 & 9 & \\ . & . & . & . & \\ . & . & . & . & \end{bmatrix} \tag{45}$$

and the set of equations (14) becomes

$$\left.\begin{array}{l} \mu(1-P_1)-\lambda(P_1-P_2)-c(1-P_1) = 0, \\ 2\{\mu(P_1-P_2)-\lambda(P_2-P_2)+c(1-P_1)-2c(P_1-P_2)\} = 0 \\ 3\{\mu(P_2-P_3)-\lambda(P_3-P_4)+2c(P_1-P_2)-3c(P_2-P_3)\} = 0 \\ \quad . \\ \quad . \\ \quad . \end{array}\right\} \tag{46}$$

Dividing the equations by 1, 2, 3,... and adding the first n equations, we obtain

$$\mu(1-P_n)-\lambda(P_1-P_{n+1})-nc(P_{n-1}-P_n) = 0.$$

When $P_i \neq 1$ (i.e. $\lambda > \mu$; see, for example, Kaplan, 1973), we shall assume not only that $P_n \to 0$, but also that $P_n \to 0$ rapidly enough for $nP_n \to 0$ (see further remarks in connection with equations (65) and (68)). It follows, as n increases, that $P_1 = \mu/\lambda$, as in the case $c = 0$ (this is no longer true if $a \neq 0$). If we define $Q_r = P_{r-1} - P_r$, and

$$\Sigma(x) = Q_1 + Q_2 x + Q_3 x^2 + \ldots,$$

it is found that the equations (46) are equivalent to

$$\mu\Sigma(x) - \lambda[\Sigma(x) - Q_1]/x + c(x-1)\partial(x\Sigma)/\partial x,$$

or

$$\Sigma(x) = \frac{\lambda Q_1}{\lambda - \mu x} - \frac{cx(1-x)}{\lambda - \mu x}\frac{\partial(x\Sigma)}{\partial x}. \tag{47}$$

The more general case may be handled similarly. The derivation becomes a little tidier if we introduce auxiliary matrices as follows. Define

$$\underset{\sim}{L} = \begin{bmatrix} 0 & 0 & 0 & 0 & .. \\ 1 & 0 & 0 & 0 & .. \\ 0 & 1 & 0 & 0 & .. \\ 0 & 0 & 1 & 0 & .. \\ . & . & . & . & \\ . & . & . & . & \end{bmatrix}, \quad \underset{\sim}{D} = \begin{bmatrix} 1 & 0 & 0 & 0 & .. \\ 0 & 2 & 0 & 0 & .. \\ 0 & 0 & 3 & 0 & .. \\ 0 & 0 & 0 & 4 & .. \\ . & . & . & . & \\ . & . & . & . & \end{bmatrix}$$

Then $\underset{\sim}{R} = -\underset{\sim}{L}\underset{\sim}{R}$, $\underset{\sim}{R}\underset{\sim}{R}_\mu = -\underset{\sim}{L}\underset{\sim}{R}_\mu^2$, $\underset{\sim}{R}_\lambda^2 = -\underset{\sim}{L}\underset{\sim}{R}_\mu\underset{\sim}{R}_\lambda$. Also $\underset{\sim}{R}_\mu = (1 - \underset{\sim}{L})\underset{\sim}{D}\underset{\sim}{L}'$, and equation (14) becomes

$$\underset{\sim}{P}'[(\mu - \lambda\underset{\sim}{L})\underset{\sim}{R}_\mu + (c - b\underset{\sim}{L})\underset{\sim}{R}_\mu^2 + (b - a\underset{\sim}{L})\underset{\sim}{R}_\mu\underset{\sim}{R}_\lambda] = 0 \tag{48}$$

or

$$\underset{\sim}{Q}'[(\mu - \lambda\underset{\sim}{L}) + (c - b\underset{\sim}{L})\underset{\sim}{D}\underset{\sim}{L}'(1 - \underset{\sim}{L}) - (b - a\underset{\sim}{L})\underset{\sim}{D}\underset{\sim}{L}'\underset{\sim}{L}(1 - \underset{\sim}{L})]\underset{\sim}{D}\underset{\sim}{L}' = 0, \tag{49}$$

where $\underset{\sim}{Q}' = \underset{\sim}{P}'(1 - \underset{\sim}{L})$. Define further

$$\underset{\sim}{X}' = (1\ x\ x^2\ ..\ ),\quad \underset{\sim}{Y}' = (1\ x\ \tfrac{1}{2}x^2\ \ \tfrac{1}{3}x^3\ ..) ;$$

then (49) becomes, if we multiply on the right by $\underset{\sim}{Y}$,

$$\underset{\sim}{Q}'[(\mu-\lambda\underset{\sim}{L}) + (c-b\underset{\sim}{L})\underset{\sim}{D}\underset{\sim}{L}'(1-\underset{\sim}{L})-(b-a\underset{\sim}{L})\underset{\sim}{D}\underset{\sim}{L}'\underset{\sim}{L}(1-\underset{\sim}{L})]x\underset{\sim}{X} = 0 . \tag{50}$$

Note that $\underset{\sim}{Q}'\underset{\sim}{X} = \Sigma(x)$ as defined above; thus in the special case $a = b = c = 0$, we have the well known result

$$\mu x\Sigma(x) - \lambda[\Sigma(x) - Q_1] = 0 \tag{51}$$

or
$$\Sigma(x) = \lambda Q_1/(\lambda-\mu x) . \tag{52}$$

(Notice that for $\mu/\lambda \geq 1$ in (52) we must conclude from putting $x = \lambda/\mu$ that Q_1 must be zero, or $P_1 = 1$, whence all $P_i = 1$; if $\mu/\lambda < 1$, the denominator is well behaved in the entire range 0 to 1 for x, whence for $x = 1$ we find $P_1 = \mu/\lambda$.) For the general case (50), note that $L'L$ is the unit matrix, and $\underset{\sim}{L}'\underset{\sim}{X} = x\underset{\sim}{X}$. We have also

$$\underset{\sim}{Q}'\underset{\sim}{D}\underset{\sim}{X} = \frac{\partial(x\Sigma)}{\partial x} , \quad \underset{\sim}{Q}'\underset{\sim}{L}\underset{\sim}{D}\underset{\sim}{X} = \frac{\partial\Sigma}{\partial x} , \quad \underset{\sim}{Q}'\underset{\sim}{D}\underset{\sim}{L}\underset{\sim}{X} = \frac{\partial[x(\Sigma-Q_1)]}{x\partial x} , \quad \underset{\sim}{Q}'\underset{\sim}{L}\underset{\sim}{D}\underset{\sim}{L}\underset{\sim}{X} = \frac{\partial(\Sigma - Q_1 - Q_2x)}{x\partial x}$$

so that finally (50) reduces to

$$(\mu x-\lambda)\Sigma(x)+\lambda Q_1+aQ_2-bQ_1+(x-1)[(cx-b)\frac{\partial(x\Sigma)}{\partial x} + (a-bx)\frac{\partial\Sigma}{\partial x}] = 0 . \tag{53}$$

Equation (53) appears at first sight to lead to an exact solution, but it is rather less tractable than expected. We consider the case $P_i \neq 1$ $(\lambda > \mu)$, so that we may put $x = 1$ in (53) to give exactly

$$\lambda Q_1 + aQ_2 - bQ_1 = \lambda - \mu . \tag{54}$$

For small a, b, c it may be adequate to expand iteratively. Thus to the first approximation in a, b, c

$$\Sigma_1(x) = \frac{\lambda - \mu}{\lambda - \mu x} + \frac{x - 1}{\lambda - \mu x}\{(cx-b)\frac{\partial(x\Sigma_o)}{\partial x} + (a-bx)\frac{\partial\Sigma_o}{\partial x}\} , \tag{55}$$

where $\Sigma_o(x) = (\lambda - \mu)/(\lambda - \mu x)$. In particular, putting $x = 0$ in (55) gives

$$(P_1)_1 = \frac{\mu}{\lambda}[1 + \frac{\lambda-\mu}{\lambda}(\frac{a}{\lambda} - \frac{b}{\mu})] . \tag{56}$$

The case b = c = 0. To illustrate the kind of difficulty that arises with the exact solution, let us consider further the case $b = c = 0$, $a \neq 0$, so that (53) reduces (for $\lambda > \mu$) to

$$\Sigma(x) = \frac{\lambda - \mu}{\lambda - \mu x} + \frac{a(x-1)}{\lambda-\mu x} \frac{\partial \Sigma(x)}{\partial x} , \tag{57}$$

with $\Sigma(x) = 1 - P_1$, when $x = 0$. Solving (57) in an orthodox manner yields

$$\partial[(x-1)^{-\xi} e^{\zeta(x-1)} \Psi(x)]/\partial x = - \zeta(x-1)^{-\xi} e^{\zeta(x-1)} , \tag{58}$$

where $\Sigma(x) - 1 \equiv \Psi(x)$, $(\lambda - \mu)/a = \xi$, $\mu/a = \zeta$. Integrating (58) from 0 to x, we obtain

$$(x-1)^{-\xi} e^{\zeta(x-1)} \Psi(x) + e^{-\zeta}(-1)^{-\xi} P_1 = -\zeta \int_0^x (x-1)^{-\xi} e^{\zeta(x-1)} dx .$$

It appears convenient to expand the exponential inside the integral and integrate term by term, whence

$$\Psi(x) = - P_1 (1-x)^{-\xi} e^{-\zeta(x-1)} \sum_{r=0}^{\infty} \frac{[\zeta(x-1)]^{r+1}}{r!(r+1-\xi)}$$

$$+ (1-x)^{-\xi} e^{-\zeta(x-1)} \sum_{r+o}^{\infty} \frac{(-\zeta)^{r+1}}{r!(r+1-\xi)} , \quad (0 \leq x < 1) \tag{59}$$

where for definiteness it is assumed that $r = 1 - \xi$ does not vanish for any r. This solution, in which the infinite sums are degenerate forms of hypergeometric functions, determines all the P_i in terms of P_1, but it is not obvious how we determine P_1. For example, the exact relation (54) (with $b = 0$) vanishes identically from (59) whatever value we assign to P_1! The difficulty is not removed by solving for, say,

$S(x) \equiv 1 + P_1 x + P_2 x^2 + \ldots$ instead of for $\Sigma(x)$ or $\Psi(x)$. The solution must agree with the curtailed case of finite n as we allow n to increase, but presumably should, as in the standard case $a = 0$, be derivable more directly. It is tempting to try to argue that, as $\xi > 0$, the terms involving the large quantity $(1-x)^{-\xi}$ as x gets near to 1 should vanish separately from the remaining term on the right of equation (59), but this does not appear valid except as an approximating solution for large ξ. For example, as $\xi \to 0$ it would give the incorrect (and impossible) answer $P_1 = 1 - e^{\zeta}$. In the case $\lambda = 100$, $\mu = 1$, $a = 3$, however, it gives $P_1 = 0.01041$, which agrees to this accuracy with the value obtained by either of the alternative methods described below.

A further difficulty with the solution (59) is the expansions in powers of $\zeta = \mu/a$, whereas we are most interested in the results for small a. This suggested a development of the approximating method in (56). Thus, if we write

$$\Sigma_s(x) = \sum_{r=0}^{s} F_s(x)a^s ,$$

then

$$\Sigma_o(x) = F_o(x) = \frac{\lambda-\mu}{\lambda-\mu x} , \quad f_s(x) = \left(\frac{x-1}{\lambda-\mu x}\frac{\partial}{\partial x}\right)^s f_o(x) . \qquad (60)$$

In particular,

$$f_1(x) = \frac{(\lambda-\mu)\mu(x-1)}{(\lambda-\mu x)^3} , \quad f_2(x) = (\lambda-\mu)\mu\left[\frac{x-1}{(\lambda-\mu x)^4} + \right.$$

$$\left. + \frac{3\mu(x-1)^2}{(\lambda-\mu x)^5}\right] , \qquad (61)$$

$$f_3(x) = (\lambda-\mu)\mu\left[\frac{x - 1}{(\lambda-\mu x)^5} + \frac{10\mu(x-1)^2}{(\lambda-\mu x)^6} + \frac{15\mu^2(x-1)^3}{(x-\mu x)^7}\right],\ldots$$

Putting $x = 0$ in these formulae yields

$$(P_1)_3 = 1-\Sigma_3(o) = \frac{\mu}{\lambda}\left[1+ \frac{a(\lambda-\mu)}{\lambda^2} + \frac{a^2(\lambda-\mu)(\lambda-3\mu)}{4}\right.$$

$$\left. + \frac{a^3(\lambda-\mu)(\lambda^2-10\lambda\mu+15\mu^2)}{\lambda^6}\right] \qquad (62)$$

To illustrate in a numerical example, let $\lambda = 10$, $\mu = 1$, $a = 2$ (cf. the values used at the end of §2) so that $\mu/\lambda = 0.1$, $\zeta = 0.5$. We obtain

$$(P_1)_3 = 0.1[1 + 0.1800 + 0.0252 + 0.0011] = 0.1206,$$

which should be reasonably accurate, and probably adequate for most practical purposes, especially if it is recalled that the 'white-noise' solution can itself be regarded as an approximation to a more realistic situation.

However, an alternative iterative method was found to yield more rapid convergence, especially for μ/λ (and a/λ) reasonably small, so that P_s decreases in value fairly rapidly as s increases. Comparably to equations (46), the equations represented by (57) may be written:

$$\left.\begin{aligned} &\mu(1-P_1) - \lambda(P_1-P_2) + a(P_1-P_2) - 2a(P_2-P_3) = 0\ , \\ &\mu(P_1-P_2) - \lambda(P_2-P_3) + 2a(P_2-P_3) - 3a(P_3-P_4) = 0\ , \\ &\mu(P_2-P_3) - \lambda(P_3-P_4) + 3a(P_3-P_4) - 4a(P_4-P_5) = 0\ , \\ &\quad\vdots \end{aligned}\right\} \qquad (63)$$

In addition to the equation (54) (with $c = 0$), we may obtain by adding the equations with omission of the first, the first two, etc.,

$$\left.\begin{aligned} &\mu - \lambda P_1 + a(P_1 - P_2) = 0, \\ &\mu P_1 - \lambda P_2 + 2a(P_2 - P_3) = 0, \\ &\mu P_2 - \lambda P_3 + 3a(P_3 - P_4) = 0 \\ &\quad\vdots \end{aligned}\right\} \qquad (64)$$

For $\lambda > \mu$, an approximation for P_{s+1} in terms of P_s, e.g. $P_{s+1} = \mu P_s/\lambda$, may be inserted in the s-th equation to give P_s in terms of P_{s-1}, and successively up to the first equation for P_1. For the set of values $\lambda = 10$, $\mu = 1$, $a = 2$, this procedure yielded the results (starting from various values of s):

$s =$	1	2	3
P_1	0.1220	0.1203	0.1206
P_2		0.0188	0.0176
P_3			0.0038

Notice that not only is the value of P_1 more rapidly obtained than from (62), but the absolute values of $P_2, P_3, \ldots$ are also available, though of course with some diminution in the relative accuracy. (Thus the cumbersome solution (59) may be dispensed with altogether.) This second numerical method has some link with the earlier remark about arriving at the solution for the infinite case as a limit of the curtailed finite n case; we are putting $P_{s+1} \sim \mu P_s/\lambda$ instead of zero for the completely absorbing boundary. Even this approximation may not be very satisfactory if a/λ is not too small; in fact, if $s = \lambda/a$, the equations give $P_{s+1} = \mu P_{s-1}/\lambda$ exactly. The value of different numerical procedures for different ranges of values of a/λ and μ/λ would no doubt repay further study.

If we suppose that asymptotically $P_s \sim \alpha s^{-\beta}\gamma^s$, the original equations (63) lead for large s to the relation:

$$(1-\gamma)\{\mu/\gamma-\lambda+sa(1-\gamma)-a\gamma+2a\beta\gamma\} = 0 \ . \tag{65}$$

The first bracket corresponds to the solution $P_s = 1$ when $\lambda \le \mu$; the second corresponds to the solution $P_{s+1} = \mu P_s/\lambda$ if $a = 0$, $\lambda > \mu$, but for $a \ne 0$ the expression does not vanish unless both $\gamma = 1$ and $\beta = \frac{1}{2}(\xi + 1)$, where $\xi = (\lambda - \mu)/a$. This rather crude argument suggests that the general assumption previously made that, if $P_s \ne 1$, $sP_s \to 0$ will require the further condition $\xi > 1$.

When $c \ne 0$ rather than a, similar manipulation of the original equations (46) enables us to by-pass the even more awkward-looking equation (47). For $sP_s \to 0$, we write down the sequence containing as its first equation the result $P_1 = \mu/\lambda$:

$$\left.\begin{aligned} \mu - \lambda P_1 &= 0 , \\ \mu P_1 - \lambda P_2 + c(1 - P_1) &= 0 , \\ \mu P_2 - \lambda P_3 + 2c(P_1 - P_2) &= 0 , \end{aligned}\right\} \qquad (66)$$

In this case the solution is now straightforward, for we may now write down *exact* results in succession from the first equation downwards, yielding for $P_2, \ldots, P_{s+1}, \ldots$

$$\left.\begin{aligned} P_2 &= (\mu/\lambda)P_1 + (c/\lambda)(1 - P_1) , \\ P_{s+1} &= (\mu/\lambda)P_s + (sc/\lambda)(P_{s-1} - P_s) , \\ &\cdot \\ &\cdot \\ &\cdot \end{aligned}\right\} \qquad (67)$$

From the original equations (46) the relation analogous to (65) is

$$(1-\gamma)\{\mu-\lambda\gamma+sc(\frac{1}{\gamma} - 1)-c+c\beta(1 + \frac{1}{\gamma})\} = 0 , \qquad (68)$$

suggesting for $\lambda > \mu$ and $c \neq 0$ that $P_s \sim \alpha s^{-\beta}$, where $2\beta = 1 + (\lambda - \mu)/c$, implying a similar condition $(\lambda - \mu)/c > 1$ to ensure $sP_s \to 0$. (To ensure stability in the P_s as s increases, we require also a large enough value of μ/c.)

The results (67) are for $a = b = 0$, but the general case of $a, b, c \neq 0$ could be handled numerically by an extension of the methods for $a \neq 0$ to include b and c.

REFERENCES

Bartlett, M.S. (1947) *Stochastic Processes* (mimeographed notes of a course given at the University of North Carolina in the Fall Quarter, 1946).

Bartlett, M.S. (1949) Some evolutionary stochastic processes, *J.R. Statist. Soc.* B 11, 211.

Bartlett, M.S. (1978) *Introduction to Stochastic Processes* (Cambridge University Press, 3rd ed.)

Bartlett, M.S. (1981) Population and community structure and interaction, in *The Mathematical Theory of the Dynamics of Biological Populations* II, ed. R.W. Hiorns and D. Cooke (Academic Press, 1981), 119-137.

Bartlett, M.S. (1982) Some stochastic models in biology (to be published).

Bartlett, M.S. and Kendall, D.G. (1951) On the use of the characteristic functional in the analysis of some stochastic processes occurring in physics and biology. *Proc. Camb. Phil. Soc.* 47, 65.

Ewens, W. (1979) *Mathematical Population Genetics* (Springer-Verlag).

Gillespie, J. (1974) Polymorphism in patchy environments. *The Amer. Nat.* 108, 145-151.

Kaplan, N. (1973). A continuous time Markov branching model with random environments. *Adv. Appl. Prob.* 5, 37-54.

Keiding, N. (1975) Extinction and exponential growth in random environments. *Theor. Pop. Biol.* 8, 49-63.

Kendall, D.G. (1948) On the generalized 'birth-and-death' process. *Ann. Math. Statist.* 19,1.

Kendall, D.G. (1949) Stochastic processes and population growth. *J.R. Statist. Soc.* B 11, 230.

May, R.M. (1973) *Stability and Complexity in Model Ecosystems* (Princeton).

Ricciardi, L.M. (1980) Stochastic equations in neurobiology and population biology, in *Vita Volterra Symposium on Mathematical Models in Biology*, ed. C. Barigozzi. Lecture Notes in Biomathematics, Vol.39 (Springer-Verlag), 248-263.

Turelli, M. (1977) Random environments and stochastic calculus. *Theor. Pop. Biol.* 12, 140-178.

ON LIMIT THEOREMS FOR OCCUPATION TIMES

N.H. Bingham and John Hawkes

§1 Local limit theory and the Darling-Kac theorem

Let F be a probability law on the line having zero.mean. If $\{X_i\}$ is a sequence of independent random variables each with law F and if $S_n = \Sigma_1^n X_i$ then (see [11, p.165]) S_n is a recurrent random walk. Suppose that F is in the domain of attraction of a stable law F_α of index α (so that $\alpha \in (1,2]$); then there exists a sequence $\{B_n\}$ of norming constants such that

$$S_n/B_n \Rightarrow F_\alpha \tag{1}$$

(here $\Rightarrow$ denotes convergence in distribution). We let F^{n*} denote the nth convolution power of F and let f_α be the density of F_α. If F is non-lattice, Stone's local limit theorem [26] gives

$$B_n F^{n*}([x,x+h]) = hf_\alpha(x/B_n) + o(1) \qquad (n \to \infty) \tag{2}$$

where $o(1)$ is uniform over x in $\mathbb{R}$ and h in compact sets. Thus in particular the expression on the left-hand side of (2) tends to $hf_\alpha(0)$. Now $f_\alpha(0)$ is a positive constant which can be absorbed into B_n and we then have

$$B_n F^{n*}(I) \to |I| \qquad (n \to \infty) \tag{3}$$

for all intervals I, where $|I|$ denotes the length of I. In the lattice case, F is supported by some arithmetic progression kh where k is an integer and h is maximal. We shall suppose for simplicity that $h = 1$. In this case (3) again holds but this time $|I|$ is the cardinality of the set I.

Let R_β and $R_\beta^{(0)}$ denote the functions that have regular variation of index β at infinity and zero respectively. Then (see [11, XVII, (5.23)]), in either the lattice or the non-lattice case, (1) implies that $B_n \in R_{1/\alpha}$. Thus, by (3), we have

$$\Sigma_0^n F^{k*}(I) \sim |I| \, n^{(\alpha-1)/\alpha} L(n) \qquad (n \to \infty) \tag{4}$$

for some slowly varying function L. Condition (4) is important in the study of occupation-times. For an interval I we let $N_n(I) = \#\{k \le n: S_k \in I\}$ be the occupation-time of I by S_k. It was shown by Darling and Kac [6] that if (4) holds then $N_n(I)/\{|I| n^{(\alpha-1)/\alpha} L(n)\}$ converges in distribution to a limit law H which has the Laplace-Stieltjes transform

$$\int_0^\infty e^{-sx} \, dH(x) = \sum_0^\infty (-s)^n / \Gamma(1+n(\alpha-1)/\alpha);$$

that is H is the Mittag-Leffler law of index $(\alpha-1)/\alpha$. Darling and Kac further show that this is the only way in which

$$N_n(I)/C_n \Rightarrow H \tag{5}$$

can hold for some norming sequence $\{C_n\}$ and some non-degenerate limit law H.

Let us call (5) the central limit theorem for $N_n(I)$ and (1) the central limit theorem for S_n. Then the condition for (5) is the Darling-Kac condition (4), while that for (1) is that F be in a domain of attraction. As noted above, if F is in a domain of attraction, the Darling-Kac condition holds; that is the central limit theorem for S_n implies that for $N_n(I)$.

The question was raised by Kesten [17] and Bretagnolle and Dacunha-Castelle [4] as to whether the converse holds. Kesten gave a positive answer in the case where F has a symmetric distribution but the general problem remains open. Of course a similar question can be asked in continuous time with the random walk being replaced by a Lévy process (that is a process with stationary and independent increments).

The purpose of this paper is two-fold. First (in §2) we extend Kesten's above-mentioned result to symmetric Lévy processes. Then (in §3) we show that both these results have verifiable analogues in the cases of left-continuous random walks and completely asymmetric Lévy processes. These two cases are of course very special. It is well known that in the symmetric case the arc-sine law also holds. We show that in the completely asymmetric case the existence of an arc-sine law is equivalent to both the central limit theorem for X_t and that for N_t.

§2 Symmetry

Our starting-point is the following important result:

THEOREM 0 (Kesten [17]). Let S_n be a symmetric random walk, $1 < \alpha \le 2$. Then the Darling-Kac condition (4) holds if and only if F is in the domain of attraction of a stable law with index α. That is, the central limit theorems for S_n and for $N_n(I)$ are equivalent.

In particular, if (4) holds for *one* I (an interval if F is non-lattice, a finite set of integers if F is lattice) it holds for *all* I.

The proof of Kesten's theorem is long and difficult, and we have no simplification to offer. We may, however, use Kesten's theorem to obtain the analogous result for Lévy processes.

THEOREM 1. Let Z be a symmetric zero-mean Lévy process, $1 < \alpha \le 2$. If I is a finite interval, the occupation-time $N_t(I)$ of Z in I up to time t has (when normed) a Mittag-Leffler law of index $1 - \frac{1}{\alpha}$ as limit-law if and only if Z is attracted to a stable process of index α. This is the only way that $N_t(I)$ can have a non-degenerate limit law.

Some notation: let Z(1) have (infinitely divisible) distribution function F and characteristic function ϕ. Write $\nu(dx)$ for the Lévy measure of Z. Then Z is compound Poisson if and only if the total mass $\nu = \nu(\mathbb{R})$ of the Lévy measure is finite. When this is the case we can set $Z(t) = S_{N(t)}$ where S_n is a random walk with step-length distribution G, where $G(dx) = \nu(dx)/\nu$ (write ψ for the characteristic function of G) and N(t) is a Poisson process of rate ν independent of $\{S_n\}$.

We prove Theorem 1 by reducing the problem to the corresponding result for random walks. There are two cases to be considered. When the Lévy measure is finite we use the fact that the process is then a random walk subordinated by a Poisson process, together with a simple summability result (Lemma 1 below). When the Lévy measure is infinite we use the fact that (for each $h > 0$) $Z(nh)$ is a random walk, together with a result due to Croft (Lemma 2 below).

LEMMA 1. Suppose that $\beta \geq 0$ and let

$$C(x) = \Sigma_0^\infty e^{-x} \frac{x^n}{n!} c_n . \qquad (6)$$

(i) If $c_n \in R_{-\beta}$, then $C(x) \in R_{-\beta}$ and $C(x) \sim c_x$ $(x \to \infty)$.

(ii) If $C(x) \in R_{-\beta}$ and c_n is asymptotic to a non-increasing sequence d_n, then $c_n \in R_{-\beta}$ and $c_n \sim C(n)$.

Proof. If X has a Poisson distribution with parameter λ and $\sigma \in (\frac{1}{2},\frac{2}{3})$, then standard estimates using Stirling's formula yield

$$P\{|X-\lambda| > \lambda^\sigma\} \sim \sqrt{\frac{2}{\pi\lambda}} \exp\{-\tfrac{1}{2}\lambda^{2\sigma-1}\} \qquad (\lambda \to \infty).$$

The summation in (6) may thus be restricted to n between $x \pm x^\sigma$ to within an exponentially small remainder term; call this restricted sum $C_1(x)$. Uniformly for n in this range, $c_n \sim c_{[x]}$; then $C_1(x) \sim c_{[x]}$ follows, and from this, $C(x) \sim c_{[x]}$.

(ii) As d_n is non-increasing,

$$d_{[x-x^\sigma]}(1+o(1)) \geq C_1(x) \geq d_{[x+x^\sigma]}(1+o(1)) \qquad (x \to \infty),$$

whence $\liminf d_n/C_1(n) \geq 1$ and $\limsup d_n/C_1(n) \leq 1$; the result follows.

LEMMA 2. Let $f: \mathbb{R} \to \mathbb{R}$ be a continuous function such that for some c

$$\lim_{n\to\infty} f(nh) = c \quad \text{for each } h > 0.$$

Then

$$\lim_{x\to\infty} f(x) = c .$$

This was proved by Croft in [5]. A more accessible reference is Kingman [19].

We also need the following lemma.

LEMMA 3. If $\nu(\mathbb{R}) = \infty$, the function $h(t) = P\{Z(t) \in I\}$ is continuous in $t > 0$ for any finite interval I.

Proof. Since $\nu(\mathbb{R}) = \infty$ the function

$$g(x) = P\{Z(s) \in I-x\}$$

is in C_0, the space of continuous functions that vanish at infinity. Now

$$\begin{aligned} h(t) &= \int P\{Z(t-s) \in dx\}\, g(x) \\ &= (P_{t-s}g)(0). \end{aligned}$$

If $\omega(f,\delta)$ denotes the modulus of continuity of f, one has $\omega(P_t f,\delta) \le \omega(f,\delta)$ so, as is well-known, Z has a Feller semigroup. In particular,

$$(P_t f)(x) \to (P_u f)(x) \quad (t \to u) \quad \text{uniformly in } x$$

Yosida [29, 232-3]). The lemma follows.

We turn now to the easier half of Theorem 1, the 'local limit' implication from Z being attracted to a stable law to $N_t(I)$ being attracted to a Mittag-Leffler law.

First, consider the case where Z is compound Poisson. Then

$$1 - \phi(t) = 1 - \exp\{-\nu[1 - \psi(t)]\} \sim \nu[1 - \psi(t)] \quad (t \to 0), \qquad (7)$$

whence G is also attracted to the same stable law (see, e.g. [13, Th. 2.6.5]). With S_n as above

$$\begin{aligned} P\{Z(t) \in I\} &= \Sigma_0^\infty P\{N_t = n\}\, P\{S_n \in I\} \\ &= \Sigma_0^\infty \frac{e^{-\nu t}(\nu t)^n}{n!} P\{S_n \in I\}. \end{aligned} \qquad (8)$$

By the argument leading from (1) to (3), $P\{S_n \in I\} \in R_{-1/\alpha}$. By Lemma 1,

$P\{Z(t) \in I\} \in R_{-1/\alpha}$, whence $\int_0^t P\{Z(s) \in I\}\ ds \in R_{1-\frac{1}{\alpha}}$. But this is the Darling-Kac condition for $N_t(I)$ to have a Mittag-Leffler limit law with index $1 - \frac{1}{\alpha}$.

In the remaining case $(\nu = \infty)$, let $B(t)$ be a norming function for Z, so that

$$Z(t)/B(t) \Rightarrow F_\alpha \qquad (t \to \infty).$$

Then $B(t) \in R_{1/\alpha}$, and we may take $B(.)$ continuous. Choose any $h > 0$; then

$$Z(nh)/B(nh) \Rightarrow F_\alpha \qquad (n \to \infty),$$

and as in §1 the local limit theorem gives

$$B(nh)\ P\{Z(nh) \in I\} \to f_\alpha(0) \qquad (n \to \infty).$$

But $B(t)\{P\ Z(t) \in I\}$ is continuous in t, by Lemma 3. Hence by Lemma 2,

$$B(t)\ P\{Z(t) \in I\} \to f_\alpha(0) \qquad (t \to \infty)$$

giving the Darling-Kac condition; the proof is finished as before. The 'local limit' part of Theorem 1 is due when Z is non-singular to Port and Stone [20, Th. 5.4]; the singular case can be obtained from their result on suitably approximating the indicator-function of I above and below by continuous functions of compact support.

For the converse 'Kesten' half of Theorem 1, we first prove a ratio limit theorem of independent interest. For the next result, compare Proposition 5.18 of Port and Stone [20], who again assumed Z non-singular (they also obtained a uniformity conclusion, which we omit as it will not be needed here).

THEOREM 2. If Z is a symmetric Lévy process,

$$\frac{P\{Z(t+s) \in I\}}{P\{Z(t) \in J\}} \to \frac{|I|}{|J|} \qquad (t \to \infty), \tag{9}$$

where I,J are intervals, $|.|$ is Lebesgue measure when Z is non-lattice, I,J are finite sets of integers and $|.|$ is counting measure when Z is lattice.

Proof. In the compound Poisson case, the result follows easily from (8) and the corresponding ratio limit theorem for random walks, due to Stone [25, Cor. 1]. In the remaining case, let

$$f(t) = P\{Z(t) \in I\}/P\{Z(t) \in J\}.$$

For each $h > 0$, Stone's result gives

$$f(nh) \to |I|/|J| \qquad (n \to \infty),$$

and, by Lemma 3, f is continuous. So by Lemma 2 again,

$$f(t) \to |I|/|J| \qquad (t \to \infty),$$

so (9) holds for $s = 0$.

We will write out the details of the general case only for Z non-lattice. Then for $h > 0$,

$$\int_0^h P\{Z(t) \in [-u,u]\}\, du \sim P\{Z(t) \in I\} \int_0^h 2u\, du/|I|$$
$$= h^2 P\{Z(t) \in I\}/|I| \qquad (t \to \infty).$$

If Z has cumulant $k(t)$ ($\phi(t) = e^{-k(t)}$; $k(.)$ is real as Z is symmetric),

$$P\{Z(t) \in I\} \sim \frac{|I|}{\pi h^2} \int_{-\infty}^{\infty} \frac{1 - \cos xh.e^{-tk(x)}}{x^2}\, dx$$

(cf. Kesten [17, (2.6)]), and the right-hand side is decreasing in t. Hence, for $s \geq 0$,

$$\limsup_{t \to \infty} P\{Z(t+s) \in I\}/P\{Z(t) \in I\} \leq 1.$$

But

$$P\{Z(t+s) \in I\} = \int P\{Z(s) \in dx\}\, P_x\{Z(t) \in I\}$$
$$= \int P\{Z(s) \in dx\}\, P\{Z(t) \in I-x\},$$

so by Fatou's lemma

$$\liminf_{t \to \infty} \frac{P\{Z(t+s) \in I\}}{P\{Z(t) \in I\}}$$

$$\geq \int P\{Z(s) \in dx\} \liminf_{t \to \infty} \frac{P\{Z(t) \in I-x\}}{P\{Z(t) \in I\}}$$

$$\geq 1 ,$$

by the $s = 0$ case. Combining, we obtain the required result for $s > 0$, and hence for $s < 0$ also.

We return to the 'Kesten' half of Theorem 1. Assume that $N_t(I)$ converges in distribution, after norming, to a Mittag-Leffler law of index $1 - \frac{1}{\alpha}$. The Darling-Kac theorem, as in §1, shows this to be equivalent to $\int_0^t P\{Z(s) \in I\} ds \in R_{1 - \frac{1}{\alpha}}$. By a 'monotone density' argument, (9) shows this is equivalent to $P\{Z(t) \in I\} \in R_{-1/\alpha}$:

$$P\{Z(t) \in I\} \sim |I|/R(t) \qquad (t \to \infty) \tag{11}$$

say. Then for each $h > 0$,

$$P\{Z(nh) \in I\} \sim |I|/R(nh) \qquad (n \to \infty) .$$

By Kesten's theorem,

$$P\{Z(nh)/R(nh) \leq x\} \to F_\alpha(cx) \qquad (n \to \infty) ,$$

writing c for $f_\alpha(0)$. When Z is not compound Poisson, as in Lemma 3 $P\{Z(t) \leq x\}$ is continuous in t. We may take $R(t)$ continuous in t, so $P\{Z(t)/R(t) \leq x\}$ is continuous in t. Now the desired conclusion

$$P\{Z(t)/R(t) \leq x\} \to F_\alpha(cx) \qquad (t \to \infty)$$

follows by Lemma 2. When Z is compound Poisson, note that $P\{S_n \in I\}$ is asymptotically non-increasing in n (Kesten [17, 282]). Referring to (11), we have by (8) and Lemma 1 that $P\{S_n \in I\} \in R_{-1/\alpha}$. By Kesten's theorem, G is in the domain of attraction of the stable law F, so the characteristic function ψ of G is such that $1-\psi \in R^{(0)}_{1/\alpha}$ ([13, Th. 2.6.5]). By

(7), $1-\phi \in R^{(0)}_{1/\alpha}$. Reversing the step above, this shows that $Z(1)$ is attracted to the stable law F_α, as required.

Remark. If the characteristic function ϕ of $Z(1)$ is monotone in a neighbourhood of the origin, a much shorter proof of Kesten's theorem is possible, using Karamata's Tauberian theorem (Kesten [17, 290]) to obtain regular variation at zero of $1-\phi$ (without the monotonicity hypothesis, we obtain merely regular variation of the *monotone rearrangement* of $1-\phi$). The monotonicity condition seems artificial, but gains in interest in view of the fact that it necessarily holds if the desired conclusion, that Z is attracted to a stable law, is true (Bretagnolle and Dacunha-Castelle [4, Prop. II-1]).

§3 Complete asymmetry

We consider next the case of a *left-continuous* random walk on the integers; that is, the possible step-lengths are restricted to the set $\{-1,0,1,2,\dots\}$ (to avoid trivialities, we assume -1 has positive probability). Such walks have many interesting properties; cf. Spitzer [23], Wendel [27], Pakes [21], Green [12]. Regarded as walks on $\mathbb{Z}$, they are called *skip-free to the left*, or walks with *continuous minimum*. We shall see that a complete analogue of Kesten's theorem holds for them. Further, one may link limit theorems for occupation-times of *intervals* on the one hand (the Darling-Kac theorem above, with Mittag-Leffler limit laws) and *half-lines* on the other. Recall the classical arc-sine laws of Spitzer [22]: let T_n denote the occupation-time of a random walk in the half-line $(-\infty,0)$ up to time n. Then T_n/n converges in law to the generalised arc-sine law with parameter $\rho \in (0,1)$ - that is, the law on $[0,1]$ with density

$$\frac{\sin \pi\rho}{\pi} \cdot \frac{1}{x^{1-\rho}(1-x)^{\rho}}$$

- if and only if

$$\frac{1}{n} \sum_0^n P\{S_k < 0\} \to \rho \qquad (n \to \infty) , \tag{12}$$

and this is the only way that T_n can have a non-degenerate limit law. We shall refer to (12) as *Spitzer's condition.*

THEOREM 3. Let S_n be a zero-mean left-continuous random walk, $1 < \alpha \leq 2$. Then the following statements are equivalent:
(i) the domain-of-attraction condition (1) holds, so S_n has a (stable) limit law,
(ii) the Darling-Kac condition (4) holds, so the occupation-time $N_n(I)$ of an interval I has a (Mittag-Leffler) limit law,
(iii) the Spitzer condition (12) holds (with $\rho = 1/\alpha$), so the occupation-time T_n of a half-line has a (generalised) arc-sine limit law.

COROLLARY. Each of the following is equivalent to each of (i)-(iii):
(iv) the first-passage time Z_k of the walk from 0 to $-k$ $(k = 1,2,\ldots)$ is attracted to a one-sided stable law of index $\rho = 1/\alpha$,
(v) the minimum $M_n = \min\{S_k: k = 0,1,\ldots,n\}$ is attracted to a Mittag-Leffler law of index ρ.

Proof. This hinges on a result of Kemperman:

$$P\{S_n = -1\} = n\, P\{Z_1 = n\} \tag{13}$$

(see Wendel [27]; we shall use a transform version of (13) below).

Consider first occupation-times of the set $\{-1\}$. The Darling-Kac condition for $\{-1\}$ is

$$U(n) = \Sigma_0^n P\{S_k = -1\} \in R_{1-\frac{1}{\alpha}}\,, \tag{14}$$

which by Karamata's Tauberian theorem is

$$u(\sigma) = \Sigma_0^\infty e^{-\sigma n}\, dU(n) = \Sigma_0^\infty e^{-\sigma n}\, P\{S_n = -1\} \in R^{(0)}_{\frac{1}{\alpha}-1}\,.$$

Write $\phi(\sigma)$ for the Laplace-Stieltjes transform of Z_1; then by (13), $u(\sigma) = -\phi'(\sigma)$. Now $u(.)$ is monotone, so by a 'monotone density' argument, (14) is equivalent to

$$1-\phi(\sigma) \in R^{(0)}_{1/\alpha} = R^{(0)}_{\rho}\,. \tag{15}$$

This is equivalent to

$$P\{Z_1 > n\} \in R_{-\rho} \tag{16}$$

and (iv) (cf. [1, XIII, §6]); thus (ii) for $I = \{-1\}$ and (iv) are equivalent.

Next, let g, G be the probability- and cumulant-generating functions of Z_1:

$$E(e^{-\sigma Z_1}) = \phi(\sigma) = g(e^{-\sigma}) = e^{-G(\sigma)}.$$

Since

$$1-\phi(\sigma) \sim -\log \phi(\sigma) = G(\sigma) \qquad (\sigma \to 0+),$$

(15) is $G \in R_\rho^{(0)}$.

Write

$$P(s) = \Sigma_{-1}^{\infty} P_k s^k = E(s^X)$$

for the probability generating function of the length X of a typical step of the random walk. Then X has Laplace-Stieltjes transform $P(e^{-s})$, which we write in the form $e^{\psi(s)}$. One may verify that the generating-function form of Kemperman's result (13) is

$$s\ P(g(s)) \equiv 1\ ,$$

which may be re-written as

$$\psi(G(s)) \equiv s \qquad (s \geq 0)$$

(the continuous-time analogue of this result, due to Kendall, Keilson, Borovkov and Zolotarev, is well-known; see e.g. [3, §4]). Since both and G are monotone, this shows that $G \in R_\rho^{(0)}$ is equivalent to $\psi \in R_\alpha^{(0)}$ (recall $\rho = 1/\alpha$). Indeed, if L is slowly varying

$$G(\sigma) \sim \sigma^\rho\ L(1/\sigma) \qquad (\alpha \to 0+)$$

if and only if

$$\psi_n(s) = n\psi(s/\{n^{\rho}L(n)\}) \to s^{\alpha} \qquad (n \to \infty) ,$$

that is,

$$E \exp\{-sS_n/(n^{\rho}L(n))\} = \exp\{\psi_n(s)\} \to \exp\{s^{\alpha}\} \qquad (n \to \infty) .$$

But for $1 < \alpha \le 2$, $\exp\{s^{\alpha}\}$ is the Laplace-Stieltjes transform of a spectrally positive stable law F_{α} of index α (see e.g. [3, 8], 748). That this is equivalent to (1) with $B_n = n^{\rho}L(n)$ follows as in [3, §9].

We now know that (i) is equivalent to (ii) for $I = \{-1\}$; it remains to pass from $\{-1\}$ to any finite set of integers. To do this, recall that $EX = 0$, so $\{S_n\}$ oscillates, so $EZ_1 = \infty$ (see [11, XII, §2]). In particular, $\Sigma s^n P\{Z_1 = n\}$ has radius of convergence 1, so $\limsup(P\{Z_1 = n\})^{1/n} = 1$, whence $\limsup(P\{S_n = -1\})^{1/n} = 1$ by (13). But this is Condition 1 of Stone [25], so the ratio limit theorem applies to S_n [25, Cor. 5]:

$$P(S_{n+k} \in x+A)/P(S_n \in y+B) \to |A|/|B| \qquad (n \to \infty) ,$$

uniformly in finite x- and y-sets, for finite sets A, B of integers and $|.|$ counting measure. We may thus replace $\{-1\}$ in the argument above by any finite set A at the cost of introducing a scale-factor $|A|$; thus (i) and (ii) are equivalent.

We have established the equivalence of (i), (ii) and (iv). The equivalence of (iv) and (v) follows as in [1] or [28], since the processes $\{Z_k\}_k$ and $\{-M_n\}_n$ are pathwise inverse to each other.

From the left-continuity property of the walk, the law of Z_k is the kth convolution power of the law of Z_1. So, since $1-\phi^k \sim k(1-\phi)$, (15), (16) are equivalent to

$$P\{Z_k > n\} \in R_{-\rho} , \tag{17}$$

or to

$$P\{-M_n < k\} \in R_{-\rho} . \tag{18}$$

But this holds if and only if the Spitzer condition (12) holds, as may be shown using Spitzer's identity (cf. [2]); thus (i) - (iv) are equivalent, as required.

Remark. Since $\Sigma_1^\infty P\{S_n = 0\}/n < \infty$ ([22]), $\frac{1}{n}\Sigma_1^n P\{S_k = 0\} \to 0$ by Kronecker's lemma. We may thus replace $(-\infty,0)$ by $(-\infty,0]$ in the result above. Since $(-\infty,c)$ and $(-\infty,0)$ differ by an interval, and Theorem 3 shows that occupation-times of intervals are $O(n^\rho) = o(n)$, we may also replace $(-\infty,0)$ by any 'left half-line', hence (replacing ρ in (12) by $1-\rho$) by any 'right half-line' also.

The property of having a continuous minimum has an analogue in continuous time. The left-continuous random walks above are analogous to the *spectrally positive* Lévy processes, whose Lévy measures give no mass to $(-\infty,0)$ (and, consequently, whose sample paths have no negative jumps); see e.g. [3, §5].

THEOREM 4. Let Z be a zero-mean spectrally positive Lévy process, let I be a finite interval, and suppose that $1 < \alpha \le 2$. Then the following three statements are equivalent:

(i) Z is in the domain of attraction of a spectrally positive stable law of index α;

(ii) N_t satisfies the Darling-Kac condition for occupation-times; and

(iii) $|\{s \le t: Z(s) \le 0\}|/t$ converges in distribution to an arc-sine law of index $1/\alpha$.

Proof. Let $U^\sigma(A) = \int_0^\infty e^{-\sigma t} P\{Z(t) \in A\}\, dt$. It follows from potential-theoretic considerations that U^σ has a density $q(\sigma,x)$ which has

$$q(\sigma,x) = q(\sigma,0)\, E^0 \exp\{-\sigma H_x\},$$

where H_x is the hitting-time of x. In [3] it is shown that q has the particular functional form

$$q(\sigma,x) = \eta'(\sigma)\, e^{x\eta(\sigma)} \qquad (\sigma > 0,\ x \le 0)$$

where

$$\eta(\sigma) = c\sigma + \int_0^\infty (1-e^{-\sigma x})\, \mu(dx)$$

is the Lévy exponent of a subordinator $\tau = \{\tau_x : x > 0\}$ (the first-passage process of the spectrally negative process $-Z$). Now suppose that $I = (a,b)$, where $a < b < 0$. Statement (i) of the theorem is equivalent to

$$\mu(x,\infty) \in R_{-1/\alpha}$$

by Theorem 10 of [3], while (ii) is equivalent to

$$\int_0^t P\{Z(s) \in I\}\, ds \in R_{1-\frac{1}{\alpha}} .$$

Using the identities

$$c + \int x\, e^{-\sigma x}\, \mu(dx) = \eta'(\sigma) ,$$

$$\int_0^\infty e^{-\sigma t}\, P\{Z(t) \in I\}\, dt = \eta'(\sigma)\, (e^{b\eta(\sigma)} - e^{a\eta(\sigma)})/\eta(\sigma)$$

$$\sim \eta'(\sigma)\, (b-a) \qquad (\sigma \to 0)$$

we find that both are equivalent to

$$\eta'(\sigma) \in R^{(0)}_{(1-\alpha)/\alpha} ;$$

thus (i) and (ii) are equivalent. For the equivalence of both with (iii), argue as above, using the continuous-time version of Spitzer's identity as in [2]. Writing $\underline{Z}(t)$ for $\inf\{Z(u) : u \in [0,t]\}$, the analogues of (17), (18) are (for fixed $x > 0$ and $T \to \infty$)

$$P\{\tau_x > t\} \in R_{-\rho} ,$$

$$P\{-\underline{Z}(t) < x\} \in R_{-\rho} .$$

For general I, choose c so large that $I-c \subset (0,\infty)$. Then

$$P\{Z(t) \in I-c\}/P\{Z(t) \in I\} \to 1 \qquad (t \to \infty).$$

To see this, use the fact that Z has zero mean, and argue as in Theorem 2. Instead of symmetry and Corollary 1 of Stone [25], use Corollary 2 to Theorem 5 of [26] ($s_0 = 0$ there, as the mean is zero). This allows us to pass from I to I-c, and the previous argument applies to I-c; the result follows.

Remarks.

1. The key to Theorems 3 and 4 is that knowledge of the law of any one of the process, its first-passage process (in the positive or negative directions) or its supremum (or infimum) process determines the laws of the others. This phenomenon occurs in other limit theorems of Darling-Kac type, due to Kasahara [15], [16], Stone [24] for diffusions, Karlin and McGregor [14] for birth-and-death processes.

2. For the continuous-time analogue of the Kemperman identity used above, see Propositions 2 and 4 of [3].

3. While the equivalence of (i) and (ii) in Theorem 3 hinges on Kemperman's identity, and so is specific to the left-continuous case, the equivalence of (i) and (iii) holds much more generally. It suffices merely that one tail of the step-length distribution be negligible with respect to the other; see Emery [9], Doney [7]. Thus the equivalence of (i) and (iii) holds 'far from symmetry'. By contrast, 'close to symmetry' it fails; see Doney [8].

§4 The general case

The 'local limit' assertion (1) ⇒ (4) in §1 is quite general, and does not require the symmetry or complete asymmetry we needed for the converse 'Kesten' assertion (4) ⇒ (1). Recall that, to within location and scale, a stable process is specified by two parameters, its index $\alpha \in (0,2]$ and its skewness (or tail-balance) parameter $\beta \in [-1,1]$. We know that (4) ⇒ (1)

(a) with symmetry (Theorems 0,1); here $\beta = 0$;

(b) with complete asymmetry (Theorems 3,4); here $\beta = \pm 1$.

Note that, while α appears explicitly in (4), does not, though in cases (a) and (b) above β is clear from context. In the general case the problem is complicated by the need to determine β, without which it is not even clear to *which* stable law of index α the desired convergence should take place. One may thus pose, in addition to the 'Kesten problem' of deriving (1) from (4), the 'modified Kesten problem' in which we assume in addition that the step-length distribution F satisfies the *tail-balance condition*

$$(1-F(x))/(1-F(x) + F(-x)) \to P \qquad (x \to \infty)$$

(and then the desired β is $p-q$ with $q = 1-p$).

REFERENCES

1. N.H. Bingham (1971) : Limit theorems for occupation-times of Markov processes. *Z. Wahrschein. 17,* 1-22.

2. N.H. Bingham (1973) : Limit theorems in fluctuation theory. *Adv. Appl. Prob. 5,* 554-569.

3. N.H. Bingham (1975) : Fluctuation theory in continuous time. *Adv. Appl. Prob. 7,* 705-766.

4. J. Bretagnolle & D. Dacunha-Castelle (1968) : Théorèmes limites à distances finies pour les marches aléatoires. *Ann. Inst. H. Poincaré 4.1,* 25-73.

5. H.T. Croft (1957): A question of limits. *Eureka 20,* 11-13.

6. D.A. Darling & M. Kac (1957) : On occupation-times of Markov processes. *Trans. Amer. Math. Soc. 84,* 444-458.

7. R.A. Doney (1977) : A note on a condition satisfied by certain random walks. *J. Appl. Prob. 14,* 843-849.

8. R.A. Doney (1980) : Spitzer's condition for asymptotically symmetric random walk. *J. Appl. Prob. 17,* 856-859.

9. D.J. Emery (1975) : On a condition satisfied by certain random walks. *Z. Wahrschein. 31,* 125-139.

10. W. Feller (1967) : On regular variation and local limit theorems. *Proc. Fifth Berkeley Sympos. Math. Stat. Prob. Vol.2 Part 1,* 373-388, Univ. Calif. Press, Berkeley, Ca.

11. W. Feller (1971) : *An introduction to probability theory and its applications, Volume 2, second ed.,* Wiley.

12. P.J. Green (1976) : The maximum and time to absorption of a left-continuous random walk. *J. Appl. Prob. 13,* 444-454.

13. I.A. Ibragimov & Yu. V. Linnik (1971) : *Independent and stationary sequences of random variables.* Wolters-Noordhoff, Groningen.

14. S. Karlin & J.L. McGregor (1961) : Occupation-time for birth-and-death processes. *Proc. Fourth Berkeley Sympos. Math. Stat. Prob. Vol. 2,* 219-273, Univ. Calif. Press, Berkeley, Ca.

15. Y. Kasahara (1975) : Spectral theory of generalised second-order differential operators and its applications to Markov processes. *Japan J. Math. 1,* 67-84.

16. Y. Kasahara (1977) : Limit theorems for occupation-times of Markov processes. *Publ. RIMS, Kyoto Univ. 12,* 801-818.

17. H. Kesten (1968) : A Tauberian theorem for random walks. *Israel J. Math. 6*, 278-294.

18. J.F.C. Kingman (1964) : A note on limits of continuous functions. *Quart. J. Math. (2) 15*, 279-282.

19. A.G. Pakes (1978) : On the maximum and absorption time of left-continuous random walk. *J. Appl. Prob. 15*, 292-299.

20. S.C. Port & C. Stone (1971) : Infinitely divisible processes and their potential theory, I, II. *Ann. Inst. Fourier (Grenoble) 21.2*, 157-275 & *21.4*, 179-265.

21. E. Seneta (1976) : Regularly varying functions. *Lecture Notes in Math. 506*, Springer.

22. F. Spitzer (1956) : A combinatorial lemma and its application to probability theory. *Trans. Amer. Math. Soc. 82*, 323-339.

23. F. Spitzer (1964) : *Principles of random walk*. Van Nostrand.

24. C. Stone (1963) : Limit theorems for random walks, birth and death processes, and diffusion processes. *Illinois J. Math. 7*, 638-660.

25. C. Stone (1966) : Ratio limit theorems for random walks on groups. *Trans. Amer. Math. Soc. 125*, 86-100.

26. C. Stone (1967) : On local and ratio limit theorems. *Proc. Fifth Berkeley Sympos. Math. Stat. Prob. Vol. 2 Part 2*, 217-224, Univ. Calif. Press, Berkeley, Ca.

27. J.G. Wendel (1975) : Left-continuous random walk and the Lagrange expansion. *Amer. Math. Monthly 82*, 494-499.

28. W. Whitt (1971) : Weak convergence of first-passage times. *J. Appl. Prob. 8*, 417-422.

29. K. Yosida (1965) : *Functional analysis*. Springer.

THE MARTIN BOUNDARY OF TWO DIMENSIONAL ORNSTEIN-UHLENBECK PROCESSES

M. Cranston, S. Orey, U. Rösler

For u a twice continuously differentiable function on R^2, B a 2 x 2 nonsingular matrix, define

$$L_B u(x) = \frac{1}{2}\Delta u(x)+(Bx,\nabla u(x)),\ x\in R^2 . \tag{0.1}$$

We are interested in a full description of the Martin boundary for this operator, including the dependence of the boundary on B. The operator L_B is the infinitesimal generator of a diffusion process X in R^2. With W the two-dimensional Wiener process, X satisfies

$$dX_t = dW_t + BX_t dt . \tag{0.2}$$

We refer to X as a two-dimensional Ornstein-Uhlenbeck process, though this may be stretching terminology, since for many choices of B (the ones that interest us chiefly) X is not recurrent.

Since Martin [6] introduced his boundary for the Laplace operator, his construction has been extended to large classes of Markov process generators, for example in Doob [4] and Kunita and T.Watanabe [5]. Explicit working out of examples is still fairly isolated. One of the first instances is the treatment of the Pólya urn scheme by Blackwell and Kendall [1].

Under the condition

$$X_0 = x \tag{0.3}$$

the solution of (0.2) can be written as

$$X_t = e^{Bt}[x + \int_0^t e^{-Bs} dW_s] . \tag{0.4}$$

Despite the fact that X is a Gaussian process, a direct calculation of the

Martin boundary does not appear easy. If both eigenvalues of B have non-positive real part, X is recurrent, and the Martin boundary consists of exactly one point. When both eigenvalues of B have positive real part, our work in [3] led us to conjecture that the minimal Martin boundary points correspond to the solutions

$$t \to e^{Bt}y, \ y \in R^2$$

of $\dot{x} = Bx$, the solution curves being curves of convergence in the Martin topology. One can indeed verify this conjecture directly. However, when B has one positive and one negative eigenvalue the situation is more complicated, and we had no obvious conjecture. It will turn out that for each such B there exists a uniquely defined $\hat{B}$ having both eigenvalues with positive real part, and such that the Martin boundary for L_B agrees with that of $L_{\hat{B}}$.

It is easier to consider the parabolic operator

$$\tilde{L}_B\, u(x,t) = \frac{\partial u}{\partial t}(x,t) + L_B u(x,t)$$

Our approach will be to find the Martin boundary for $\tilde{L}_B$ (parabolic Martin boundary) in Section 1, and then study in Section 2 how this can be "collapsed" to obtain the Martin boundary for L_B. This is a natural and very useful construction; it was mentioned already in H. Cohn [2].

The operator $\tilde{L}_B$ is the generator of the space-time diffusion $\tilde{X}$ corresponding to X. $\tilde{X}$ moves on $R^2 \times (-\infty,\infty)$ with $\tilde{X}_t = (X_t, \xi_t)$, where $\xi_t = \xi_0 + t$.

Our work in Section 1 is simplified by the observation that for B with eigenvalues of opposite sign there exists a positive parabolic function h (i.e. solution of $\tilde{L}_B h = 0$) so that the corresponding h-transformation takes $\tilde{L}_B$ into $\tilde{L}_{\hat{B}}$, where $\hat{B}$ is a matrix both of whose eigenvalues are positive. Since h can be explicitly found, everything reduces to the case of eigenvalues with positive real part, treated in Subsection 1.1. If one considers only the operators L_B they can not in general be transformed into $L_{\hat{B}}$ by an h-transformation. In any case the Martin boundary for the parabolic operators is easier to study, since the potential kernel has simpler form.

In [7] Rösler extends the work to a more general class of time-dependent Ornstein-Uhlenbeck processes. Even in the time independent case that work covers some aspects not treated here. In [3] the tail σ-field and invariant σ-field of X were studied by probabilistic methods. Some connections with this paper are explained in Section 3.

1. The parabolic Martin boundary. We proceed to the identification of the Martin boundary for $\tilde{L}_B$. By a *parabolic function* we mean a solution of $\tilde{L}_B h = 0$ in the entire space $R^2 \times R^1$. Martin's results [6] deal only with the Laplace operator. We could apply the general theory of [5], but, as explained below, our case is so simple that we can just follow Martin [6] with some minor modifications. (For L_B we could refer to Shur [8], but $\tilde{L}_B$ involves some slight differences).

The operator $\tilde{L}$ is the generator of $\tilde{X}_t = (X_t, \xi_t)$, as explained above, and X has Gaussian transition probability densities,

$$p(x,s;\ y,s+t)dy = P[X_t \in dy | X_s = x] \quad .$$

In the parabolic case the densities function as potential kernels, and it is this fact, that time is not integrated out, which makes computations involving the Martin boundary simpler than in the elliptic case. Further simplification can be achieved in our case by introducing the function $\psi(x,t) = (e^{-Bt}x, t)$ and the process $\tilde{Z}_t = \psi(\tilde{X}_t) = (Z_t, \xi_t)$. Note

$$Z_t = e^{-B\xi_t} X_t = e^{-B\xi_0}[X_0 + \int_0^t e^{-Bs} dW_s] \tag{1.1}$$

by (0.4). Observe that if both eigenvalues of B have positive real part, the integral converges as t approaches infinity. The process Z_t is a diffusion with non-stationary transition probabilities; of course $\tilde{Z}_t$ has stationary transition probabilities. Since ψ is a one-one transformation, the Martin boundary for $\tilde{X}$ is obtained immediately from that for $\tilde{Z}$. Denote the transition densities for Z by

$$q(w,s;z,t)dz = P[Z_{t-s} \in dz | Z_0 = w, \xi_0 = s], \qquad -\infty < s \leq t = \infty \ , \ w \in R^2, z \in R^2 \quad .$$

Then $P[Z_{t-s} \in \cdot | Z_0 = w, \xi_0 = s]$ is a Gaussian distribution with mean w and covariance C_s^t given by

$$C_s^t = \int_s^t e^{-B\tau} e^{-B^*\tau} d\tau \ , \ -\infty < s \leq t < \infty$$

so that

$$q(x,s;y,t) = \frac{1}{2\pi|\det C_s^t|^{1/2}} \exp\{-\frac{1}{2}(y-x)^*(C_s^t)^{-1}(y-x)\} \quad (1.2)$$

and

$$p(x,s;y,t) + q(e^{-Bs}x,s;e^{-Bt}y,t)|\det e^{-Bt}| \quad . \quad (1.3)$$

Note that

$$C_s^u = C_s^t + C_t^u \ , \ C_s^t = e^{-Bs}C_0^{t-s}\ e^{-B^*s} \ , \ s \le t \le u \quad . \quad (1.4)$$

Finally introduce the Martin kernel for $\tilde{Z}$

$$M(x,s;y,t) = \frac{q(x,s;y,t)}{q(0,0;y,t)} =$$

$$\left|\frac{\det C_0^t}{\det C_s^t}\right|^{1/2} \exp\{-[\frac{1}{2}\ (y-x)^*\ (C_s^t)^{-1}(y-x)-y^*(C_0^t)^{-1}y]\} \quad (1.5)$$

This expression will be defined on $S^+ \times S^+$, where $S^+ = (R^2 \times (0,\infty)) \cup \{(0,0)\}$. In case $t>s$ and $t>0$, the expression is well defined. For $(x,s) = (y,t)$ interpret the quotient as $+\infty$, provided $(y,t) \neq (0,0)$, and for $(x,s) = (y,t) = (0,0)$, the value is one; for $t<s$ the value of M is zero.

1.1 Both eigenvalues have positive real part. Let λ_1 and λ_2 be the eigenvalues of B. In this sub-section we assume that $Re(\lambda_1)>0$, $Re(\lambda_2)>0$.

The kernel M was defined only on $S^+ \times S^+$, and we begin by considering $\tilde{Z}$ only on this space. Later we extend the construction to obtain the Martin boundary for $\tilde{Z}$ on $R^2 \times (-\infty,\infty)$.

Following Martin [6] a sequence of points (y_n,t_n) of S^+ with no accumulation point will be called a *fundamental sequence* if $M(x,s;y_n,t_n)$ converges as n approaches infinity to a limit $h(x,s)$, for all $(x,s) \in S^+$. In Martin's setup any sequence without accumulation points has a subsequence which is fundamental; and for fundamental sequences the convergence turns out to be not just pointwise convergence, but uniform on compact sets. We shall see that the first part of the last sentence carries over to our present situation, and the second part does almost. A function will be called *normalized* if it assumes the value 1 at the origin. In the present situation, if (y_n, t_n) is a fundamental sequence the corresponding limit function $h(x,s)$ is normalized, and, as we shall see, it is either positive and parabolic on S^+, or equal to h_o defined by

$$h_o(x,s) = \begin{cases} 0, & s>0 \\ 1 & (x,s) = (0,0) \end{cases}$$

A positive parabolic function h is called *minimal* if for any other positive parabolic function $k, k \leq h$ implies that k is a constant multiple of h. Two fundamental sequences are called *equivalent* if they correspond to the same limiting h, and an equivalence class of fundamental sequences is a Martin boundary point. As in Martin [6], these points can be obtained by compactifying S^+ with respect to a suitable metric. A Martin boundary point is *minimal* if the corresponding h is a minimal parabolic function. By a *Martin* sequence we mean a fundamental sequence corresponding to a minimal boundary point. Sometimes, instead of moving out to infinity along a discrete sequence (y_n, t_n) we prefer to move along a curve $(y(s), t(s))$, $s \to \infty$, or $(y(t), t)$, $t \to \infty$, and we shall speak of fundamental curves, *Martin curves*.

From the fact that λ_1 and λ_2 both have positive real part one sees easily that

$$\lim_{t\to\infty} C_s^t = C_s^\infty \text{ exists, and } C_s^\infty \text{ is invertible .}$$

This will allow us to see that every sequence (y_n, t_n) of points of S^+ without an accumulation point has a fundamental sub-sequence. Consider first $t_n \to \infty$, $y_n \to y$. Then (1.4) and (1.5) imply

$$\begin{aligned} M(x,s;y_n,t_n) &\to e^{(\lambda_1+\lambda_2)s} \exp\{-\tfrac{1}{2}[(y-x)^* e^{B^* s}(C_0^\infty)^{-1} e^{Bs}(y-x) - y^*(C_0^\infty)^{-1} y] \\ &=: M(x,s;[y]) \end{aligned} \qquad (1.6)$$

where we denote the corresponding boundary point by $[y]$. In this case the convergence is uniform for (x,s) in a compact set, and $M(x,s;[y])$ is parabolic.

Now consider (y_n, t_n) with $|y_n| \to \infty$. By going over to a sub-sequence one may suppose $t_n \to T \leq \infty$. From (1.4), $C_0^t > C_s^t$ (i.e. $C_0^t - C_s^t$ is positive definite) for $0<s<t$, and so $(C_s^t)^{-1} > (C_0^t)^{-1}$ and $(y_n-x)^*(C_s^{t_n})^{-1}(y_n-x)-y_n^*(C_0^{t_n})^{-1}y_n\to\infty$, showing

$$M(x,s;y_n,t_n)\to h_0(x,s)$$

as $n\to\infty$. Thus this class of sequences corresponds to one Martin boundary point, which is not minimal.

The results of Martin [6] now carry over to our situation. In particular we have the Martin *representation theorem*: the class of minimal boundary points Δ_1 is a Borel set in the compactified space, and for every normalized, positive parabolic function h, there exists a unique probability measure μ on Δ_1 such that

$$h(x,s) = \int_{\Delta_1} M(x,s;[y])d\mu \ .$$

For $y\in R^2$, $[y]$ is a candidate for minimal Martin boundary point: it will be shown that $[y]$ is indeed minimal, and $y \neq z$ implies $[y] \neq [z]$. As a function of $x, M(x,s;[y])$ attains its maximum at $x = y$ only, and this shows that the functions $M(x,s;[y])$ and $M(x,s;[z])$ are distinct for $y \neq z$. Next note that

$$\lim_{s\to\infty} e^{-(\lambda_1+\lambda_2)s} M(x,s;[y]) \to \begin{cases} 0, & x \neq y \\ \exp\{\frac{1}{2}\, y^*(C_0^\infty)^{-1}\}y\ , & x = y \end{cases}$$

and the convergence is bounded pointwise convergence. From Martin's representation theorem,

$$e^{-(\lambda_1+\lambda_2)s} M(x,s,[y]) = e^{-(\lambda_1+\lambda_2)s}\int_{\Delta_1} M(x,s,[z])\mu(dz)\ .$$

Setting $x = y$ and letting $s\to\infty$ gives

$$\exp\{\tfrac{1}{2}\, y^*(C_0^\infty)^{-1}\, y\} = \exp\{\tfrac{1}{2}\, y^*(C_0^\infty)^{-1}\, y\}\cdot\mu(\{[y]\})$$

proving $\mu(\{[y]\}) = 1$, and $[y]$ minimal.

Evidently, instead of normalizing at the origin $(0,0)$ one could normalize at $(-t_o,0)$, with $t_o>0$, and find the Martin boundary for $\tilde{Z}$ on $S^+_{t_o} =: [R^2\times(-t_o,\infty)]\cup\{(0,-t_o)\}$. Proceeding as above, the minimal normalized parabolic functions would correspond to $y\in R^2$ via

$$M_{t_o}(x,s;[y]) = \lim_{t\to\infty} \frac{q(x,s;y,t)}{q(0,-t_o;y,t)} , \quad (x,s)\in S^+_{t_o}$$

and for $s>0$,

$$M_{t_o}(x,s;[y]) = M(x,s;[y])M_{t_o}(0,0;[y]) .$$

Dividing through by the second factor on the right one sees that $M(x,s;[y])$ can be extended to $R^2 \times (-t_o,\infty)$, hence, since t_o is arbitrary, to $R^2 \times (-\infty,\infty)$, the extended function still being a minimal positive harmonic function. Not too surprisingly, the formula (1.6), which is meaningful for all values of $s,-\infty<s<\infty$ gives the correct expression for $M(x,s,[y])$.

Recalling that $\tilde{Z}_t = \psi(\tilde{X}_t)$ we obtain at once the Martin boundary for $\tilde{Z}$. We summarize.

In case B has eigenvalues λ_1 and λ_2 with positive real part, the minimal positive normalized parabolic functions for $\tilde{\mathbf{X}}$ are in 1-1 correspondence with R^2, y in R^2 corresponding to

$$\begin{aligned} K_B(x,s;[y]) &= M(e^{-Bs}x,s;[y]) \\ &= e^{(\lambda_1+\lambda_2)s} \exp\{-\tfrac{1}{2}[(e^{Bs}y-x)^*(C_0^\infty)^{-1}(e^{Bs}y-x)-y^*(C_0^\infty)^{-1}y]\} \end{aligned} \tag{1.7}$$

A Martin curve corresponding to the minimal boundary point y is given by $t\to(e^{Bt}y,t)$, that is a solution of

$$\dot{x} = Bx, \quad x(0) = y$$

In addition there is exactly one other point in the Martin boundary, but it does not correspond to a parabolic function.

1.2 One eigenvalue negative. Suppose now that B has one negative and one positive eigenvalue. It will be shown how to reduce this case to the case when both eigenvalues are positive.

If A is the generator of a diffusion, and h a positive solution of $Ah = 0$, a new generator is defined by

$$A^h g = \frac{1}{h} A(hg) \quad \text{on} \quad \{g: hg \in \text{domain of } A\} \quad .$$

For further information and historical references we refer to Doob [4]. If A corresponds to a process with stationary transition probability densities $p(t,x,y)$, then A^h will correspond to a process with stationary transition probability densities $h^{-1}(x)p(t,x,y)h(y)$, and it is easy to see that A and A^h will have identical Martin boundaries. In particular, if h is a positive parabolic function for $\tilde{L}_B$ one obtains

$$(\tilde{L}_B)^h g = \frac{\partial g}{\partial t} + \frac{1}{2} \Delta g + (Bx, \nabla g) + \frac{1}{h}(\nabla h, \nabla g) \quad .$$

In case h has the form

$$h(x,t) = e^{\lambda t + x^* Q x} \tag{1.8}$$

where Q is a 2x2 matrix this becomes

$$(\tilde{L}_B)^h = \frac{\partial}{\partial t} + \frac{1}{2}\Delta + (Bx + (Q+Q^*)x, \nabla) = \tilde{L}_{\hat{B}}$$

where

$$\hat{B} = B + (Q+Q^*) \quad .$$

If h is given by (1.8) and $Q^s = \frac{1}{2}(Q+Q^*)$

$$\tilde{L}_B h = h[\lambda + \mathrm{Tr}(Q) + 2(Q^s x, Q^s x) + 2(Q^s x, Bx)]$$

and a sufficient condition for h to be parabolic is given by

$$\lambda + \mathrm{Tr}(Q) = 0 \qquad Q^s[Q^s + B] = 0 \; . \tag{1.9}$$

The first relation determines λ in terms of Q. We now show that if the eigenvalues of B are $\lambda_2<0<\lambda_1$, (1.9) has a solution and the resulting B has eigenvalues $-\lambda_2$, and λ_1. According to (1.9) we may as well assume that Q is symmetric, so $Q = Q^s$. By introducing an orthogonal change of coordinates B can be put in subtriangular form, and since such a change of coordinates preserves symmetry, we may as well assume

$$B = \begin{bmatrix} \lambda_2 & 0 \\ b & \lambda_1 \end{bmatrix}$$

and a solution for (1.9) is provided by

$$Q = Q^s = \begin{bmatrix} -\lambda_2 & 0 \\ 0 & 0 \end{bmatrix} .$$

Note that the resulting h in (18) becomes

$$h_B(x,t) = e^{\lambda_2 t + x^* Q x} \tag{1.10}$$

and

$$\hat{B} = \begin{bmatrix} -\lambda_2 & 0 \\ b & \lambda_1 \end{bmatrix}, \quad \lambda_1>0,\ -\lambda_2>0 \ . \tag{1.11}$$

Thus $\tilde{L}_{\hat{B}}$ has the same Martin topology as $\tilde{L}_B$ and one has

$$K_B(x,s;[y]) = h_B(x,s)K_{\hat{B}}(x,s;[y]) \tag{1.12}$$

with $K_{\hat{B}}$ as given in the previous subsection.

If both eigenvalues of B have negative real part the minimal points of the parabolic boundary still correspond to R^2; see [7]. Here we will comment only on the case that the two eigenvalues λ_1 and λ_2 of B are negative. One can proceed as above to reduce this to the case of $\hat{B}$ with positive eigen-values: first construct an h-transform $\tilde{L}_{B'}$ of $\tilde{L}_B$

with B' having eigenvalues $-\lambda_2$ and λ_1. Then construct an h-transform $\tilde{L}_{\hat{B}}$ of $\tilde{L}_{B'}$, having eigenvalues $-\lambda_2$ and $-\lambda_1$. So in two steps one finds Q such that

$$h_B(x,t) = e^{(\lambda_1+\lambda_2)t-x^*Qx} \tag{1.13}$$

is parabolic, and $\hat{B}$ has positive eigenvalues, and once again (1.12) holds.

2. <u>The Martin boundary</u> In the previous section the boundary for $\tilde{L}_B$ was obtained. Now it will be shown how one can obtain from this the boundary for L_B.

A solution of $L_B u = 0$ will be called a *harmonic* function. Again such a function is normalized if it assumes the value one at the origin. A *minimal* positive harmonic function, and all the terminology of Martin boundaries is now defined just as in Subsection 1.1.

If both eigenvalues of B have non-positive real part, X is recurrent and the Martin boundary reduces to a single point.

Assume then that either both eigenvalues of B have positive real part, or that B has one negative and one positive eigenvalue. In the second case we showed in Subsection 1.2 how to obtain $\hat{B}$ with positive eigenvalues such that $\tilde{L}_B$ and $\tilde{L}_{\hat{B}}$ have the same Martin boundaries. In the first case set $\hat{B} = B$.

2.1 <u>Collapsing the parabolic boundary</u> Each minimal Martin boundary point η of X can be identified with a corresponding Martin curve $t \to (\eta(t),t)$ such that

$$\frac{p(x,s;\eta(t),t)}{p(0,0;\eta(t),t)} \to K(x,s;\eta) \quad \text{as} \quad t \to \infty \quad .$$

If $\tilde{X}$ has generator $\tilde{L}_B$, $K = \tilde{K}_B$, of course. Note that though $\tilde{L}_B$ and $\tilde{L}_{\hat{B}}$ have the same Martin boundaries, $K_B \neq K_{\hat{B}}$ if $B \neq \hat{B}$.

If v is any real number and

$$\eta_v(t) = \eta(t+v) \quad , \quad -\infty<t<\infty$$

then $t \to (\eta_v(t),t)$ is again a Martin path corresponding to a minimal boundary point η_v and

$$K(x,s;\eta_v) = \frac{K(x,s+v;\eta)}{K(0,v;\eta)} \quad . \tag{2.1}$$

Note $\eta_0 = \eta$. Three cases can arise: (i) $\eta_0 = \eta_v$ only for $v = 0$, (ii) $\eta_v = \eta_0$ for all v, (iii) η_v is periodic but not constant.

Each minimal boundary point has the form $[y], y \in R^2$, with corresponding Martin curve $T \to (e^{Bt}y,t)$ and we see that $[y]_v = [e^{\hat{B}v}y]$. Hence we are in case (i) if $y \neq 0$, and in case (ii) if $y = 0$. Now put $y \sim z$ or $[y] \sim [z]$ in case $[y]_v = [z]$ for some reall v. Then $\sim$ is an equivalence relation with y and z equivalent if and only if they lie on the same trajectory of $\dot{x} = \hat{B}x$. Except for the trajectory resting at the origin, each trajectory intersects $S^1 = \{y: |y| = 1\}$ exactly once, and so the equivalence classes can be identified with

$$S^1 \cup \{\text{origin}\}$$

and the minimal Martin boundary points become identified with

$$S^1 \times (-\infty,\infty) \cup \{\text{origin}\} \quad .$$

(That is, if $y \in S^1$, $-\infty<v<\infty$ then (y,v) corresponds to $[y]_v = [e^{\hat{B}v}y]$). For h a positive, normalized parabolic function, Martin's representation theorem gives a unique representing measure μ, and using (2.1) one obtains

$$\begin{aligned} h(x,s+v) &= \int_{R^2} K(x,s+v;[y])d\mu = \int K(x,s;[y]_v)K(0,v;[y])d\mu \\ &= \int_{R^2} K(x,s;[y])K(0,v;[y]_{-v})\mu(d[y]_{-v}) \quad . \end{aligned} \tag{2.2}$$

Suppose now that h is harmonic, i.e. $h(x,s) \equiv h(x,0) \equiv: h(x)$, and $L_B h(x) \equiv 0$. Then the uniqueness part of the Martin representation theorem implies

$$\mu(d[y]) = K(0,v,[y]_{-v})\mu(d[y]_{-v}) \quad . \tag{2.3}$$

Now we may consider μ to be a measure on the set given in (2.2). Since $[0]_v = [0]$ for all v, but $K[0,v,[0])$ does depend on v, the relation (2.3) implies that μ can place no mass at the origin. So μ is on $S^1 \times (-\infty,\infty)$ and we can "disintegrate" the measure as

$$\mu(dz \times d\tau) = \mu_2(d\tau|z)\mu_1(dz), \quad \mu_1 \text{ on } S^1, \quad \mu_2(\cdot|z) \text{ on } (-\infty,\infty) .$$

Relation (2.3) becomes

$$\mu_2(d\tau|z)\mu_1(dz) = K(0,v,[z]_{\tau-v})\mu_2(d(\tau-v)|z)\mu_1(dz) .$$

Hence for μ_1-a.e. z

$$\mu_2(d\tau|z) = K(0,v,[z]_{\tau-v})\mu_2(d(\tau-v)|z) . \tag{2.4}$$

The probability measure $\mu_2(\cdot|z)$ will satisfy this relationship if and only if it has a density $f(\tau)$ such that

$$f(\tau + v) = K(0,v,[z]_\tau)f(\tau)$$

for all real numbers τ and v. Because of (2.1), the relation will hold for all τ and v if it holds for $\tau = 0$ and all v. Setting

$$c_z = \int_{-\infty}^{\infty} K(0,v,[z])dv$$

it appears that (2.4) has a solution if and only if $c_z < \infty$, and then

$$\mu_2(d\tau|z) = c_z^{-1}K(0,\tau,[z])d\tau .$$

This identifies the minimal harmonic functions as

$$\bar{K}_B(x,[z]) = c_z^{-1}\int_{-\infty}^{\infty} K_B(x,\tau;[z])d\tau, \; z\in S^1, \; c_z < \infty . \tag{2.5}$$

Observe now that if the second factor in the last member of (1.7) is denoted by $k(x,s;y)$ then

$$\int_{-\infty}^{\infty} e^{\mu s}k(x,s;y)ds < \infty$$

provided $\mu > 0$. This suffices to give us

$$c_z < \infty \text{ for all } z \in S^1 . \tag{2.6}$$

If both eigenvalues of B have negative real part we already know that all positive harmonic functions are constants. It is instructive to see how this follows from collapsing boundaries, at least if both eigenvalues are negative. By the observation at the end of Subsection 1.2, (1.12) again holds, but with h_B given by (1.13). In this case $c_z = \infty$ for all $z \in S^1$. On the other hand $K_B(x,s,[0]) \equiv 1$ in this case, and this is the unique minimal, normalized, positive harmonic function.

2.2 The Martin topology In the previous subsection we have identified the positive normalized harmonic functions. However the Martin topology, that is the fundamental sequences, or Martin curves have not been obtained. It is natural to expect that these are obtained from parabolic Martin curves in $R^2 \times (-\infty,\infty)$ by projecting onto R^2. Suppose that $t \to (z(t),t)$ is a parabolic Martin curve corresponding to the minimal boundary point $[z]$, so that

$$\lim_{t\to\infty} K(x,s;z(t),t) = K(x,s;[z])$$

and suppose further that

$$c_z = \int_{-\infty}^{\infty} K(x,t;[z])dt < \infty \quad .$$

Then by (2.1),

$$\int_{-\infty}^{\infty} K(x,t;[z])dt = \int_{-\infty}^{\infty} \frac{K(x,0;[z]_t)}{K(0,-t;[z]_t)} dt$$

$$= \int_{-\infty}^{\infty} \lim_{\tau\to\infty} \frac{p(x,s;z(\tau),\tau-t)/p(0,0;z(\tau),\tau-t)}{p(0,-t;z(\tau),\tau-t)/p(0,0;z(\tau),\tau-t)} dt$$

$$= \int_{-\infty}^{\infty} \lim_{\tau\to\infty} \frac{p(x,t;z(\tau),\tau)}{p(0,0;z(\tau),\tau)} dt$$

and provided the interchange of integration can be justified the last member will equal the first member of the following identities:

$$\lim_{\tau\to\infty} \frac{\int_{-\infty}^{\infty} p(x,t;z(\tau),\tau)dt}{p(0,0;z(\tau),\tau)} = \lim_{\tau\to\infty} \frac{g(x,z(\tau))}{p(0,0;z(\tau),\tau)}$$

$$= \lim_{\tau\to\infty} \frac{g(x,z(\tau)}{g(0,z(\tau))} \frac{g(0,z(\tau))}{p(0,0;z\ (\tau),\tau)}$$

where

$$g(x,z) = \int_0^\infty p(x,0;z,t)dt$$

is the Green function for the elliptic operator. Setting $x = 0$ reveals then that

$$c_z = \lim_{\tau\to\infty} \frac{g(0,z(\tau)}{p(0,0,z(\tau),\tau)}$$

and then

$$\bar{K}(x,[z]) = c_z^{-1} \int_{-\infty}^{\infty} K(x,t,[z])dt = \lim_{\tau\to\infty} \frac{g(0,z(\tau))}{g(0,z(\tau))}$$

and $t \to z(t)$ does indeed represent a Martin curve. For the particular kernel K_B obtained in Section 1, and the Martin curves $t \to (e^{Bt}z,t)$ or $t \to (e^{\hat{B}t}z,t)$, $z \neq 0$, the necessary interchange of limits is easy to justify. It suffices of course to consider $z \in S^1$, since z and $e^{Bs}z$ (or $e^{\hat{B}s}z$) will give rise to the same projected curve. (We feel the interchange of limits should be valid "in general", not just for our specific problem, but for that we have found no justification.)

3. <u>Bounded solutions</u> The Martin boundary provides information on the non-negative solutions of $L_B u = 0$. Sometimes it is of interest to know the bounded solutions. Indeed this question is well known to be intimately connected with probabilistic properties of the associated diffusion. It was investigated from that viewpoint in [3]. We now recall some of the results.

In case both eigenvalues of B have positive real part, Z_t converges to a limiting random variable Z_∞, which has a two dimensional Gaussian distribution, and it was shown that Z_∞ generates the tail σ-field of X. Hence all bounded parabolic functions have the form $h(x) = E^x g(Z_\infty)$, where g is a bounded Borel function on R^2. In case B has one negative eigenvalue λ_2 and one positive one λ_1, we may assume

$$B = \begin{pmatrix} \lambda_1 & 0 \\ \alpha & \lambda_2 \end{pmatrix}$$

(making an orthogonal change of coordinates to achieve this). In this case one finds that if X^1 denotes the first coordinate of X,

$e^{-\lambda_1 t} X_t^1 \to \hat{Z}_\infty$ and $\hat{Z}_\infty$ is a one-dimensional Gaussian random variable: it generates the tail σ-field in this case. Going from the tail σ-field of X (which is the invariant σ-field of $\tilde{X}$) to the invariant σ-field of X involves a procedure analogous to the "collapsing" explained in Subsection 2.1. Consider first the case in which both eigenvalues of B have positive real part. As remarked earlier for every $x \in R^2\setminus\{0\}$ there is a unique point $\theta(x)$ on the unit circle, and unique real number S such that $e^{Bs}\theta(x) = x$. One finds readily that $\theta(Z_t) \to \theta(Z_\infty) \equiv \Theta$, and this limiting random variable generates the invariant σ-field of X. Θ has a distribution function concentrated on the unit circle, and equivalent to Lebesgue measure on the circle. Thus the bounded harmonic functions have the form

$$h(x) = E^x g(\Theta)$$

with g a bounded Borel function on S^1. None of these functions are minimal harmonic functions. Returning to the case $\lambda_2<0<\lambda_1$ one finds, with $\hat{\theta}(s) = 1$ for $s\geq 0, \hat{\theta}(s) = -1$ for $s<0$ that

$$\hat{\theta}(e^{-\lambda_1 t} X_t^1) \to \hat{\theta}(\hat{Z}_\infty) = \hat{\Theta}$$

and $\hat{\Theta}$ generates the invariant σ-field. The probability distribution of $\hat{\Theta}$ is concentrated on $\{-1,+1\}$. Thus if

$$h_1(x) = 2P^x[\hat{\Theta} = 1], \quad h_{-1}(x) = 2P^x[\hat{\Theta} = -1]$$

then $1 = \frac{1}{2}(h_1 + h_{-1})$, h_1 and h_{-1} are minimal, bounded, positive harmonic functions, and every bounded harmonic function is a linear combination of h_1 and h_{-1}.

Let us show the connection with the present paper. First, examining the integral in (2.5), under the assumption that both eigenvalues of B have positive real part, we find that $\bar{K}_B(x,z)$ is not bounded as a function of x, in fact for $x = e^{Bt}z$, the function approaches infinity with t. So there are no bounded minimal harmonic functions.

Now assume B has eigenvalues $\lambda_2<0<\lambda_1$. Then 1 should be the average of two minimal positive normalized harmonic functions, and we proceed to identify these. The minimal harmonic functions for this case were given in (2.5) (with K_B given in (1.12). Each $z \in S^1$ may be considered to be a unit vector, and choosing u to be the unit eigenvector for B corresponding to the eigenvalue λ_1, one finds, after some manipulation, that

$$1 = \frac{1}{2}(\bar{K}_B(\cdot,u) + \bar{K}_B(\cdot,-u)) .$$

where $\bar{K}$ is defined in (2.5) .

It is natural to consider the Dirichlet problem $L_B u = 0$ in R^2, with u assuming assigned values on the Martin boundary. From the remarks of this section one can deduce that in the case both eigenvalues of B have positive real part, all points on the boundary are regular points. On the other hand, when $\lambda_2 < 0 < \lambda_1$, only the two boundary points corresponding to bounded harmonic functions are regular.

References

[1] Blackwell, D., Kendall, D.G., The Martin boundary for Pólya's urn scheme and an application to stochastic population growth, *J. Appl. Probability 1* (1964), 184-296.

[2] Cohn, H., Harmonic functions for a class of Markov chains, *J. Austral. Math. Soc.* (Series A), *28* (1979), 413-422.

[3] Cranston, M., Orey, S., Rösler, U., Exterior Dirichlet problems and the asymptotic behaviour of diffusions, Lecture Notes in Control and Information Sciences, 15, Springer, 1980, Berlin, Edited by B.Grigelionis, 207-220.

[4] Doob, J.L., Discrete potential theory and boundaries, *J. Math. Mech. 8* (1956), 433-458.

[5] Kunita, H., Watanabe, T., Markov processes and Martin boundaries I, *Ill. J. Math. 9* (1965), 485-526.

[6] Martin, R.S., Minimal positive harmonic functions, *TAMS 49* (1941), 137-172.

[7] Rösler, U., The Martin boundary for time dependent Ornstein-Uhlenbeck processes, preprint.

[8] Shur, M.G., The Martin boundary for a linear elliptic second order operator. *Izv. Abad. Nauk SSSR Ser. Mat. 27* (1963), 45-60; *AMS Translation Series 2* 56, 19-35.

GREEN'S AND DIRICHLET SPACES FOR A SYMMETRIC MARKOV TRANSITION FUNCTION*

E.B. Dynkin

1. Introduction

1.1. This is the first paper in a series devoted to Green's and Dirichlet spaces. In the next publications we shall study the spaces associated with fine Markov processes and with a certain class of multiparameter processes.

For the Brownian motion with exponential killing, the Dirichlet space is Sobolev's space H_1 and Green's space is the dual space H_{-1}. Both spaces are widely used in the theory of the free field (arising in quantum field theory). General Dirichlet and Green's spaces can be applied in an analogous way to Gaussian random fields associated with Markov processes [2].

1.2. Axiomatic theory of Dirichlet spaces was developed by Beurling and Deny [1]. Silverstein [5] and Fukushima [3] investigated the relation between Dirichlet spaces and Markov processes.

We start from a symmetric Markov transition function and we deal simultaneously with a pair: the Dirichlet space H and Green's space K. They are in a natural duality and they play symmetric roles but, in some respects, K is simpler than H. We consider several models for K and H. In particular, we represent them by L-valued functions of time t where L is a functional Hilbert space. We get the conventional representation of H by passage to the limit as $t \to \infty$. Analogously, letting $t \to 0$, we arrive at a representation of K by distributions (generalized functions).

We do not use any topology in the state space. This makes it easier to apply the theory to processes in functional spaces.

1.3. A Markov transition function $p_t(x,B)$ in a measurable space $(E,\mathcal{B})$ is symmetric with respect to a measure m if the expression $\int_A m(dx)p_t(x,B)$ is symmetric in A and B. Operators

*Research supported by NSF Grant No. MCS 77-03543

$$T_t f(x) = \int_E p_t(x,dy) f(y) \tag{1.1}$$

act in $L^2(m)$ and preserve the subspace $\hat{L}$ which consists of f such that $T_t f \to f$ as $t \to 0$, $T_t f \to 0$ as $t \to \infty$. Green's operator G and the infinitesimal operator A are defined by the formulas

$$Gf = \lim_{t \to \infty} \int_0^t T_u f \, du \, ,$$

$$Af = \lim_{t \to 0} t^{-1}(T_t f - f) \quad .$$

Their domains D_G and D_A consist of all $f \in \hat{L}$ for which the corresponding limits exist in $L^2(m)$. D_G with the inner product $(Gf,\tilde{f})$ and D_A with the inner product $-(Af,\tilde{f})$ are pre-Hilbert spaces. Green's and Dirichlet's spaces can be defined as their completions. However we wish to construct these spaces in a more explicit way.

1.4. We say that an $L^2(m)$-valued function ϕ_t, $0 < t < \infty$ is a K-function if $T_s \phi_t = \phi_{s+t}$ for all $s,t > 0$. We put

$$(\phi,\tilde{\phi})_K = \int_0^\infty \int_E \phi_{t/2}(x)\tilde{\phi}_{t/2}(x) \, dt \, m(dx). \tag{1.2}$$

Green's space K is the space of all K-functions ϕ for which $(\phi,\phi)_K < \infty$ with the inner product (1.2).

We say that an $L^2(m)$-valued function $h_t, 0 \le t \le \infty$ is an H-function if $h_0 = 0$ and $h_s + T_s h_t = h_{s+t}$ for all $s,t > 0$. We put $h \in H$ if the derivative

$$h'_t = \lim_{u \to t} \frac{h_u - h_t}{u - t} \tag{1.3}$$

exists in $L^2(m)$ for all $t > 0$ and if h' belongs to K. The Dirichlet space is H with the inner product

$$(h,\tilde{h})_H = (h',\tilde{h}')_K = \int_0^\infty \int_E h'_{t/2}(x) \, \tilde{h}'_{t/2}(x) \, dt \, m(dx) \quad . \tag{1.4}$$

To every $f \in D_G$ there corresponds a $\phi \in K$ defined by the formula

$$\phi_t = T_t f$$

and to every $F \in D_A$ there corresponds an $h \in H$ such that

$$h_t = F - T_t F .$$

We have $(\phi,\phi)_K = (Gf,f)$ and $(h,h)_H = -(AF,F)$, and D_G, D_A can be identified with everywhere dense subsets of K and H respectively.

The spaces K,H are in duality with respect to the form

$$\langle\phi,h\rangle = \int_0^\infty \int_E \phi_{t/2}(x)\, h'_{t/2}(x)\, dt\, m(dx) . \qquad (1.5)$$

1.5. The case of a general symmetric Markov transition function can be reduced to two extreme cases: transient (dissipative) and recurrent (conservative)(see section 4).

In the transient case, the Dirichlet space can be represented by functions on E.

We say that a function f represents $h \in H$ if $h_{t_K} \to f$ m-a.e. for a sequence $t_k \to \infty$. It turns out that, for every $h \in H$, there exists a unique, up to m-equivalence, function f representing h. Moreover

$$(h,h)_H = \lim_{t\to 0} \phi_t(f)$$

where

$$\phi_t(f) = (2t)^{-1} \int_E \int_E m(dx)p_t(x,dy)|f(x)-f(y)|^2 + t^{-1} \int_E m(dx)(1-p_t(x,E))f(x)^2 . \qquad (1.6)$$

Hence the Dirichlet space can be realized as a space $\mathcal{H}$ of functions on E considered up to m-equivalence with the inner product

$$(f,f)_{\mathcal{H}} = \lim_{t\to 0} \phi_t(f) . \qquad (1.7)$$

(Silverstein and Fukushima call $\mathcal{H}$ the extended Dirichlet space.)

Green's space can be realized as the space $\mathcal{K}$ of all continuous linear functionals on $\mathcal{H}$: if $k \in \mathcal{K}$ corresponds to $\phi \in K$, then, in a certain sense,

$$k = \lim_{\varepsilon \to 0} \phi_\varepsilon .$$

2. Operators in $L^2(m)$ associated with a symmetric transition function

2.1.A. We denote by R^+ the open positive half-line $(0,\infty)$ and by $\mathcal{B}_{R^+}$ the Borel σ-algebra in R^+.

A (stationary) *transition function* in a measurable space $(E,\mathcal{B})$ is a function $p_t(x,B)$, $t > 0$, $x \in E$, $B \in \mathcal{B}$ with the following properties:

2.1.A. For every $t,x,p_t(x,\cdot)$ is a measure on $\mathcal{B}$.

2.1.B. For every $B \in \mathcal{B}$, $p_t(x,B)$ is $\mathcal{B}_{R^+} \times \mathcal{B}$-measurable in t,x.

2.1.C. $p_t(x,E) \leq 1$ for all t,x and it tends to 1 as $t \to 0$.

2.1.D. For all s,t,x,B

$$\int_E p_s(x,dy)p_t(y,B) = p_{s+t}(x,B)$$

The corresponding operators T_t act on functions by the formula (1.1).

Two important examples of transition functions in the Euclidean space R^n are

$$p_t(x,dy) = (2\pi t)^{-n/2} \exp\left(\frac{-|y-x|^2}{2t}\right) dy \tag{2.1}$$

corresponding to the Brownian motion (BM) and

$$p_t(x,dy) = (2\pi(1-e^{-t}))^{-n/2} \exp\left(\frac{-|y-e^{-t/2}x|^2}{2(1-e^{-t})}\right) dy \tag{2.2}$$

corresponding to the Ornstein-Uhlenbeck process (OUP). The corresponding operators T_t are given by the formulas

$$T_t f(x) = \int f(x + \sqrt{t}\, z)\gamma(dz) \quad \text{for BM} \tag{2.3}$$

$$T_t f(x) = \int f(e^{-t/2}x + \sqrt{1-e^{-t}}\, z)\gamma(dz) \quad \text{for OUP} \tag{2.4}$$

where

$$\gamma(dz) = (2\pi z)^{-n/2} \exp(-|z|^2/2)\, dz . \tag{2.5}$$

2.2 Let m be a σ-finite measure on $(E, \mathcal{B})$. A transition function p is called *symmetric relative to* m if

$$\int_A m(dx)p_t(x,B) = \int_B m(dx)p_t(x,A) \quad \text{for all } A,B \in \mathcal{B} . \tag{2.6}$$

The transition function (2.1) is symmetric relative to the Lebesgue measure $m(dx) = dx$ and the transition function (2.2) is symmetric relative to the measure γ given by (2.5).

It follows from (2.6) that for every $f \geq 0$

$$m(T_t f) \leq m(f) \quad \text{and} \quad m(T_t f) \to m(f) \quad \text{as} \quad t \to 0 \tag{2.7}$$

(i.e., m is an excessive measure). We put $(f,g) = m(fg)$ and $\|f\| = (f,f)^{1/2}$.

Proposition 2.1. *Operators* T_t *act in* $L^2(m)$ *and have the following properties:*

2.2.A. $\|T_t f\| \leq \|f\|$.

2.2.B. $T_s T_t = T_{s+t}$.

2.2.C. $(T_t f,g) = (f,T_t g)$

2.2.D. $\|(T_t f,f) = \|T_{t/2}f\|^2$ *is a positive monotone decreasing function of* t.

Proof. By the Schwarz inequality, for every t,x, $(T_t f(x))^2 \leq T_t(f^2)(x)$. Integrating relative to m and using (2.7), we get 2.2.A. It follows from 2.2.A that T_t preserves m-equivalence of functions and that it acts in $L^2(m)$. 2.2.B follows from 2.1.D, 2.2.C follows from (2.6), and 2.2.D is an implication of 2.2.A,B,C.

Proposition 2.2. *For every* $f \in L^2(m)$, *the function*

$$\phi_t = T_t f, \; 0 < t < \infty \tag{2.8}$$

is strongly continuous and has limits in $L^2(m)$ *as* t *tends to* 0 *and to* ∞. *The operators*

$$T_0 f = \lim_{t\to 0} T_t f, \quad T_\infty f = \lim_{t\to\infty} T_t f \tag{2.9}$$

satisfy the relations

$$T_0 T_t = T_t T_0 = T_t, \; T_\infty T_t = T_t T_\infty = T \quad . \tag{2.10}$$

T_0 *is the orthogonal projection on the subspace* $L_0 = \{f : T_0 f = f\}$ *and* T_∞ *is the orthogonal projection on the subspace* $L_\infty = \{f : T_\infty f = f\}$.

Proof. By 2.2.D the function $a_t = \| f_t \|^2$ has one-sided limits at every $t \in [0,\infty]$. It follows from the equation

$$\| f_u - f_s \|^2 = a_u + a_s - a_{(u+s)/2}$$

that f_t has one-sided strong limits in $L^2(m)$ at each $t \in [0,\infty]$. By 2.2.A $\| f_{t+} - f_{t-} \| \le \| f_{s+} - f_{s-} \|$ for $s < t$. Since f_{s+} and f_{s-} tend to f_{0+} as $s \to 0$, $f_{t+} = f_{t-}$ for all t $(0,\infty)$. The rest of Proposition 2 is obvious.

For both BM and OUP, $L_0 = L^2(m); L_\infty$ contains only 0 for BM and consists of constants for OUP.

2.3. *Proposition* 2.3. *Formulas*

$$G_t f(x) = \int_0^t T_u f(x)\,du, \tag{2.11}$$

$$A_s f(x) = \frac{T_s f(x) - f(x)}{s} \tag{2.12}$$

define bounded self-adjoint linear operators in $L^2(m)$ *with the following properties:*

$$T_s G_t f = G_t T_s f = \int_s^{t+s} T_v f \, dv = G_{t+s} f - G_s f \; ; \tag{2.13}$$

$$\begin{aligned} s\, A_s G_t f = s G_t A_s f &= \int_s^{t+s} T_u f \, du - \int_0^s T_u f \, du \\ &= G_{s+t} f - G_s f - G_t f \; . \end{aligned} \tag{2.14}$$

Proof. By 2.1.B, $T_u f(x)$ is measurable in u, x. Using Fubini's theorem, 2.2.C,B and A, we get that $\| \int_0^t T_u |f| du \|^2 \le \| f \|^2 t^2$.

Hence the integral (2.11) converges absolutely and defines an element of $L^2(m)$. Similar arguments plus change of variables in integrals are sufficient to prove (2.13) and (2.14).

2.4. We put $f \in \hat{L}$ if $f \in L_0$ and $T_\infty f = 0$. *Green's operator* is defined by the formula

$$Gf = \lim_{t\to\infty} G_t f \tag{2.15}$$

in the domain D_G which consists of all $f \in \hat{L}$ such that the limit (2.15) exists in $L^2(m)$. Analogously we define *the infinitesimal operator*

$$Af = \lim_{s\to 0} A_s f \ . \tag{2.16}$$

in the domain $D_A = \{f : f \in \hat{L}, \text{limit (2.16) exists in } L^2(m)\}$.

All functions $f \in C^2$ with compact support belong to D_A for BM and for OUP and

$$Af(x) = \frac{1}{2}\Delta f(x) \text{ for BM; } Af(x) = \frac{1}{2}(\Delta f(x) - x\cdot\nabla f(x)) \text{ for OUP,} \tag{2.17}$$

where Δ is the Laplacian and ∇ is the gradient. Hence

$$\begin{aligned} -(Af,f) &= \frac{1}{2}\int |\nabla f(x)|^2 dx \quad \text{for BM,} \\ -(Af,f) &= \frac{1}{2}\int |\nabla f(x)|^2 \gamma(dx) \quad \text{for OUP.} \end{aligned} \tag{2.18}$$

By passing to the limit in (2.14) we prove the following

Proposition 2.4. *If* $f \in D_A$, *then*

$$G_t Af = tA_t f, \quad Af \in D_G \quad \textit{and} \quad -GAf = f \ . \tag{2.19}$$

If $f \in D_G$, *then*

$$-sA_s Gf = G_s f, \ Gf \in D_A \quad \textit{and} \quad -AGf = f \ . \tag{2.20}$$

If $f \in \hat{L}$, *then*

$$G_t f \in D_A \qquad AG_t f = tA_t f \ ; \tag{2.21}$$

$$A_s f \in D_G \qquad sGA_s f = -G_s f \tag{2.22}$$

Proposition 5. If $f \in D_A$, *then* $T_t f \in D_A$ *and*

$$\frac{dT_t f}{dt} = T_t A f = A T_t f \tag{2.23}$$

(the derivative in $L^2(m)$*).*

This follows from the relations $\delta^{-1}(T_t f - T_{t+\delta} f) = T_t[\delta^{-1}(f - T_\delta f)] = \delta^{-1}(T_t f - T_\delta T_t f)$ and 2.2.A.

3. The spaces K and H.

3.1. A real-valued function $\phi_t(x)$, $t > 0$, $x \in E$ is called a *K-function* if $m(\phi_t^2) < \infty$ for all t and $T_s \phi_t(x) = \phi_{s+t}(x)$ for all s,t,x.

Proposition 3.1. If ϕ *is a K-function, then:*

3.1.A. ϕ is $\mathcal{B}_{R^+} \times \mathcal{B}$*-measurable in* t,x.

3.1.B. $\|\phi_t\|$ *decreases.*

3.1.C. ϕ_t *is continuous in* $L^2(m)$ *and tends to a limit* ϕ_∞ as $t \to \infty$.

3.1.D. $\phi_0 = \lim_{t \to 0} \phi_t$ *exists if and only if* $\|\phi_t\|$ *is bounded.*

3.1.E. *Formulas*

$$\phi_t = T_t f , \tag{3.1}$$

$$f = \lim_{t \to 0} \phi_t \tag{3.2}$$

establish a 1-1 correspondence between $\hat{\mathcal{L}}$ *and a set of K-functions.*

All these properties follow easily from Propositions 2.1 and 2.2.

3.2. *Green's space* K consists of all K-functions ϕ such that $\int_{R^+} \|\phi_t\|^2 dt < \infty$ (two functions ϕ and ψ are identified if $\phi_t = \psi_t$ m-a.e. for each t).

Proposition 3.2. The set K *with the inner product* (1.2) *is a Hilbert space.*

If $f \in D_G$, *then the K-function* (3.1) *belongs to* K *and*

$(\phi,\phi)_K = (Gf,f)$.

For every $\phi \in K$ *for all* $u > s > 0$, $\phi_s - \phi_u \in D_G$ *and*

$$\phi = \lim_{s \to 0, u \to \infty} \psi^{s,u} \quad \textit{in} \quad K \tag{3.3}$$

where $\psi^{s,u}$ *corresponds to* $\phi_s - \phi_u$ *by formula* (3.1).

Proof. We consider the product measure $\nu(dt,dx) = dt\, m(dx)$ on $R^+ \times E$ and we observe that $K \subset L^2(\nu)$. Suppose that $\phi^n \in K$ converge in $L^2(\nu)$ to $\psi \in L^2(\nu)$. We have $\|\phi^n - \phi^m\|^2_{L^2(\nu)} \geq t\|\phi^n_t - \phi^m_t\|$ for every t by 3.1.B. Hence $\phi^n_t \to \psi_t$ in $L^2(m)$ for every t, and $T_s\psi_t = \psi_{s+t}$ m-a.e. for all s,t. It is easy to construct a K-function ϕ such that $\phi_t = \psi_t$ m-a.e. for every t. Hence K is closed in $L^2(\nu)$ and the first statement of Proposition 3.2 is proved. The second statement is obvious.

Let us prove the third statement. If $\phi \in K$, then $\phi_s \in \hat{L}$ for every s. Indeed $T_t\phi_s = \phi_{s+t} \to \phi_s$ as $t \to 0$ by 3.1.C and $T_t\phi_s \to 0$ as $t \to \infty$ by 3.1.B since $\|\phi_{s+t}\|^2$ is integrable. For every $0 < s < u$,

$$\phi_s - \phi_u = \phi_s - T_{t-s}\phi_s = -(u-s)A_{u-s}\phi_s \in D_G$$

by (2.22). We have

$$\psi^{s,u}_t = T_t(\phi_s - \phi_u) = \phi_{s+t} - \phi_{u+t}$$

and

$$\|\psi^{s,u} - \phi\|^2_K = 2\int_0^\infty \|\phi_{s+t} - \phi_{u+t} - \phi_t\|^2 dt$$

By 3.1.B,C and the dominated convergence theorem, this expression tends to 0 as $s \to 0$, $u \to \infty$.

3.3. A real-valued function $h_{s,u}(x), 0 \leq s < u, x \in E$ is called an H-*function* if $m(h^2_{s,u}) < \infty$ for all s, u, $h_{s,t} + h_{t,u} = h_{s,u}$ for all $s < t < u$ and all x and $T_t h_{s,u} = h_{s+t,u+t}$ for all s,u,t,x. (This is equivalent to the definition of an H-function given in Subsection 1.4. An

1-1 correspondence between h_t and $h_{s,u}$ is established by the formulas $h_{s,u} = h_s - h_u$, $h_u = h_{0,u}$.) We say that h is differentiable if the limit

$$h'_t = \lim_{s\uparrow t, u\downarrow t} \frac{h_{s,u}}{u-s}$$

(which is equal to the limit (1.3)) exists in $L^2(m)$ for all $t \in R^+$. Obviously h' is a K-function.

We put $h \in H$ if h is a differentiable H-function and if $h' \in K$. The inner product in H is given by the formula (1.4).

Theorem 3.1. *We have the following commutative diagrams*

$$\begin{array}{ccc} D_G & \xrightarrow{\delta} & K \\ G\downarrow & & \downarrow\alpha \\ D_A & \xrightarrow{\beta} & H \end{array} \qquad\qquad \begin{array}{ccc} D_G & \xrightarrow{\delta} & K \\ -A\uparrow & & \uparrow\alpha \\ D_A & \xrightarrow{\beta} & H \end{array} \tag{3.4}$$

where the mappings G, A, δ *are defined by the formulas* (2.15),(2.16) *and* (3.1) *and*

$$\beta \ : \ h_{s,u} = T_s f - T_u f \ ,$$

$$\alpha \ : \ \phi_t = h'_t \ ,$$

$$\gamma \ : \ h_{s,u} = \int_s^u \phi_t \, dt \ .$$

All these mappings are isometries. $-A$ *is the inverse of* G, α *is the inverse of* γ. *For every* $\phi \in K$, $\phi_s - \phi_u \in D_G$ *and*

$$\phi = \lim_{s\to 0, u\to\infty} \delta(\phi_s - \phi_u) \quad \textit{in} \quad K. \tag{3.5}$$

For every $h \in H$, $h_{s,u} \in D_A$ *and*

$$h = \lim_{s\to 0, u\to\infty} \beta(h_{s,u}) \quad \textit{in} \quad H\ . \tag{3.6}$$

For every $h \in H$, $\phi \in K$,

$$(\alpha h,\phi)_K = (h,\gamma\phi)_H = \langle h,\phi\rangle \qquad (3.7)$$

where $\langle h,\phi\rangle$ *is defined by* (1.5).

Proof. The statement on A and G follows from (2.19) and (2.20). The statements on α and γ are obvious. Formula (3.5) is equivalent to (3.3).

To prove that the second diagram (3.4) is commutative, we put $\phi = -\delta Af$, $\psi = \alpha\beta f$. We have $\phi_t = -T_tAf$, $\psi_t = h'_t$ where $h_{s,u} = T_sf - T_uf$. By (2.23),

$$\psi_t = -T_tAf = \phi_t.$$

Since the second diagram (3.4) is commutative, so is the first one.

To prove (3.6) we note that, if $h = \gamma\phi$, then

$$G(\phi_s-\phi_u) = \int_s^u \phi_t dt = h_{s,u}.$$

Hence $h_{s,u} \in D_A$ and

$$\beta h_{s,u} = \beta G(\phi_s-\phi_u) = \gamma\delta(\phi_s-\phi_u) \to \gamma\phi = h \text{ in } H.$$

3.4. We denote by K^+ the set of all $\phi \in K$ such that $\phi_t \geq 0$ m-a.e. and by H^+ the set of all $h \in H$ such that $h_{s,u} \geq 0$ m-a.e. Obviously $H^+ = \gamma(K^+)$ and $K^+ = \alpha(H^+)$.

3.5. Operators T_t act on K and H by formulas

$$(T_t\phi)_s = \phi_{t+s}\ ; \quad (T_th)_{s,u} = h_{s+t,u+t}.$$

It is easy to see that T_t commute with α, β, γ, δ and they form contraction self-adjoint semi-groups in K and H. Moreover, for every $\phi \in K$,

$$\lim_{t\to 0} T_t\phi = \phi \text{ in } K \qquad (3.8)$$

and, for every $h \in H$,

$$\lim_{t\to 0} T_th = h \text{ in } H. \qquad (3.9)$$

Using this, we prove that for every $f \in D_A$

$$-(A_s f,f) \leq -(Af,f) \tag{3.10}$$

Indeed, let $h = \beta f$. By (2.23) and 2.2.C, $(AT_u f, T_u f) = (T_{2u} Af, f)$. Hence, $-(T_t Af,f) = \|T_{t/2}h\|_H^2 \leq \|h\|_H^2 = -(Af,f)$. We get (3.10) by integrating over $(0,s)$ and taking into account (2.19).

4. Conservative and dissipative semi-groups

4.1. For every measurable function $f \geq 0$, we put

$$\|f\|_G = \left(\int_0^\infty (T_t f,f)dt\right)^{1/2} .$$

We note that $\|f\|_G = \|\phi\|_k$ where ϕ is a K-function defined by (3.1). Hence $\|f+\tilde{f}\|_G \leq \|f\|_G + \|\tilde{f}\|_G$.

Obviously $\|f\|_G = 0$ if $f = 0$ m-a.e. We say that T_t is *conservative* if there exist no other functions $f \geq 0$ with $\|f\|_G < \infty$. We say that T_t is *dissipative* if $\|f\|_G < \infty$ for some strictly positive function f. The general situation can be reduced to these two extreme cases. The proof is based on the following lemma.

Lemma 4.1. *Let* m *be a* σ*-finite measure on* $(E,\mathcal{B})$ *and let* $\tilde{\mathcal{B}} \subset \mathcal{B}$ *be closed under countable union. Then* $\tilde{\mathcal{B}}$ *contains an element* C *which is maximal* mod m *that is* $m(B \cap (E\setminus C)) = 0$ *for every* $B \in \tilde{\mathcal{B}}$.

Proof. There exists a finite measure equivalent to m and therefore we can assume without any loss of generality that m is finite. Let a be the supremum of $m(B)$ over all $B \in \tilde{\mathcal{B}}$. For every n there exists $B_n \in \tilde{\mathcal{B}}$ such that $m(B_n) > a-1/n$. The element $C = B_1 \cup \ldots \cup B_n \cup \ldots$ has the property described in our lemma.

4.2. *Theorem* 4.1. *Let* p *be a symmetric transition function on* $(E,\mathcal{B})$. *There exists a partition* $E=E_c \cup E_d$ *of* E *into disjoint parts* $E_c, E_d \in \mathcal{B}$ *such that:*

(i) $p_t(x,E_d) = 0$ *m-a.e. on* E_c ;

(ii) $p_t(x,E_c) = 0$ *m-a.e. on* E_d ;

(iii) T_t *is conservative on* E_c *and dissipative on* E_d.

Proof. We put $B \in \tilde{\mathcal{B}}$ if there exists $f \geq 0$ such that $\|f\|_G < \infty$ and $B = \{f > 0\}$. We note that $\tilde{\mathcal{B}}$ contains all sets B for which $m(B) = 0$. Without changing the set $\{f > 0\}$, we can modify f in such a way to satisfy the conditions $\|f\|_G \leq 1$ and $\|f\| \leq 1$.

Let $B_n = \{f_n > 0\}$ with $\|f_n\|_G \le 1$ and $\|f_n\| \le 1$. Then $\bigcup_1^\infty B_n = \{f > 0\}$ where $f = \Sigma 2^{-n} f_n$ satisfies the analogous conditions $\|f\|_G \le 1$, $\|f\| \le 1$. Hence $\tilde{\mathcal{B}}$ is closed under countable union. By Lemma 4.1, $\tilde{\mathcal{B}}$ contains an element $E_d = \{\rho > 0\}$ which is maximal mod m. Here $\rho \ge 0$, $\|\rho\|_G < \infty$. Put $E_c = E \setminus E_d$.

We have $f = 0$ m a.e. on E_c if $f \ge 0$, $\|f\|_G < \infty$.

Since $\|T_t \rho\|_G \le \|\rho\|_G < \infty$, we have

$$\int_E p_t(x,dy)\rho(y) = 0 \quad \text{m-a.e. on } E_c$$

which implies (i).

We can choose $q \in L^2(m)$ such that $q > 0$ on E_c and $q = 0$ on E_d. By symmetry, $(T_t q, \rho) = (q, T_t \rho) = 0$. Hence $T_t q = 0$ m-a.e. on E_d and we get (ii). Property (iii) is obvious.

4.3. We list some properties of conservative and dissipative semi-groups.

4.3.A. T_t is conservative if and only if $K^+ = H^+ = \{0\}$.

4.3.B. Suppose that $p_t(x,dy) = p_t(x,y)m(dy)$ with a $\mathcal{B}_{R^+} \times \mathcal{B} \times \mathcal{B}$-measurable $p_t(x,y)$. Then T_t is dissipative if and only if

$$g(x,y) = \int_0^\infty p_t(x,y)\,dt$$

is finite for $m \times m$ - almost all x,y.

4.3.C. The following conditions are equivalent:

(i) for every t, $p_t(x,E) = 1$ m-a.e.;

(ii) $mT_t = m$ for all t.

Each of them implies that T_t is conservative if $m(E) < \infty$.

4.3.D. OUP is conservative and BM in R^n is conservative for $n = 1,2$ and is dissipative for $n > 2$.

4.3.E. If T_t is dissipative, then $T_\infty f = 0$ m-a.e. for every $f \in L^2(m)$ and $\phi_\infty = 0$ m-a.e. for every $\phi \in K$.

5. The spaces H and K.

5.1. *In this section we consider only dissipative semi-groups* T_t.

Lemma 5.1. *If* $f \in D_A$ *and* $T_\infty |f| = 0$, *then for every* $\rho \ge 0$

$$m^{\rho}|f| \le c_{\rho}(-Af,f)^{1/2} \tag{5.1}$$

where $c_{\rho} = 2\|\rho\|_G$ *and* $m^{\rho}(dx) = \rho(x)m(dx)$.

Proof. 1°. $|f| \in \hat{L}$.

Indeed $\|f\| \ge \|T_s|f|\| \ge \|T_s f\|$. Hence $\|T_s|f|\| \to \||f|\|$ as $s \to 0$ and $\|T_s|f| - |f|\|^2 \to 0$.

2°. If $F^s = s^{-1}G_s|f|$, then $-(AF^s,F^s) \le -2(Af,f)$.

Indeed, by (2.22) $-sAf^s = -sA_s|f| = |f| - T_s|f|$. Hence $-(sAF^s,T_u|f|) = (|f| - T_s|f|,T_u|f|) \le (f,f) - (T_{s+u}f,f) = -(s+u)(A_{s+u}f,f)$ and, by (3.10),

$$-(Af^s,F^s) \le -s^{-1}\int_0^s \frac{s+u}{s}(A_{s+u}f,f)du \le -2(Af,f).$$

3 . Ny the Schwarz inequality and 2°,

$$(F^s,\rho) = -(GAF^s,\rho) \le (GAF^s,AF^s)^{1/2}\|\rho\|_G = (F^s,-AF^s)^{1/2}\|\rho\|_G$$
$$\le c_{\rho}(-Af,f)^{1/2}. \tag{5.3}$$

By 1° $F^s \to |f|$ in $L^2(m)$ as $s \to 0$, and (5.1) follows from (5.3) by Fatou's lemma.

5.2. We say that *a function* f *represents an element* h *of* H if there exist sequences $s_n \to 0$ and $u_n \to \infty$ such that $h_{s_n,u_n} \to f$ m-a.e. This implies $h_{0,u_n} \to f$ m-a.e. Indeed, by Theorem 3.1, $h_{0,s} = \int_0^s h'_t\,dt$ and, by the Schwarz inequality $|h_{0,s}| \le \|h'\|_K\, s$.

Let $\|\rho\|_G < \infty$. Since $L^1(m^{\rho})$ is a complete space, it follows from (3.6) and (5.1) that there exists, for every $h \in H$, a function f such that

$$m^{\rho}|f-h_{s,u}| \to 0 \quad \text{as} \quad s \to 0,\ u \to \infty.$$

This implies: for every sequences $s_n \to 0$, $u_n \to \infty$, there exist subsequences s_{n_k}, u_{n_k} such that $h_{s_{n_k},u_{n_k}} \to f$ m^{ρ}-a.e. Hence if $\rho > 0$, then f represents h and is the only representative of h.

We note that:

5.2.A. If f represents h, then $h_{s,u} = T_s f - T_u f$ for all $0 \le s < u$.

5.2.B. An element f of D_A represents $h = \beta f$.

To prove 5.2.A, we fix a function $\rho > 0$ such that $\|\rho\|_G < \infty$ and we observe that $\|T_s\rho\|_G \le \|\rho\|_G < \infty$. Hence $(|f - h_{0,u}|, T_s\rho) \to 0$ as $u \to \infty$. But $(|T_s f - h_{s,u+s}|, \rho) \le (|f - h_{0,u}|, T_s\rho)$. Therefore $h_{s,u+s} \to T_s f$ in $L^1(m^\rho)$. On the other hand $h_{0,u+s} \to f$ in $L^1(m^\rho)$. We conclude that $f - T_s f = \lim(h_{0,u+s} - h_{s,u+s}) = h_{0,s}$.

5.2.B follows from 5.2.A and the definition of β.

5.3. Denote by $\mathcal{H}$ the set of all functions f which represent elements of h (we do not distinguish m-equivalent functions). Consider an inner product in $\mathcal{H}$ defined by the formula: $(f,\tilde{f})_{\mathcal{H}} = (h,\tilde{h})_H$ if f represents h and $\tilde{f}$ represents $\tilde{h}$. Put $\|f\|_A = (-Af,f)^{1/2}$, $\|f\|_t = (-A_t f,f)^{1/2}$. Note that formula (1.6) is meaningful for every $\mathcal{B}$-measurable function f and that

$$\Phi_t(f) = \|f\|_t^2 \quad \text{for all} \quad f \in L^2(m) . \tag{5.5}$$

Theorem 5.1. *If* $f \in \mathcal{H}$, *then*

$$\|f\|_{\mathcal{H}}^2 = \lim_{t\to 0} \Phi_t(f) . \tag{5.6}$$

Proof. There exists an element h of H and sequences $s_n \to 0$, $u_n \to \infty$ such that $f_n - h_{s_n,u_n} \to f$ m-a.e. By (3.6)

$$\varepsilon_n^2 = \lim_{m\to\infty}\|f_m - f_n\|_A = \|h - \beta f_n\|_H \to 0 \quad \text{as} \quad n \to \infty \tag{5.7}$$

and

$$\|f_n\|_A = \|\beta f_n\|_H \to \|h\|_H = \|f\|_{\mathcal{H}} . \tag{5.8}$$

By Fatou's lemma, (5.5) and (3.10),

$$\Phi_t(f - f_n) \le \liminf \Phi_t(f_m - f_n) \le \varepsilon_n^2 .$$

Since Φ_t is a positive semi-definite quadratic form,

$$|\Phi_t(f)^{1/2} - \Phi_t(f_n)^{1/2}| \leq \Phi_t(f-f_n)^{1/2} \leq \varepsilon_n . \tag{5.9}$$

Since $f_n \in D_A$, we have $A_t f_n \to Af_n$ in $L^2(m)$ and $\Phi_t(f)^{1/2} = \|f_n\|_t \to \|f_n\|_A$ as $t \to 0$. By (5.9) every limit point of $\Phi_t(f)^{1/2}$ as $t \to 0$ lies between $\|f_n\|_A - \varepsilon_n$ and $\|f_n\|_A + \varepsilon_n$ for all n. Hence (5.6) follows from (5.7) and (5.8).

Remark 1. It follows from (5.9) and (5.7) that

$$\Phi_t(f) = \lim_{n\to\infty} \Phi_t(f_n). \tag{5.10}$$

Remark 2. We have

$$\Phi_t(f_n) = t^{-1} \int_0^t \|T_{u/2} f_n\|_A^2 du$$

(cf. the proof of (3.10)). Since T_t is a contraction in H, $\Phi_t(f_n)$ is a decreasing function of t and, by (5.10), so is $\Phi_t(f)$. Hence

$$\Phi_t(f) \leq \|f\| \quad \text{for } f \in \mathcal{H}. \tag{5.11}$$

5.4. It follows from 5.2.B that D_A is everywhere dense in $\mathcal{H}$ and that $\|f\|_{\mathcal{H}} = \|f\|_A$ for $f \in D_A$. By (5.4) and (5.1)

$$m^\rho |f| \leq c_\rho \|f\| \quad \text{for all } f \in \mathcal{H} . \tag{5.12}$$

Hence if $f_n \to f$ in $\mathcal{H}$, then a subsequence f_{n_k} converges to f m-a.e.

Lemma 5.2. *If a sequence* f_n *bounded in* $\mathcal{H}$ *converges m-a.e. to a function* f, *then* $f \in \mathcal{H}$.

Proof. It is well-known (see e.g. [4], Section 38) that if f_n is a bounded sequence in an arbitrary Hilbert space $\mathcal{H}$, then there exists a subsequence f_{n_k} such that $F_k = n_k^{-1}(f_{n_1} + \ldots + f_{n_k})$ converges in $\mathcal{H}$. In our case the relation $F_k \to F$ in $\mathcal{H}$ implies that a subsequence F_{k_j} converges to F m-a.e. Since $f_n \to f$ m-a.e., we have $F_{k_j} \to f$ m-a.e. and

$f = F$ m-a.e. Therefore $f \in H$.

Corollary. A function f *belongs to* H *if and only if there exists a sequence* $f_n \in D_A$ *such that* $\sup\|f_n\|_A < \infty$ *and* $f_n \to f$ *m-a.e.*

5.5. *Theorem* 5.2. *A function* $f \in L^2(m)$ *belongs to* H *if and only if* $\Phi_t(f)$ *is bounded.*

Proof. The "only if" part follows from Theorem 5.1. Suppose that $\Phi_s(f) \le c$ for all s. By (5.5), $(f,f)-(T_s f,f) = s\Phi_s(f) \le cs$. Cons quently $(f-T_s f, f-T_s f) \to 0$ as $s \to 0$. Therefore $F^s = s^{-1}G_s f \to f$ in $L^2(m)$ as $s \to 0$ and $F^{s_k} \to f$ m-a.e. for some $s_k \to 0$.

We have

$$-(AF^s,F^s) \le -s^{-1}\int_0^s \frac{s+u}{u}(A_{s+u}f,f)du$$

cf. (5.2)). Hence $\|F^s\|_H^2 = \|F^s\|_A^2 \le 2c$ and $f \in H$ by Lemma 5.2.

5.6. We list some properties of the functional Φ_t. We denote by $a \wedge b$ the smallest of two numbers a and b and we put $x^+ = 2^{-1}(x+|x|)$.

5.6.A. $\Phi_t(f\wedge g) \le \Phi_t(f) + \Phi_t(g)$.

This follows from inequalities $(a\wedge b - a'\wedge b')^2 \le (a-a')^2+(b-b')^2$, $(a\wedge a')^2 \le a^2+(a')^2$ which hold for all real numbers.

We say that $\tilde{f}$ is *a normal contraction* of f if

$$|\tilde{f}(x)| \le |f(x)| \quad \text{and} \quad |\tilde{f}(x)-\tilde{f}(y)| \le |f(x)-f(y)| \quad \text{for all } x,y.$$

5.6.B. If $\tilde{f}$ is a normal contraction of f, then $\Phi_t(\tilde{f}) \le \Phi_t(f)$.

This is an immediate implication of the expression (1.6).

5.6.C. $\Phi_t|f| \le \Phi_t(f)$ and $\Phi_t(f^+) \le \Phi_t(f)$.

Indeed $|f|$ and f^+ are normal contraction of f.

5.6.D. If $g \ge 0$ m-a.e., then $\Phi_t(f\wedge g) \le \Phi_t(f)$.

This is because $f\wedge g$ is a normal contraction of f.

Theorem 5.3. *If* $f \in H$ *and if* $\tilde{f}$ *is a normal contraction of* f, *then* $\tilde{f} \in H$ *and* $\|\tilde{f}\|_H \le \|f\|_H$.

Proof. We consider a sequence $f_n \in D_A$ such that $\|f_n\|_H \le c < \infty$ for all n and $f_n \to f$ m-a.e. Put $\tilde{f}_n = (|f_n|\wedge f)^+$. By 5.6.A,B,C, $\Phi_t(\tilde{f}_n) \le 2c$ for all t and n. Besides $\tilde{f}_n \in L^2(m)$. By Theorem 5.2, $\tilde{f}_n \in H$. By Theorem 5.1, $\|\tilde{f}_n\|_H \le 2c$ and, by Lemma 5.2, $f^+ = \lim \tilde{f}_n$ belongs to H. Analogously $(-f)^+ \in H$. Hence $\tilde{f} = \tilde{f}^+ - (\tilde{-f})^+ \in H$. The inequality $\|\tilde{f}\|_H \le \|f\|_H$ follows from 5.6.B and Theorem 5.1.

Corollary 1. *If* $f \in H$, *then* $|f| \in H$ *and* $\||f|\|_H \le \|f\|_H$.

Corollary 2. *If* $f \in H$, *and* $\rho \ge 0$, *then*

$$m^\rho |f| \le \|f\|_H \|\rho\|_G . \tag{5.13}$$

(This is an improved version of the estimate (5.12).)

To prove (5.13), we choose $f_n \in D_A$ such that $f_n \to |f|$ in H. We have $(f_n,\rho) = -(GAf_n,\rho) \le \|-Af_n\|_G\|\rho\|_G$. This implies (5.13) because $\|-Af_n\|_G = \|f_n\|_H \to \||f|\|_H \le \|f\|_H$ and $f_n \to |f|$ in $L^1(m^\rho)$ by (5.12).

5.7. We say that f is *an almost excessive function* if

$$f \ge 0 \text{ m-a.e. and } T_t f \le f \text{ m-a.e. for all } t. \tag{5.14}$$

Lemma 5.3. *If* f *and* g *are almost excessive functions and if* $f \le g$ m-a.e. *then* $\Phi_t(f) \le \Phi_t(g)$.

Proof. For an almost excessive function f,

$$\Phi_t(f) = (f-T_t f, f)t^{-1} \tag{5.15}$$

(cf. (5.6)). Hence $\Phi_t(f) \le (f-T_t f,g) = (f,g-T_t g) \le \Phi_t(g)$.

We denote by H^+ the set of all functions representing elements of H^+.

Theorem 5.4. *A function* f *belongs to* H^+ *if and only if it is almost excessive and* $\Phi_t(f)$ *is bounded.*

Proof. Suppose f represents $h \in H^+$. Then obviously $f \ge 0$ m-a.e. and f is almost excessive by 5.2.A. $\Phi_t(f)$ is bounded by Theorem 5.1.

Now let f be almost excessive and let $\Phi_t(f) \le c$ for all t. Take $\rho > 0$ such that $\|\rho\|_G < \infty$. The function $q = G\rho$ is almost excessive and belongs to H. All functions $f_n = (nq)\wedge f$ are almost excessive. Since f_n is a normal contraction of nq, $f_n \in H$ by Theorem 5.3. By 5.6.D, $\Phi_t(f_n) \le \Phi_t(f) \le c$. By Theorem 5.1, $\|f_n\|_H \le c$, and f belongs to H by Lemma 5.2. It belongs to H^+ by 5.2.A.

Lemma 5.4. *Put* $f \in H'$ *if* $f \in H$ *and* $f \ge 0$ m-a.e. *A function* $f \in H'$ *belongs to* H^+ *if and only if*

$$(f,\tilde{f})_{\mathcal{H}} \geq 0 \quad \textit{for all} \quad f \in \mathcal{H}' . \tag{5.16}$$

Proof. Let $f \in \mathcal{H}^+$. By (5.15) and (5.6), $(f-T_t f,\tilde{f})t^{-1} \to (f,\tilde{f})_{\mathcal{H}}$ as $t \to 0$ for all $\tilde{f} \in \mathcal{H}'$. Thus (5.16) holds.

Let $\rho \geq 0$ and $||\rho||_G < \infty$. Then $G\rho$ and $GT_t\rho$ belong to $\mathcal{H}$. Hence $\tilde{f} = G\rho - GT_t\rho \in \mathcal{H}$. Suppose that (5.16) holds. Using (5.6), we show that $(f,G\rho)_{\mathcal{H}} = (f,\rho)$ for all $f \in \mathcal{H}$. Therefore $(f-T_t f,\rho) = (f,\rho-T_t\rho) = (f,G\rho) - (f,GT_t\rho)_{\mathcal{H}} = (f,\tilde{f})_{\mathcal{H}} \geq 0$ and $f \in \mathcal{H}^+$ by Theorem 5.4.

5.8. Let $\mathcal{K}$ stand for the space of all (continuous) linear functionals on $\mathcal{H}$ and let $\mathcal{K}^+$ denote the set of positive functionals, i.e. set of $k \in \mathcal{K}$ such that $k(f) \geq 0$ if $f \geq 0$ m-a.e. Using the duality $\langle h,\phi\rangle$, we identify $\mathcal{K}$ with K: the functional corresponding to $\phi \in K$ is given by the formula $k(f) = \langle h,\phi\rangle$ where h is the element of H represented by f.

Theorem 5.4. *An element* k *of* $\mathcal{K}$ *belongs to* $\mathcal{K}^+$ *if and only if the corresponding element* ϕ *of* K *belongs to* K^+. *In this case, for all* $f \in H$,

$$k(f) = \lim_{\varepsilon\to 0}(\phi_\varepsilon,f) . \tag{5.17}$$

Proof. By (3.7), $k(f) = \langle h,\phi\rangle = (h,\gamma\phi)_H$ and, by Lemma 5.4, $k \in \mathcal{K}^+$ if and only if $\gamma\phi \in H^+$ which is equivalent to the condition $\phi \in K^+$.

For every $\varepsilon, t > 0$, $(\phi_{(t+\varepsilon)/2},h'_{(t+\varepsilon)/2}) = (\phi_\varepsilon,h'_t)$. Therefore

$$\int_\varepsilon^\infty (\phi_{t/2}, h'_{t/2})dt = \int_0^\infty (\phi_\varepsilon,h'_t)dt . \tag{5.18}$$

By Fubini's theorem, for every $u > s > 0$,

$$\int_s^u (\phi_\varepsilon,h'_t)dt = (\phi_\varepsilon,h_{s,u}) .$$

By (5.4), $h_{s,u} \to f$ in $L^1(m^{\phi_\varepsilon})$. Thus the right side of (5.15) is equal to (ϕ_ε,f). Now (5.17) follows from (5.18) and (1.2).

Remark. (5.17) holds for all $k \in \mathcal{K}$ if $f \in \mathcal{H} \cap L^2(m)$.

REFERENCES

1. A. Beurling and J. Deny, Dirichlet spaces, *Proc. Nat. Acad. Sci. U.S.A.*, *45* (1959), 208-215.

2. E.B. Dynkin, Markov processes and random fields, *Bull. Amer. Math. Soc.*, *3* (1980), 975-999.

3. M. Fukushima, *Dirichlet forms and Markov processes*, North-Holland, Amsterdam/Oxford/New York, 1980. Kodashira, Tokyo.

4. F. Riesz and B. Sz. Nagy, *Functional analysis*, Unger, New York, 1955.

5. M.L. Silverstein, *Symmetric Markov processes*, *Lecture Notes in Math.*, *vol. 426*, Springer-Verlag, Berlin/Heidelberg/New York, 1974.

ON A THEOREM OF KABANOV, LIPTSER AND ŠIRJAEV

G.K. Eagleson and R.F. Gundy

The "Theorem" referred to in the title is an extension of Kakutani's classical theorem on the absolute continuity/singularity dichotomy of product measures. Here we present a new proof of this extension. The new proof is more general than the original, providing, as a bonus, information about the contiguity/complete separability of two sequences of measures.

1. INTRODUCTION

Recently Kabanov, Liptser and Širjaev (1977) have published an elegant and useful generalization of Kakutani's Theorem on the absolute continuity/singularity dichotomy of product measures. Suppose that P and Q are two measures on some measure space (Ω, F) and let $\{F_n; n \geq 0\}$ be an increasing sequence of sub-σ-fields of F with $\bigvee_{n=0}^{\infty} F_n = F$. Denote the restrictions of P and Q to F_n by P_n and Q_n respectively. If $Q_n << P_n$ for all n, one is interested in conditions which ensure that $Q << P$. In this case, denote the Radon-Nikodym derivative of Q_n with respect to P_n by $L_n = dQ_n/dP_n$. For $n \geq 1$, set

$$\alpha_n = L_n \tilde{L}_{n-1},$$

where $\tilde{L}_n = L_n^{-1}$ if $L_n \neq 0$ and zero otherwise; α_n is the density of Q_n with respect to P_n, conditional on F_{n-1}. The result of Kabanov, Liptser and Širjaev, given as a Theorem for easy reference, is

__Theorem 0.__ The measure Q is absolutely continuous with respect to P if and only if

$$Q[\sum_{j=1}^{\infty} E_P(1 - \sqrt{\alpha_j} \,\|\, F_{j-1}) < \infty] = 1,$$

whereas $Q \perp P$ if and only if

$$Q\left[\sum_{j=1}^{\infty} E_P(1 - \sqrt{\alpha_j} \| F_{j-1}) = \infty\right] = 1 .$$

When P and Q are both product measures and the F_n are product σ-fields, the α_n are independent random variables and the above theorem coincides with Kakutani's classical dichotomy.

We remark that the above theorem also has a "martingale" statement:

Given a nonnegative (F_n, P) martingale L_n, $n \geq 1$, let $\alpha_n = L_n \tilde{L}_{n-1}$ where $\tilde{L}_n$ is defined above. By the martingale property, $[L_{n-1} = 0] \subset [L_n = 0]$ and by the martingale convergence theorem, $\lim L_n$ exists (almost surely), so that $\lim \alpha_n$ exists and is either one or zero. Let us measure the rapidity of this convergence by the series $A = \sum_{j=1}^{\infty} E(1 - \sqrt{\alpha_j} \| F_{j-1})$. It turns out that each term of the series is nonnegative so that A is a nonnegative random variable, possibly infinite on a set of positive probability. Let us agree to say the sequence of random variables α_n, $n \geq 0$, converges rapidly to one if A is finite. The theorem of Kabanov, Liptser and Širjaev says that, neglecting a set of probability zero, *the set where* $L_\infty > 0$ *is precisely the set where* α_n, $n \geq 0$, *converges rapidly to one. That is,*

$$[L_\infty = 0] \overset{P\text{-a.s.}}{=} \left[\sum_{j=1}^{\infty} E(1 - \sqrt{\alpha_j} \| F_{j-1}) = \infty\right]$$

and

$$[L_\infty > 0] \overset{P\text{-a.s.}}{=} \left[\sum_{j=1}^{\infty} E(1 - \sqrt{\alpha_j} \| F_{j-1}) < \infty\right] .$$

Here we present an alternative proof of the Kabanov, Liptser and Širjaev result in its "martingale" form, the search for a different proof being motivated by the desire to have a simpler and more "natural" derivation.

But how does one judge "naturalness"? In the theory of testing statistical hypotheses, Le Cam has introduced two concepts, contiguity and complete separability, which are analogues of the absolute continuity and singularity of two measures. (See Le Cam (1960) and (1977).) They are concepts which relate to sequences of measures which are not necessarily connected with each other.

Consider a sequence of measure spaces $(\Omega_n, \mathcal{A}_n)$, each endowed with two measures, $P^{(n)}$ and $Q^{(n)}$. The sequence of measures $\{Q^{(n)}\}$ is said to be *contiguous* to $\{P^{(n)}\}$ if, whenever

$$\lim_{n\to\infty} P^{(n)}(A_n) = 0$$

for some sequence of events $A_n \in \mathcal{A}_n$, then

$$\lim_{n\to\infty} Q^{(n)}(A_n) = 0$$

also. When $\Omega_n = \Omega$ for all n, the $\mathcal{A}_n$ are increasing and the measures $P^{(n)}$ and $Q^{(n)}$ are obtained from measures P and Q by restriction to $\mathcal{A}_n$, contiguity is equivalent to the absolute continuity of Q with respect to P on $\bigvee_n \mathcal{A}_n$.

On the other hand, the sequences of measures $\{P^{(n)}\}$ and $\{Q^{(n)}\}$ are said to *separate entirely* if there exist a subsequence n' and sets $A_{n'} \in \mathcal{A}_{n'}$ such that

$$\lim_{n'\to\infty} P^{(n')}(A_{n'}) = 0$$

while

$$\lim_{n'\to\infty} Q^{(n')}(A_{n'}) = 1 .$$

Again, if $P^{(n)}$ and $Q^{(n)}$ are obtained by restricting two measures P and Q on a single measure space to an increasing sequence of σ-fields, then the entire separation of $\{P^{(n)}\}$ and $\{Q^{(n)}\}$ is equivalent to the mutual singularity of P and Q.

Recently it has been shown that a number of classical results about the absolute continuity/singularity of measures have analogues in terms of contiguity/complete separability. Thus, for example, Theorem 1 of Oosterhoff and van Zwet (1979) can be seen as a generalization of Kakutani's theorem to sequences of product measures whereas Eagleson (1981) extended the classical dichotomy between Gaussian measures to sequences of such measures. It therefore seems reasonable to judge a proof of the Kabanov, Liptser and Širjaev Theorem to be "natural" if it generalizes to give analogous results about the contiguity of two sequences of measures. The proof presented here does precisely that.

2. SEQUENCES OF RADON-NIKODYM DERIVATIVES

We shall generalize the martingale version of the theorem of Kabanov, Lipster and Širjaev so as to include the result of Oosterhoff and van Zwet. First we establish some notation. Suppose for each n=1,2,..., $P^{(n)}$ and $Q^{(n)}$ are two measures on (Ω_n, A_n).

Set $L_n = dQ^{(n)}/dP^{(n)}$ when it exists and $L_n = \infty$ otherwise. Let $\{F_{nj};\ j=1,2,\dots,n\}$ be an increasing sequence of sub-σ-fields of A_n with $F_{nn} = A_n$, n=1,2,... . Set

$$P_{nj} = P^{(n)}|F_{nj},\quad Q_{nj} = Q^{(n)}|F_{nj}$$

and define $L_{nj} = dQ_{nj}/dP_{nj}$, which is well-defined P_{nj}-a.s.. Now write

$$\alpha_{nj} = L_{nj}.\ \tilde{L}_{n,j-1}$$

where $\tilde{L}_{n,j-1} = L_{n,j-1}^{-1}$ if $L_{n,j-1} \neq 0$, and zero otherwise. Note that $\{L_{nj},\ j=1,2,\dots,n\}$ is an $(F_{nj}, P^{(n)})$-supermartingale, so that $L_{nj} = 0$ on $[L_{n,j-1} = 0](P^{(n)}$-a.s.), or $\alpha_{n,j+k} = 0$, k=1,2,... as soon as $\alpha_{nj} = 0$. Also, we have

$$E_{P^{(n)}}(\alpha_{nj} \| F_{n,j-1}) \leq 1 \tag{1}$$

for all n and j. (Where no confusion should arise, we shall suppress the subscript on the expectation operator.) Finally, recall that a sequence of random variables f_n, n=1,2,..., is *tight* for a sequence of measures $P^{(n)}$ if

$$P^{(n)}(|f_n| > N) = o(1)$$

as N tends to infinity, uniformly in n.

Theorem 1. Let B_n denote a sequence of A_n-measurable sets with $P^{(n)}(B_n)$ bounded away from zero, and $P^{(n)}_{B_n}(.)$ denote the conditional probability $P^{(n)}(.|B_n)$. If for all $k > 0$,

$$\lim_{n\to\infty} P^{(n)}_{B_n}\Big(\sum_{j=1}^{n} E(1 - \sqrt{\alpha_{nj}} \,\|\, F_{n,j-1}) > k\Big) = 1 \ , \tag{2}$$

then for all $\varepsilon > 0$,

$$\lim_{n\to\infty} P^{(n)}_{B_n}(L_n > \varepsilon) = 0 \ . \tag{3}$$

On the other hand, if

$$\sum_{j=1} E(1 - \sqrt{\alpha_{nj}} \,\|\, F_{n,j-1}) \quad \text{is tight} \quad (P^{(n)}_{B_n}) \tag{4}$$

as well as

$$\max_{j\le n} \alpha_{nj}^{-\frac{1}{2}} \quad \text{being tight} \quad (P^{(n)}_{B_n}) \ , \tag{5}$$

then

$$\log \ L_n \quad \text{is tight} \quad (P^{(n)}_{B_n}) \ . \tag{6}$$

The basic ideas of our proof are simple, in fact, simpler (in our opinion) than the original proof of Kabanov, et al. To illustrate this point, let us outline our approach as it applies to their result before giving a proof of the Theorem. Let $\Omega_n = \Omega$, $P^{(n)} = P$ and assume that $E(L_n) \equiv 1$, $n=1,2,...,$ so that the nonnegative martingale L_n, $n=1,2,...$ is viewed as a sequence of Radon-Nikodym derivatives of measures Q_n, $n=1,2,...$ Let $B_n = B = [\sum_{j=1}^{\infty} E(1 - \sqrt{\alpha_j} \,\|\, F_{j-1}) = \infty]$. We shall show that

$$B \subset [L_\infty = 0] \quad P\text{-a.s.} \quad .$$

(i) Let $B = B_1 \cup B_2$ where

$$B_1 = B \cap \{\alpha_j \equiv 0 \text{ for } j \geq n \text{ for some } n\} \quad .$$

In this case, $L_j \equiv 0$ for the same indices, and obviously, $B_1 \subset [L_\infty = 0]$.
(ii) We now consider

$$B_2 = B \cap \{\alpha_j > 0, j=1,2,\ldots\} \quad .$$

As $\alpha_j = 0$ P-a.s. on $[E(\sqrt{\alpha_j} \| F_{j-1}) = 0]$, $E(\sqrt{\alpha_j} \| F_{j-1}) > 0$ P-a.s. on B_2. Hence on B_2 we have the identity, valid for each n,

$$\sqrt{L_n} = \prod_{j=1}^{n} (\sqrt{\alpha_j}/E(\sqrt{\alpha_j} \| F_{j-1}))\, E(\sqrt{\alpha_j} \| F_{j-1}) = M_n \cdot D_n \;,$$

where M_n, $n=1,2,\ldots$ is a non-negative martingale and D_n, $n=1,2,\ldots$ is the sequence of products $D_n = \prod_{j=1}^{n} E(\sqrt{\alpha_j} \| F_{j-1})$.

From (1), $E(\sqrt{\alpha_j} \| F_{j-1}) \leq 1$ and so

$$D_n \leq \exp(-\sum_{j=1}^{n} E(1 - \sqrt{\alpha_j} \| F_{j-1})) \quad .$$

It follows that $L_\infty = 0$ on B_2, which proves the inclusion

$$B_2 \subset [L_\infty = 0] \quad P\text{-a.s.} \quad .$$

Now consider the other inclusion. Let $B_n = C = [\sum_{j=1}^{\infty} E(1 - \sqrt{\alpha_j} \| F_{j-1}) < \infty]$ and let us prove $C \subset [L_\infty > 0]$ P-a.s. .
We note that

$$0 \leq E(1 - \sqrt{\alpha_j} \| F_{j-1}) \leq 1$$

and

$$E((1 - \sqrt{\alpha_j})^2 \| \mathcal{F}_{j-1}) \leq 2E(1 - \sqrt{\alpha_j} \| \mathcal{F}_{j-1}) .$$

These two inequalities, together with a stopping time argument, given as a Lemma in the body of the proof of the Theorem, imply that

$$\sum_{j=1}^{\infty} (1 - \sqrt{\alpha_j})^2 < \infty \quad \text{P-a.s. on C}$$

and $\sum_{j=1}^{\infty} (1 - \sqrt{\alpha_j})$ exists and is finite P-a.s. on C.

With these facts in hand we may estimate $\log L_n$, using the inequalities

$$-c \sum_{j=1}^{\infty} (1 - \sqrt{\alpha_j})^2 \leq \log L_n + \sum_{j=1}^{n} (1 - \sqrt{\alpha_j}) \leq d \sum_{j=1}^{n} (1 - \sqrt{\alpha_j})^2 .$$

to conclude $C \subset [L_\infty > 0]$ P-a.s. .

Proof of the Theorem. First, let us prove that (2) implies (3). Condition (2) implies that a sequence of constants k_n can be chosen tending to ∞ such that

$$\lim_{n\to\infty} P_{B_n}^{(n)} \left(\sum_{j=1}^{n} E(1 - \sqrt{\alpha_{nj}} \| \mathcal{F}_{n,j-1}) > k_n \right) = 1 \tag{2'}$$

Each set B_n may be decomposed as $B_n = B_{n1} \cup B_{n2}$ where $B_{n1} = B_n \cap \{\alpha_{n,j+k} = 0$ for some $j < n$ and $k = 1,2,\ldots,n-j\}$. (Notice that for each n, $\{L_{nj}, j=1,\ldots,m\}$ is an $(\mathcal{F}_{nj}, P^{(n)})$ supermartingale so that $L_{n,j+k} = 0$ on $[L_{nj} = 0]$ for $k = 1,2,\ldots$ $P^{(n)}$-a.s. .) We have $L_n = 0$ on B_{n1}. Therefore, it suffices to restrict attention to the set $B_{n2} = B_n \cap \{\alpha_{nj} > 0$ for $j = 1,2,\ldots,n\}$. As before, $E(\sqrt{\alpha_{nj}} \| \mathcal{F}_{n,j-1}) > 0$ on B_{n2} and so we have the identity

$$\sqrt{L_n} = \prod_{j=1}^{n} (\sqrt{\alpha_{nj}}/E(\sqrt{\alpha_{nj}} \| \mathcal{F}_{j,j-1}))\, E(\sqrt{\alpha_{nj}} \| \mathcal{F}_{n,j-1}) = M_n \cdot D_n .$$

Here M_n are positive random variables with $E_{P^{(n)}}(M_n) \equiv 1$ for all n. Hence M_n $n = 1,2,\ldots$ is tight $P^{(n)}$ and so $I_{B_{n2}} M_n$ is also tight. On the other hand

$$D_n \le \exp(-\sum_{j=1}^{n} E(1 - \sqrt{\alpha_{nj}} \,\|\, F_{n,j-1})) ,$$

and condition (2′) implies that for some sequence of constants $\varepsilon_n \downarrow 0$,

$$\lim_{n\to\infty} P^{(n)}(I_{B_{n2}} D_n > \varepsilon_n) = 0 .$$

This, together with the tightness of M_n, implies that for all $\varepsilon > 0$

$$\lim_{n\to\infty} P^{(n)}(I_{B_{n2}} L_n > \varepsilon) = 0$$

and hence that (3) holds.

To prove the second part of the Theorem, we state the following facts in the form of a lemma

Lemma Suppose that (4) holds: $\sum_{j=1}^{n} E_{P^{(n)}}(1 - \sqrt{\alpha_{nj}} \,\|\, F_{n,j-1})$ is tight $(P_{B_n}^{(n)})$.

Then the following sequences are also tight:

(i) $\sum_{j=1}^{n} E_{P^{(n)}}((1 - \sqrt{\alpha_{nj}})^2 \,\|\, F_{n,j-1})$, $n=1,2,\ldots$;

(ii) $\sum_{j=1}^{n} (1 - \sqrt{\alpha_{nj}})^2$, $n=1,2,\ldots$;

(iii) $\sum_{j=1}^{n} (1 - \sqrt{\alpha_{nj}})$, $n=1,2,\ldots$.

Let us accept this lemma for the moment and complete the proof of the Theorem. Thus, condition (4) implies that $\sum_{j=1}^{n} (1 - \sqrt{\alpha_{nj}})^2$, $n=1,2,\ldots$ is tight $(P_{B_n}^{(n)})$. which implies that $\max_{j\le n} \sqrt{\alpha_{nj}}$ is also tight $(P_{B_n}^{(n)})$. Condition (4) also implies that $\sum_{j=1}^{n} (1 - \sqrt{\alpha_{nj}})$, $n=1,2,\ldots$ is tight $(P_{B_n}^{(n)})$. Finally, condition (5) gives control on how *small* the sequence α_{nj}, $j=1,2,\ldots,n$; $n=1,2,\ldots$ is allowed to be. Together these conditions imply that for all $\varepsilon > 0$ there exist positive constants c and d, not depending on n, such that

$$P^{(n)}_{B_n}\{-c\sum_{j=1}^{n}(1-\sqrt{\alpha_{nj}})^2 \le \log L_n + \sum_{j=1}^{n}(1-\sqrt{\alpha_{nj}})$$

$$\le d\sum_{j=1}^{n}(1-\sqrt{\alpha_{nj}})^2\} > 1-\varepsilon.$$

Therefore, condition (6) holds.

Proof of the Lemma (cf. Proposition VII-2-3(c) of Neveu (1972)). Notice that

$$\sum_{j=1}^{n} E_{P^{(n)}}((1-\sqrt{\alpha_{nj}})^2 \| F_{n,j-1}) \le 2\sum_{j=1}^{n} E_{P^{(n)}}(1-\sqrt{\alpha_{nj}} \| F_{n,j}) .$$

so that part (i) is verified. For each n and $c > 0$, define

$$\nu_c^{(n)} = \inf\{k: \sum_{j=1}^{k+1} E_{P^{(n)}}((1-\sqrt{\alpha_{nj}})^2 \| F_{n,j-1}) > c\}$$

if such a k exists; otherwise set $\nu_c^{(n)} = n$. Then $\nu_c^{(n)}$ is a predictable stopping time with respect to F_{nj}, $j=1,2,\ldots$ and

$$E_{P^{(n)}}[\sum_{j=1}^{\nu_c^{(n)}}(1-\sqrt{\alpha_{nj}})^2] \le c.$$

This implies that the sequence $\sum_{j=1}^{\nu_c^{(n)}}(1-\sqrt{\alpha_{nj}})^2$, $n=1,2,\ldots$ is tight $(P^{(n)}_{B_n})$. Since

$$P^{(n)}_{B_n}\{\sum_{j=1}^{n}(1-\sqrt{\alpha_{nj}})^2 \ne \sum_{j=1}^{\nu_c^{(n)}}(1-\sqrt{\alpha_{nj}})^2\}$$

$$= P^{(n)}_{B_n}\{\sum_{j=1}^{n} E_{P^{(n)}}((1-\sqrt{\alpha_{nj}})^2 \| F_{n,j-1}) > c\}$$

which tends to zero, uniformly in n, as c tends to infinity, we see that $\sum_{j=1}^{n} (1 - \sqrt{\alpha_{nj}})^2$, $n = 1,2,\ldots$ is tight also.

In order to prove (iii), consider the sum

$\sum_{j=1}^{\nu_c(n)} [1 - \sqrt{\alpha_{nj}} - E(1 - \sqrt{\alpha_{nj}} \| F_{n,j-1})]$. This time we estimate the L_2-norms

$$E_{P^{(n)}} \left| \sum_{j=1}^{\nu_c^{(n)}} [1 - \sqrt{\alpha_{nj}} - E(1 - \sqrt{\alpha_{nj}} \| F_{n,j-1})] \right|^2$$

$$= E_{P^{(n)}} \{ \sum_{j=1}^{\nu_c^{(n)}} [1 - \sqrt{\alpha_{nj}} - E(1 - \sqrt{\alpha_{nj}} \| F_{n,j-1})]^2 \}$$

$$\leq 2E_{P^{(n)}} \{ \sum_{j=1}^{\nu_c^{(n)}} E((1 - \sqrt{\alpha_{nj}})^2 \| F_{n,j-1}) \}$$

$$\leq 2c.$$

This inequality implies that the sequence

$$\sum_{j=1}^{\nu_c^{(n)}} \left[1 - \sqrt{\alpha_{nj}} - E(1 - \sqrt{\alpha_{nj}} \| F_{n,j-1}) \right] \quad n = 1,2,\ldots$$

is tight ($P_{B_n}^{(n)}$ - a.s.). This, together with the assumption of the Lemma implies that $\sum_{j=1}^{\nu_c^{(n)}} (1 - \sqrt{\alpha_{nj}})$, $n = 1,2,\ldots$ is tight ($P_{B_n}^{(n)}$ - a.s.). But this implies the conclusion (iii) by the argument given above.

<u>Corollary</u> (Oosterhoff and van Zwet (1979))

For each fixed $n = 1,2,\ldots,$ Let $(\mathcal{X}_{n1},\mathcal{A}_{n1}),\ldots,(\mathcal{X}_{nn},\mathcal{A}_{nn})$ be arbitrary measurable spaces and P_{nj}, Q_{nj} probability measures defined on $(\mathcal{X}_{nj},\mathcal{A}_{nj})$, $j = 1,\ldots,n$. Let μ_{nj} be any σ-finite measure dominating $P_{nj} + Q_{nj}$ and set

$$\alpha_{nj} = (dQ_{nj}/d\mu_{nj})/(dP_{nj}/d\mu_{nj}) \ .$$

Let $\Omega_n = \underset{j=1}{\overset{n}{\times}} \mathcal{X}_{nj}$ and $\mathcal{A}_n = \mathcal{A}_{n1} \otimes \cdots \otimes \mathcal{A}_{nn}$. The sequence of product measures on $(\Omega_n,\mathcal{A}_n)$, $\{Q_n = \prod_{j=1}^{n} Q_{nj}\}$ is contiguous to the product measures

$\{P_n = \prod_{j=1}^{n} P_{nj}\}$ if, and only if,

$$\limsup_{n\to\infty} \sum_{j=1}^{n} E_{P_{nj}} (1 - \sqrt{\alpha_{nj}})^2 < \infty \tag{7}$$

and

$$\lim_{n\to\infty} \sum_{j=1}^{n} Q_{nj}(\alpha_{nj} \geq c_n) = 0 \tag{8}$$

whenever $c_n \to \infty$.

<u>Proof</u> As we are here concerned with product measures,

$$E_{P_{nj}}(1 - \sqrt{\alpha_{nj}} \,\|\, \mathcal{F}_{n,j-1}) = E_{P_{nj}}(1 - \sqrt{\alpha_{nj}})$$

and

$$E_{P_{nj}}\left[(1 - \sqrt{\alpha_{nj}})^2 \,\|\, \mathcal{F}_{n,j-1}\right] = E_{P_{nj}}(1 - \sqrt{\alpha_{nj}})^2 \ .$$

Now

$$E_{P_{nj}}(\alpha_{nj}) = E_{P_{nj}}(\alpha_{nj} I(\alpha_{nj} < \infty))$$
$$= Q_{nj}[\alpha_{nj} < \infty],$$

so that conditions (1) and (8) ensure that

$$2 \sum_{j=1}^{n} E_{P_{nj}} (1 - \sqrt{\alpha_{nj}}) - \sum_{j=1}^{n} E_{P_{nj}} (1 - \sqrt{\alpha_{nj}})^2$$

$$= \sum_{j=1}^{n} Q_{nj}[\alpha_{nj} = \infty] \to 0 \text{ as } n \to \infty .$$

It follows that (7) implies that

$$\limsup_{n\to\infty} \sum_{j=1}^{n} E_{P_{nj}} (1 - \sqrt{\alpha_{nj}}) < \infty . \tag{9}$$

Set $B_n = \Omega_n$ for all n. Reversing the roles of $P^{(n)}$ and $Q^{(n)}$ in the theorem (in which case α_{nj} must be replaced by $1/\alpha_{nj}$) and noting that

$$E_{P_{nj}} (1 - \sqrt{\alpha_{nj}}) = E_{Q_{nj}} (1 - 1/\sqrt{\alpha_{nj}}) ,$$

we see that (9) and (8) imply (4) and (5) and hence that $\{\log L_n\}$ is tight (Q_n). But this is a well-known sufficient condition for $\{Q^{(n)}\}$ to be contiguous to $\{P^{(n)}\}$ (see Proposition 4 of Hall and Loynes (1977)).

On the other hand, if

$$\limsup \sum_{j=1}^{n} E_{P_{nj}} (1 - \sqrt{\alpha_{nj}}) = \infty ,$$

the first part of the Theorem shows that $\{Q^{(n)}\}$ and $\{P^{(n)}\}$ are completely separable. Hence if $\{Q^{(n)}\}$ is contiguous to $\{P^{(n)}\}$, then (9) and hence (7) must hold.

The condition (8) now follows from the fact that

$$\lim_{n\to\infty} \sum_{j=1}^{n} P_{nj} ((1 - \sqrt{\alpha_{nj}})^2 > c_n) \le \lim_{n\to\infty} c_n^{-1} \sum_{j=1}^{n} E_{P_{nj}} (1 - \sqrt{\alpha_{nj}})^2 = 0$$

and, as noted in Oosterhoff and van Zwet (1979), because one is dealing with product measures, for any collection of measurable sets $\{A_{nj}\}$, when $\{Q^{(n)}\}$ is contiguous to $\{P^{(n)}\}$

$$\lim_{n\to\infty} \sum_{j=1}^{n} P_{nj}(A_{nj}) = 0$$

implies that

$$\lim_{n\to\infty} \sum_{j=1}^{n} Q_{nj}(A_{nj}) = 0 \quad \text{also.}$$

3. REFERENCES

Eagleson, G.K. An extended dichotomy theorem for sequences of pairs of Gaussian measures. *Ann. Prob.* *9* (1981), 453-459.

Hall, W.J. and Loynes, R.M. On the concept of contiguity. *Ann. Prob.* *5* (1977), 278-282.

Kabanov, Yu M., Liptser, R.S. and Širjaev, A.N. On the question of the absolute continuity and singularity of probability measures (Russian). *Mat. Sbornik* *104* (146) (1977), 227-247.

Kakutani, S. On equivalence of infinite product measures. *Ann. Math.* *49* (1948), 214-224.

Le Cam, L. Local asymptotically normal families of distributions. *Univ. Calif. Publ. Statist.* *3* (1960), 37-98.

Le Cam, L. On the asymptotic normality of estimates. *Proceedings of the Symposium to Honour Jerzey Neyman* (Warsaw 1974). Panstw. Wydawn Nauk. Warsaw, (1977), 203-217.

Neveu, J. *Martingales à temps discret.* Masson et Cie, Paris. (1972).

Oosterhoff, J. and van Zwet, W.R. A note on contiguity and Hellinger distance. *Contributions to Statistics - Jaroslav Hájek* Memorial Volume (Ed. J. Jureckavá). Academia, Prague 1979.

Oxford commemoration ball

By J. M. Hammersley

1. Question for a demy.

I hope this salute to my old friend David Kendall will remind him of his Oxford days, for it commemorates an excellent scholarship question that he once set in the 1950s when he was the mathematics tutor at Magdalen. The gist of the question ran as follows. A spherical ball of unit radius rests on an infinite horizontal table. You may imagine that it is a globe with a map of the world painted on its surface to distinguish its spatial orientations. The *state* of the ball is specified by specifying both its spatial orientation and its position on the table. You have to transfer the ball from a given initial state to an arbitrary final state via a sequence of *moves*. Each move consists of rolling the ball along some straight line on the table: the length and direction of any move are at your disposal, but the rolling must be pure in the sense that the axis of rotation must be horizontal and there must be no slipping between ball and table. How many moves, N, will be necessary and sufficient to reach any final state?

The original version of the question, set for 18–year–old schoolboys, invited candidates to investigate how two moves, each of length π, would change the ball's orientation; and to deduce in the first place that $N \leqslant 11$, and in the second place that $N \leqslant 7$. Candidates scored bonus marks for any improvement on 7 moves. When he first set the question, Kendall knew that $N \leqslant 5$; but, interest being aroused amongst professional mathematicians at Oxford, he and others soon discovered that the answer must be either $N = 3$ or $N = 4$. But in the 1950s nobody could decide between these two possibilities. There was renewed interest in the 1970s, and not only amongst professional mathematicians: for example, the then President of Trinity (a distinguished biochemist) spent some time rolling a ball around his drawing room floor in search of empirical insight. In 1978, while delivering the opening address to the first Australasian Mathematical Convention, I posed the problem to mathematicians down under; but I have not subsequently received a solution from them. So this is an opportunity to publish the solution. However there are one or two surprises in store; so I shall not reveal until much nearer the end of this paper whether $N = 3$ or 4, and in the meantime the reader may care to ponder which horse to back.

Any mathematical paper ought to raise more unsolved problems than it resolves. So in §15, I shall mention some variants and generalizations of Kendall's problem: most of these are very difficult and some may be quite beyond the reach of contemporary mathematical techniques. Accordingly I call them problems for the twenty–first century.

At the other end of the mathematical scale, §2 will deal with some nineteenth century mathematics for the benefit of young mathematicians —undergraduates, clever sixth–formers (for whom Kendall set the problem originally) — and schoolteachers. Kendall's problem raises some educational issues, belonging particularly to the twentieth century; for, both at school and university, the pursuit of abstraction and generalization have come to vie for time and attention against the development of manipulative skills.To handle rotations, there are two main tools: orthogonal matrices and quaternions. The former generalize to n dimensions, but the latter win hands down in the particular case $n = 3$ when it comes to manipulations of the sort needed in Kendall's problem. Although undergraduate lectures devote much time to vector analysis in mechanics, scarcely any newly–fledged honours graduate nowadays knows what is meant by the quaternion product of two vectors or how to represent rotations by quaternions. Indeed, if undergraduates have met quaternions at all, they may only have met them as a passing illustration of a non–commutative division ring; and that is a sad reflection on modern education. So I do not hesitate to describe quaternions in §2, in a thoroughly old–fashioned style, as a valuable manipulative tool.

The first paragraph of §2 summarizes the conventions, notations, and results that will be used in this paper; and readers, who are familiar with quaternions, will not need to read the remainder of §2. Other readers, who wish for an introduction to quaternions, should not be deterred by that first paragraph; for they will find explanations, definitions, and proofs in the remainder of §2. I shall assume that all readers are familiar with the elementary facts of vector analysis; and these prerequisites appear in the second paragraph of §2 for the benefit of clever sixth–formers and schoolteachers. Anyone, who understands the second paragraph of §2 (supplemented, if need be, by a textbook on vectors – for example the first nine pages of D.E. Rutherford's *Vector Methods* (Oliver and Boyd, 1939)), ought to be able to understand most of this paper, except perhaps §§13 and 14 which are more advanced.

2. Nineteenth century prolegomenon on quaternions.

We shall work in three–dimensional Euclidean space referred to a right–handed Cartesian co–ordinate system, with mutually perpendicular axes Ox, Oy, Oz of which Ox and Oy are horizontal and Oz is vertically upwards. Bold–face letters denote vectors in this space, and the corresponding italic letters denote the length of these vectors: thus v is the length of the vector $\mathbf{v}$. We write $\mathbf{i}$, $\mathbf{j}$, $\mathbf{k}$ for unit vectors along the axes Ox, Oy, Oz respectively. The letter $\mathbf{u}$ (with or without suffices) is reserved to denote a unit vector; the letter $\mathbf{h}$ (with or without suffices) is reserved to denote a horizontal unit vector; the letter q (with or without suffices) is reserved to denote a quaternion. The quaternion product of two vectors $\mathbf{v}$ and $\mathbf{w}$ is written $\mathbf{vw}$ (without a dot or cross, to distinguish it from the scalar product $\mathbf{v}.\mathbf{w}$ and the vector product $\mathbf{v}\times\mathbf{w}$). The *angle* θ *between two vectors* $\mathbf{v}$ and $\mathbf{w}$ is the angle between their directions, and is always understood to satisfy $0 \leqslant \theta \leqslant \pi$, where $\theta = \pi$ if and only if $\mathbf{v}$ and $\mathbf{w}$ are in opposite directions. If $\mathbf{u}$ is perpendicular to both $\mathbf{v}$ and $\mathbf{w}$, the *angle from* $\mathbf{v}$ *to* $\mathbf{w}$ *along* $\mathbf{u}$ is the angle from the direction of $\mathbf{v}$ to the direction of $\mathbf{w}$ in the plane of $\mathbf{v}$ and $\mathbf{w}$ measured clockwise when looking along the forward direction of $\mathbf{u}$, such an angle being interpreted modulo 2π; and, conversely, use of the phrase "from $\mathbf{v}$ to $\mathbf{w}$ along $\mathbf{u}$" will automatically imply (without further explicit mention) that $\mathbf{u}$ is perpendicular to both $\mathbf{v}$ and $\mathbf{w}$. A rotation *through* φ *about* $\mathbf{u}$ means a rotation about an axis $\mathbf{u}$ with the angle of rotation φ measured clockwise when looking along the forward direction of $\mathbf{u}$. This rotation is represented by the quaternion pair

$$\pm q = \pm e^{\frac{1}{2}\varphi\mathbf{u}} = \pm(\cos\tfrac{1}{2}\varphi + \mathbf{u}\sin\tfrac{1}{2}\varphi) = \pm\mathbf{u}_2\mathbf{u}_1 , \tag{2.1}$$

where the angle from $\mathbf{u}_1$ to $\mathbf{u}_2$ along $\mathbf{u}$ is $\frac{1}{2}\varphi$. Angles of rotation may be interpreted modulo 2π when the context permits. Conversely, if $\pm q_1$ and $\pm q_2$ represent rotations whose axes are perpendicular to $\mathbf{u}$, then

$$\pm 1 \neq \pm e^{\frac{1}{2}\varphi\mathbf{u}} = \pm q_2 q_1 \tag{2.2}$$

implies that $\pm q_p = \pm\mathbf{u}_p$ $(p = 1, 2)$, where $\frac{1}{2}\varphi$ is the angle from $\mathbf{u}_1$ to $\mathbf{u}_2$ along $\mathbf{u}$. A sequence of rotations, represented by the pairs $\pm q_1, \pm q_2, \ldots, \pm q_n$ and performed in that order, yield a resultant rotation represented by

$$\pm q = \pm q_n \cdots q_2 q_1 . \tag{2.3}$$

We shall assume the following prerequisite knowledge about vectors. We shall always use the noun *scalar* to mean a real number. The vector $\mathbf{v} = (x,y,z)$ is the directed line from the origin to the point having real Cartesian co-ordinates x,y,z; and the length of $\mathbf{v}$ is $v = \sqrt{(x^2 + y^2 + z^2)} \geqslant 0$. The vector $\mathbf{v}$ is a *unit* vector if $v = 1$. Let $\mathbf{w} = (\xi,\eta,\zeta)$ be another typical vector. Vector addition is commutative: $\mathbf{v}+\mathbf{w} = \mathbf{w}+\mathbf{v} = (x+\xi, y+\eta, z+\zeta)$. A vector can be multiplied commutatively by a scalar: $a\mathbf{v} = \mathbf{v}a = (ax, ay, az)$. The *scalar product* of $\mathbf{v}$ and $\mathbf{w}$ is $\mathbf{v}.\mathbf{w} = \mathbf{w}.\mathbf{v} = x\xi + y\eta + z\zeta = vw\cos\theta$, where θ is the angle between $\mathbf{v}$ and $\mathbf{w}$. The *vector product* of $\mathbf{v}$ and $\mathbf{w}$ is $\mathbf{v}\times\mathbf{w} = (y\zeta - z\eta, z\xi - x\zeta, x\eta - y\xi)$, which is a vector of length $vw\sin\theta$ perpendicular to $\mathbf{v}$ and $\mathbf{w}$ and has the direction of $\mathbf{u}$ when θ is the angle from $\mathbf{v}$ to $\mathbf{w}$ along $\mathbf{u}$. Thus $\mathbf{v}\times\mathbf{w} = -\mathbf{w}\times\mathbf{v}$. All the ordinary laws of algebra apply to the addition and multiplication of vectors, *provided* that due care is taken over the order of the factors in any vector product *and* over the bracketing of terms in the triple products

$$(\mathbf{v}\times\mathbf{w}).\mathbf{t} = \mathbf{v}.(\mathbf{w}\times\mathbf{t}) , \tag{2.4}$$

$$\mathbf{v}\times(\mathbf{w}\times\mathbf{t}) = (\mathbf{v}.\mathbf{t})\mathbf{w} - (\mathbf{v}.\mathbf{w})\mathbf{t} , \tag{2.5}$$

$$(\mathbf{v}\times\mathbf{w})\times\mathbf{t} = (\mathbf{v}.\mathbf{t})\mathbf{w} - (\mathbf{w}.\mathbf{t})\mathbf{v} . \tag{2.6}$$

With the aid of the foregoing preliminaries, we can now define quaternions and derive their properties. For conciseness, we shall not repeat the conventions and definitions above, to which the reader may refer as and when he needs them. A *quaternion* is a combination of a scalar and vector; and, for reasons of manipulative convenience, we write this combination in terms of commutative addition:

$$q = a + \mathbf{v} = \mathbf{v} + a \, . \tag{2.7}$$

Here a is called the *scalar component* of q, and $\mathbf{v}$ is the *vector component* of q. Two quaternions $a + \mathbf{v}$ and $b + \mathbf{w}$ are equal if and only if $a = b$ and $\mathbf{v} = \mathbf{w}$. Scalars and vectors can be considered as particular sorts of quaternion by taking $\mathbf{v} = \mathbf{0}$ or $a = 0$ respectively, giving $q = a$ or $q = \mathbf{v}$. The addition of two quaternions is defined by the addition of their components:

$$(a + \mathbf{v}) + (b + \mathbf{w}) = (a + b) + (\mathbf{v} + \mathbf{w}) \, . \tag{2.8}$$

Thus quaternion addition is commutative. The *quaternion product* of two vectors $\mathbf{v}$ and $\mathbf{w}$ is the quaternion with $-\mathbf{v}.\mathbf{w}$ for its scalar component and $\mathbf{v} \times \mathbf{w}$ for its vector component; and it is denoted by

$$\mathbf{vw} = -\mathbf{v}.\mathbf{w} + \mathbf{v} \times \mathbf{w} \, . \tag{2.9}$$

More generally the product of two quaternions is defined by

$$\begin{aligned}(a + \mathbf{v})(b + \mathbf{w}) &= ab + a\mathbf{w} + b\mathbf{v} + \mathbf{vw} \\ &= (ab - \mathbf{v}.\mathbf{w}) + (a\mathbf{w} + b\mathbf{v} + \mathbf{v} \times \mathbf{w}).\end{aligned} \tag{2.10}$$

It follows, by a straightforward calculation from (2.4), (2.5), (2.6), (2.9) and (2.10), that the quaternion product of vectors is associative, $(\mathbf{vw})\mathbf{t} = \mathbf{v}(\mathbf{wt})$; and hence from (2.10) that quaternion multiplication is associative, $(q_1q_2)q_3 = q_1(q_2q_3)$. (Note that vector multiplication, as instanced by (2.5) and (2.6), is *not* associative: to this extent, quaternion algebra is easier to handle than vector algebra.) Quaternion multiplication is non–commutative, because $\mathbf{v} \times \mathbf{w} = -\mathbf{w} \times \mathbf{v}$. However, according to (2.10), any scalar $q = a$ commutes with every quaternion under multiplication. The *conjugate* of the quaternion $q = a + \mathbf{v}$ is defined and denoted by $\bar{q} = a - \mathbf{v}$. Hence

$$\mathbf{wv} = -\mathbf{v}.\mathbf{w} - \mathbf{v} \times \mathbf{w} = \overline{\mathbf{vw}} \, ; \tag{2.11}$$

and (2.10) and (2.11) now show that

$$\overline{q_1q_2} = \bar{q}_2\bar{q}_1 \, . \tag{2.12}$$

Moreover, if $q = a + \mathbf{v}$, then

$$q\bar{q} = \bar{q}q = a^2 + v^2 \geqslant 0 \, , \tag{2.13}$$

with equality if and only if $q = 0$. Since $q\bar{q}$ is a scalar, any non–zero quaternion q has a reciprocal $q^{-1} = \bar{q}/(q\bar{q})$ such that $qq^{-1} = q^{-1}q = 1$. Since scalars commute with quaternions, (2.12) gives

$$(q_1q_2)\overline{(q_1q_2)} = q_1(q_2\bar{q}_2)\bar{q}_1 = (q_1\bar{q}_1)(q_2\bar{q}_2) \, . \tag{2.14}$$

The quaternion q is a *unit quaternion* if $q\bar{q} = 1$, in which case $q^{-1} = \bar{q}$. According to (2.14), the product of two unit quaternions is a unit quaternion. Hence the unit quaternions form a non–commutative group under multiplication. All the ordinary laws of algebra, including division by any non–zero quaternion, hold for the manipulation of quaternions, provided only that the order of the factors in any multiplication or division is respected: in technical language we express this by saying that the quaternions form a *non–commutative division ring Q*. In this system, each unit vector is a square root of -1, because $\mathbf{u}^2 = -u^2 = -1$, by (2.9); and the converse holds as well, since $q^2 = (a + \mathbf{v})^2 = a^2 - v^2 + 2a\mathbf{v} = -1$ implies $v^2 = a^2 + 1 \geqslant 1$ and $2a\mathbf{v} = \mathbf{0}$, whence $\mathbf{v} \neq \mathbf{0}$ and $a = 0$ and $v^2 = 1$ and $q = \mathbf{v}$ must be a unit vector. As a consequence of (2.13) and the fact that scalars are real, we may write any unit quaternion in the form

$$q = \cos\tfrac{1}{2}\varphi + \mathbf{u}\sin\tfrac{1}{2}\varphi = e^{\frac{1}{2}\varphi\mathbf{u}}, \qquad (2.15)$$

for some angle φ and some unit vector $\mathbf{u}$, upon expanding the exponential formally as a power series and using $\mathbf{u}^2 = -1$. This exponential is to be thought of firstly as a convenient shorthand; and, since quaternions do not commute, we *cannot* write $e^{\mathbf{v}}e^{\mathbf{w}} = e^{\mathbf{v}+\mathbf{w}}$ unless $\mathbf{v}$ and $\mathbf{w}$ are parallel. However, the notation (2.15) is much more than a shorthand: it is a powerful manipulative asset in more advanced situations, such as the solution of quaternion differential equations. The reason for inserting the fraction $\frac{1}{2}$ in (2.15) will become apparent presently, when we consider rotations.

We now identify the quaternion q in (2.15) with rotation through φ about $\mathbf{u}$, as defined in the first paragraph of §2. But, since this is the same as a rotation through $\varphi + 2\pi$ about $\mathbf{u}$, the quaternion $-q = \cos\frac{1}{2}(\varphi + 2\pi) + \mathbf{u}\sin\frac{1}{2}(\varphi + 2\pi)$ is also identified with this same rotation. Accordingly we say that the quaternion pair $\pm q = \pm e^{\frac{1}{2}\varphi\mathbf{u}}$ represents a rotation through φ about $\mathbf{u}$, as stated in (2.1). To prove the remainder of (2.1), consider any two unit vectors $\mathbf{u}_1$ and $\mathbf{u}_2$, each perpendicular to $\mathbf{u}$, such that (in the language defined in the first paragraph of §2) the angle from $\mathbf{u}_1$ to $\mathbf{u}_2$ along $\mathbf{u}$ is $\frac{1}{2}\varphi$. By (2.9)

$$\begin{aligned} \mathbf{u}_2\mathbf{u}_1 &= -\mathbf{u}_2 . \mathbf{u}_1 + \mathbf{u}_2 \times \mathbf{u}_1 \\ &= -(\mathbf{u}_1 . \mathbf{u}_2 + \mathbf{u}_1 \times \mathbf{u}_2) = -(\cos\tfrac{1}{2}\varphi + \mathbf{u}\sin\tfrac{1}{2}\varphi), \end{aligned} \qquad (2.16)$$

and accordingly $\pm\mathbf{u}_2\mathbf{u}_1$ is the representative pair of quaternions for the rotation through φ about $\mathbf{u}$.

On the other hand, given the foregoing vectors $\mathbf{u}$, $\mathbf{u}_1$, and $\mathbf{u}_2$, consider a sphere S freely pivoted about the origin O and subjected first to a rotation through π about $\mathbf{u}_1$ and then to a rotation through π about $\mathbf{u}_2$. These rotations have representative pairs $\pm\mathbf{u}_1$ and $\pm\mathbf{u}_2$, as we see on putting $\varphi = \pi$ in (2.15). Let C be the great circle on S passing through the ends of $\mathbf{u}_1$ and $\mathbf{u}_2$ (or any such circle if $\mathbf{u}_1 = \pm\mathbf{u}_2$). Each of the rotations is a rotation through π about a diameter of C; and therefore each point of C is transformed into some other point of C after both rotations are complete. So the resultant rotation is a rotation through some angle

θ about $\mathbf{u}$. To see what θ is, consider the point of C initially coinciding with the end of $\mathbf{u}_1$. It remains fixed under the first rotation, and under the second rotation it moves to its reflected image in $\mathbf{u}_2$. Hence $\theta = 2(\frac{1}{2}\varphi) = \varphi$. Thus the resultant rotation is represented by the pair $\pm q = \pm \mathbf{u}_2\mathbf{u}_1$. In other words, we have proved the particular case of (2.3) in which $n = 2$ and $q_1 = \mathbf{u}_1$ and $q_2 = \mathbf{u}_2$.

Next consider the somewhat more general case of (2.3) in which $n = 2$ and also $q_1 = e^{\frac{1}{2}\varphi_1\mathbf{u}_1}$ and $q_2 = e^{\frac{1}{2}\varphi_2\mathbf{u}_2}$. Let $\mathbf{u}_3$ be a unit vector perpendicular to both $\mathbf{u}_1$ and $\mathbf{u}_2$. According to (2.1), we can find vectors $\mathbf{u}_4$ and $\mathbf{u}_5$, such that $\frac{1}{2}\varphi_1$ is the angle from $\mathbf{u}_4$ to $\mathbf{u}_3$ along $\mathbf{u}_1$ and $\frac{1}{2}\varphi_2$ is the angle from $\mathbf{u}_3$ to $\mathbf{u}_5$ along $\mathbf{u}_2$; whereupon $\pm q_1 = \pm \mathbf{u}_3\mathbf{u}_4$ and $\pm q_2 = \pm \mathbf{u}_5\mathbf{u}_3$. Let $\pm q = \pm e^{\frac{1}{2}\varphi\mathbf{u}}$ be the resultant rotation from performing the rotations represented by $\pm q_1$ and $\pm q_2$ in that order. From the special case proved in the previous paragraph, this is the resultant rotation produced by rotations through π about $\mathbf{u}_4, \mathbf{u}_3, \mathbf{u}_3, \mathbf{u}_5$ in that order. Since two successive rotations through π about $\mathbf{u}_3$ together preserve the orientation of S, the net resultant rotation represented by $\pm q$ is the resultant of rotations through π about $\mathbf{u}_4$ and $\mathbf{u}_5$ in that order, which by a further appeal to the special case proved in the previous paragraph, is represented by $\pm q = \pm \mathbf{u}_5\mathbf{u}_4$. However $\mathbf{u}_3^2 = -1$ commutes with any quaternion. Hence $\pm q = \mp \mathbf{u}_5\mathbf{u}_3\mathbf{u}_3\mathbf{u}_4 = \pm q_2 q_1$; and this proves (2.3) for $n = 2$ and arbitrary rotations represented by $\pm q_1, \pm q_2$.

We can now prove the general case of (2.3) from the particular case $n = 2$. Since quaternion multiplication is associative, we put $\pm q = \pm q_n(q_{n-1} \ldots q_2 q_1)$ and, by induction on n, use the special case to combine the rotation represented by $\pm q_{n-1} \ldots q_2 q_1$ with its successor represented by $\pm q_n$. This completes the proof of (2.3).

Finally, to prove (2.2), we may write it in the form

$$\pm 1 \neq \pm(c + s\mathbf{u}) = \pm(c_2 + s_2\mathbf{u}_2)(c_1 + s_1\mathbf{u}_1), \tag{2.17}$$

in which $c = \cos\frac{1}{2}\varphi$, $s = \sin\frac{1}{2}\varphi$, $c_p = \cos\frac{1}{2}\varphi_p$, $s_p = \sin\frac{1}{2}\varphi_p$ $(p=1,2)$. We cannot have $s = 0$, for otherwise $c = \pm 1$. So, equating the vector components on each side of (2.17) we have

$$s\mathbf{u} = \pm(c_2 s_1 \mathbf{u}_1 + c_1 s_2 \mathbf{u}_2 + s_1 s_2 \mathbf{u}_2 \times \mathbf{u}_1) \ . \tag{2.18}$$

By hypothesis $\mathbf{u}_1$ and $\mathbf{u}_2$ are perpendicular to $\mathbf{u}$. Resolution of (2.18) along $\mathbf{u}$ and perpendicular to $\mathbf{u}$ gives

$$s\mathbf{u} = s_1 s_2 \mathbf{u}_2 \times \mathbf{u}_1 \ , \qquad c_2 s_1 \mathbf{u}_1 + c_1 s_2 \mathbf{u}_2 = \mathbf{0} \ . \tag{2.19}$$

Thus $s_1 \neq 0$ and $s_2 \neq 0$, for otherwise $s = 0$. Hence $s_1\mathbf{u}_1$ and $s_2\mathbf{u}_2$ are linearly independent vectors; and the second equation in (2.19) now implies $c_1 = c_2 = 0$. Hence $s_1 = \pm 1$, and $s_2 = \pm 1$. So $\pm q_p = \pm \mathbf{u}_p$ $(p=1,2)$, as required.

Quaternions were used quite extensively in nineteenth century research: for example, Maxwell originally wrote his equations of electromagnetic radiation in terms of them.

3. Necessary and sufficient conditions for an N-move solution.

We may suppose that, in its initial state, the centre of the ball is at the origin O of our fixed co–ordinate system; and, in its final state, the centre of the ball lies at the end of a prescribed vector $\mathbf{v}$. Since $\mathbf{v}$ is horizontal, it is perpendicular to the unit vector $\mathbf{k}$ in the upward direction; and hence $-\mathbf{vk} = \mathbf{k} \times \mathbf{v}$ is also a horizontal vector, which we can write as

$$-\mathbf{vk} = \theta\mathbf{h} . \tag{3.1}$$

For the pth move ($p = 1,2,...,N$) we suppose that the ball is rolled through an angle φ_p about an axis that has the direction $\mathbf{h}_p$. Here φ_p may have any real value, positive, negative, or zero; and $|\varphi_p|$ will be the length of the pth move. The vector $\varphi_p\mathbf{h}_p$ completely represents this move; and, for brevity, we shall simply call it " the move $\varphi_p\mathbf{h}_p$ ", although actually it will shift the centre of the ball by a displacement vector $\varphi_p\mathbf{h}_p \times \mathbf{k} = \varphi_p\mathbf{h}_p\mathbf{k}$. To bring the centre of the ball to its desired final position, we must have

$$\sum_{p=1}^{N} \varphi_p\mathbf{h}_p\mathbf{k} = \mathbf{v} ; \tag{3.2}$$

and multiplying this equation on the right by $-\mathbf{k}$, we get

$$\sum_{p=1}^{N} \varphi_p\mathbf{h}_p = \theta\mathbf{h} . \tag{3.3}$$

To attain the desired final orientation, the ball must undergo a prescribed rotation, represented by $\pm e^{\frac{1}{2}\varphi\mathbf{u}}$ say. Hence

$$e^{\frac{1}{2}\varphi_N\mathbf{h}_N} \ldots e^{\frac{1}{2}\varphi_2\mathbf{h}_2} e^{\frac{1}{2}\varphi_1\mathbf{h}_1} = \pm e^{\frac{1}{2}\varphi\mathbf{u}} . \tag{3.4}$$

To achieve an N–move solution, it is necessary and sufficient to be able to solve (3.3) and (3.4) for the N horizontal vectors $\varphi_p\mathbf{h}_p$, given any prescribed vector $\varphi\mathbf{u}$ (not necessarily horizontal) and any prescribed horizontal vector $\theta\mathbf{h}$. This includes the possibility that fewer than N moves might be sufficient in some particular case: for example, if the initial and final states coincided, no move would be required and we could satisfy (3.3) and (3.4) by taking $\varphi_p\mathbf{h}_p = \mathbf{0}$ for all p.

We say that the pth move is *integral* (or *even* or *odd*) if φ_p is an integral (or even or odd) multiple of π; and we say that we have the *integral* (or *even* or *odd*) *version* of Kendall's problem if we require at least one move to be integral (or even or odd). We write N_{integral}, N_{even}, N_{odd} for the necessary and sufficient number of moves in these respective cases. Notice that a null move (i.e. one with $\varphi_p = 0$) counts as an even move.

Generally, if $N_{\text{prescribed}}$ is the necessary and sufficient number of moves when at least one move (independently of other moves) must be of some arbitrarily prescribed kind, we have

$$N \leqslant N_{\text{prescribed}} \leqslant N+1 ; \tag{3.5}$$

for we could take the first move to be of the prescribed kind, after which only N more moves would be needed to attain the desired final state. In particular, we have

$$N \leqslant N_{\text{integral}} \leqslant N_{\text{even}} \leqslant N+1, \quad N \leqslant N_{\text{integral}} \leqslant N_{\text{odd}} \leqslant N+1. \quad (3.6)$$

4. Elementary arguments.

We shall now prove that

$$3 \leqslant N \leqslant N_{\text{even}} \leqslant 4\,, \quad 3 \leqslant N \leqslant N_{\text{odd}} \leqslant 4\,. \quad (4.1)$$

To prove that 2 moves are insufficient, consider the particular case $\theta = 0$. The first move rolls the ball away from the origin; and then the second move must roll it back to the origin to satisfy (3.3). At the end of the second move the ball will have its original orientation; and therefore 2 moves cannot satisfy the case $\theta = 0$, $\varphi = \pi$. So $N \geqslant 3$.

Next we notice that an even move does not change the ball's orientation. We can translate the ball from any given position T_1 on the table to any other position T_2 on the table in two even moves of equal length; for we have only to choose an integer n such that $4n\pi$ exceeds the distance from T_1 to T_2, and pick a point T on the table at an equal distance $2n\pi$ from T_1 and T_2. Then the move from T_1 to T, followed by the move from T to T_2, will shift the ball from T_1 to T_2 without changing its orientation.

We shall say that the ball has a *spin discrepancy* φ_0 if it requires a rotation $\pm e^{\frac{1}{2}\varphi_0 \mathbf{k}}$ to bring it to its desired final orientation. If $\mathbf{h}_1$ and $\mathbf{h}_2$ are any two horizontal vectors such that the angle from $\mathbf{h}_1$ to $\mathbf{h}_2$ along $\mathbf{k}$ is $\frac{1}{2}\varphi_0$, the two odd moves $\pi\mathbf{h}_1$ followed by $\pi\mathbf{h}_2$ will correct a spin disrepancy φ_0 and bring the ball into its final orientation. This follows from (2.1).

Let B be the point (fixed on the surface of the ball) that will be in contact with the table when the ball is in its final state. Starting from the initial state, let us roll the ball until B comes into contact with the table. Now this will be a move of the form $(\varphi_1 - \pi)\mathbf{h}_1$ for some angle φ_1 and some vector $\mathbf{h}_1$. Let φ_0 be the spin discrepancy when B touches the table. Continue this move for a further distance π in the same direction, giving a complete first move $\varphi_1\mathbf{h}_1$. Then take an odd move $\varphi_2\mathbf{h}_2 = \pi\mathbf{h}_2$, where $\frac{1}{2}\varphi_0$ is the angle from $\mathbf{h}_1$ to $\mathbf{h}_2$ along $\mathbf{k}$. The ball now has the correct final orientation at some point T_1, and it can then be brought to its final state by two even moves. This gives a 4–move solution, the second move being odd, and the third and fourth being even. This completes the proof of (4.1).

5. Heuristic considerations.

A quaternion $q = a + \mathbf{v}$ has four independent co–ordinates, namely a and the co–ordinates x,y,z of $\mathbf{v}$. But a unit quaternion satisfies $a^2 + x^2 + y^2 + z^2 = 1$, and so only has three independent co–ordinates. Equation (3.4) has unit quaternions on each side, and is therefore equivalent to three scalar equations. Equation (3.3) is an equation involving horizontal vectors, and is equivalent to two scalar equations. So we have 5 equations to satisfy. Each of the horizontal vectors $\varphi_p\mathbf{h}_p$ has 2 degrees of freedom; and so we have $2N$ degrees of freedom available for the solution of 5 simultaneous equations. We would therefore expect, on heuristic grounds, to be able to solve them when $N = 3$ with some room to spare. Indeed, we should still expect a solution if we sacrificed one of the $2N = 6$ degreees of freedom by imposing some sort of restriction on the solution, for example by insisting that the first move should have a prescribed direction. A little more generously, we might insist only that one of the moves should have a length restricted to some countable set of values, leaving us with $5+0$ degrees of freedom (in a manner of speaking) for 5 equations. Thus the even version of Kendall's problem has $5+0$ degrees of freedom and ought to have a 3–move solution.

This conjecture can be supported (though not rigorously proved) by the following rough and ready topological argument, which we sketch very briefly leaving details to the reader. Consider the solution used in §4. Instead of extending the first move a distance π beyond the point when B first touches the table, we might have extended it by any odd multiple of π; and we might also have taken $n_2\pi\mathbf{h}_2$ with any odd integer n_2 for the second move. Also, for any positive $\varepsilon > 0$, we might have taken the angle from $\mathbf{h}_1$ to $\mathbf{h}_2$ along $\mathbf{k}$ to be $\frac{1}{2}\varphi_0^*$, for any φ_0^* such that $0 \leqslant \varphi_0 - \varphi_0^* \leqslant \varepsilon$. Let T_2 denote the point of the table reached by the ball at the end of the second move. The available locus of T_2 under all these choices, is a set of circular arcs, described when the points of a skew integral lattice are subjected to an angular distortion ε in the fundamental lattice angle. Some point, on one of these arcs sufficiently far from the desired final position T^* of the ball on the table, will be such that the distance T_2T^* is an even multiple of π. Consequently, we can replace the third and fourth even moves by a single third even move. Of course, the ball will then arrive at T^* in three moves with a spin discrepancy not exceeding ε. However, since $\varepsilon > 0$ is arbitrary, this means that we can obtain, in 3 moves of the integral version, an approximate final state as close to the desired final state as we please. An appeal to continuity, accompanied by a certain amount of topological hand–waving, therefore suggests that

$$N_{\text{even}} = N_{\text{odd}} = N_{\text{integral}} = 3 \ . \tag{5.1}$$

Let us accordingly see whether or not we can substantiate (5.1) by a rigorous argument.

6. The even version of Kendall's problem.

We have to consider equations (3.3) and (3.4) with $N = 3$ and one of the φ_p equal to an even multiple of π. The corresponding quaternion $e^{\frac{1}{2}\varphi_p\mathbf{h}_p}$ will be ± 1, which commutes with the other quaternions on the left–hand side of (3.4). The addition on the left–hand side of (3.3) is of course commutative. Hence there is no loss of generality in assuming that the even move is the third move. So (3.3) and (3.4) become

$$\varphi_1\mathbf{h}_1 + \varphi_2\mathbf{h}_2 = \theta\mathbf{h} - n_3\pi\mathbf{h}_3 \ , \tag{6.1}$$

$$e^{\frac{1}{2}\varphi_2\mathbf{h}_2} e^{\frac{1}{2}\varphi_1\mathbf{h}_1} = \pm e^{\frac{1}{2}\varphi\mathbf{u}} \ , \tag{6.2}$$

where n_3 may be any even integer. Here the prescribed angle φ can be interpreted modulo 2π; so we assume $0 \leqslant \varphi < 2\pi$. If $\varphi = 0$, the initial and final orientations are the same; we can reach the final state in two even moves, as in §4; so we have a solution with $n_3 = 0$. Thus we can assume hereafter that

$$0 < \varphi < 2\pi \ , \quad -1 < c = \cos\tfrac{1}{2}\varphi < 1 \ , \quad s = \sin\tfrac{1}{2}\varphi > 0 \ . \tag{6.3}$$

Let ψ be the angle between $\mathbf{u}$ and $\mathbf{k}$. If we replace $\mathbf{u}$ by $-\mathbf{u}$ and at the same time φ by $2\pi - \varphi$, the right–hand side of (6.2) is not changed. So there is no loss of generality in assuming that $0 \leqslant \psi \leqslant \frac{1}{2}\pi$. Similarly, we can suppose $\theta \geqslant 0$, since if necessary we can replace θ by $-\theta$ and $\mathbf{h}$ by $-\mathbf{h}$ on the right–hand side of (6.1). If $\psi = \frac{1}{2}\pi$, then $\mathbf{u}$ is horizontal; so, by taking $\varphi_1\mathbf{h}_1 = \varphi\mathbf{u}$ for the first move, we correctly orient the ball at the end of the first move; whereafter we can reach the final state in two even moves, as in §4. So there is a 3–move solution when $\psi = \frac{1}{2}\pi$, and we may hereafter suppose that

$$0 \leqslant \psi < \tfrac{1}{2}\pi \ . \tag{6.4}$$

This leaves us with three mutually exclusive and exhaustive cases:

Case 1. $0 < \psi < \frac{1}{2}\pi \ , \quad \theta \geqslant 0$;
Case 2. $\psi = 0 \ , \quad \theta > 0$;
Case 3. $\psi = \theta = 0$.

Before embarking upon details, we give an outline of the plan for handling (6.1) and (6.2). We introduce a parameter λ which can vary over $0 \leqslant \lambda \leqslant 2\pi$; and we obtain, for each such value of λ, a solution $\varphi_p\mathbf{h}_p$ $(p=1,2)$ of (6.2). We write P_λ for the point at the end of the vector $\varphi_1\mathbf{h}_1 + \varphi_2\mathbf{h}_2$; and we show that, as λ varies over $0 \leqslant \lambda \leqslant 2\pi$, P_λ describes a continuous closed curve Γ in the horizontal plane Oxy. As $\mathbf{h}_3$ varies over all horizontal directions, the right–hand side of (6.1) represents a circle of radius $|n_3|\pi$ centred at $\theta\mathbf{h}$. If we can show, for a suitable choice of n_3, that this circle intersects the curve Γ, then (6.1) will be satisfied as well as (6.2) and there will be a 3–move solution of the even version of Kendall's problem.

7. Even version: case 1.

Let $\mathbf{h}_0$ be a fixed horizontal vector that is perpendicular to the prescribed vector $\mathbf{u}$; and let $\mathbf{u}_1$ and $\mathbf{u}_2$ be perpendicular to $\mathbf{u}$, such that $\lambda - \frac{1}{4}\varphi$ is the angle from $\mathbf{h}_0$ to $\mathbf{u}_1$ along $\mathbf{u}$ and $\lambda + \frac{1}{4}\varphi$ is the angle from $\mathbf{h}_0$ to $\mathbf{u}_2$ along $\mathbf{u}$. As λ varies over $0 \leqslant \lambda \leqslant 2\pi$, the ends of $\mathbf{u}_1$ and $\mathbf{u}_2$ move round a circle C, whose plane is perpendicular to $\mathbf{u}$; but the angle from $\mathbf{u}_1$ to $\mathbf{u}_2$ along $\mathbf{u}$ remains fixed at $\frac{1}{2}\varphi$, and (2.1) yields

$$\pm\, e^{\frac{1}{2}\varphi\mathbf{u}} = \pm\, \mathbf{u}_2\mathbf{u}_1 = \mp\, (\mathbf{u}_2\mathbf{k})(\mathbf{k}\mathbf{u}_1) \ , \tag{7.1}$$

since $\mathbf{k}^2 = -1$ commutes with $\mathbf{u}_2$. Write $\frac{1}{2}\varphi_p^*$ $(p=1,2)$ for the angle between $\mathbf{u}_p$ and $\mathbf{k}$. Clearly, φ_p^* is a continuous function of λ satisfying $0 < \frac{1}{2}\varphi_p^* < \pi$ with strict inequality, because otherwise $\mathbf{u}$ would be horizontal in violation of the hypothesis $0 < \psi < \frac{1}{2}\pi$ made for case 1. Define γ by

$$0 < \gamma = \cos\psi < 1 \ ; \tag{7.2}$$

and choose a fixed integer n such that

$$n > \left(\frac{5}{2} + \frac{\theta}{2\pi}\right) \Big/ \{(1+\gamma^2 \tan^2 \tfrac{1}{4}\varphi)^{-\frac{1}{2}} - (1+\gamma^{-2}\tan^2\tfrac{1}{4}\varphi)^{-\frac{1}{2}}\}. \tag{7.3}$$

This is possible because of (6.3) and (7.2). Then define

$$\varphi_1 = 2n\pi - \varphi_1^* \ , \quad \varphi_2 = 2n\pi + \varphi_2^* \ , \tag{7.4}$$

and note that φ_1 and φ_2 are continuous functions of λ. For $p = 1,2$, the vectors $\mathbf{k} \times \mathbf{u}_p$ are non–zero, because $0 < \frac{1}{2}\varphi_p^* < \pi$, and horizontal because $\mathbf{k}$ is vertical; and hence we can define uniquely a horizontal unit vector $\mathbf{h}_p$ having the same direction as $\mathbf{k} \times \mathbf{u}_p$. Now, by (7.4),

$$\begin{aligned} \pm e^{\frac{1}{2}\varphi_1 \mathbf{h}_1} &= \pm(\cos\tfrac{1}{2}\varphi_1 + \mathbf{h}_1 \sin\tfrac{1}{2}\varphi_1) = \cos\tfrac{1}{2}\varphi_1^* - \mathbf{h}_1 \sin\tfrac{1}{2}\varphi_1^* \\ &= \mathbf{k}.\mathbf{u}_1 - \mathbf{k} \times \mathbf{u}_1 = -\mathbf{k}\mathbf{u}_1 \ , \end{aligned} \tag{7.5}$$

and

$$\begin{aligned} \pm e^{\frac{1}{2}\varphi_2 \mathbf{h}_2} &= \pm(\cos\tfrac{1}{2}\varphi_2 + \mathbf{h}_2 \sin\tfrac{1}{2}\varphi_2) = \cos\tfrac{1}{2}\varphi_2^* + \mathbf{h}_2 \sin\tfrac{1}{2}\varphi_2^* \\ &= \mathbf{k}.\mathbf{u}_2 + \mathbf{k} \times \mathbf{u}_2 = -\mathbf{u}_2\mathbf{k} \ . \end{aligned} \tag{7.6}$$

Assembly of (7.1), (7.5), and (7.6) shows that $\varphi_1\mathbf{h}_1$ and $\varphi_2\mathbf{h}_2$ satisfy (6.2).

The directions of $\mathbf{h}_p$ $(p=1,2)$, that is to say the directions of $\mathbf{k} \times \mathbf{u}_p$, are continuous functions of λ for $0 \leqslant \lambda \leqslant 2\pi$. Hence the point P_λ, at the end of the vector $\varphi_1\mathbf{h}_1 + \varphi_2\mathbf{h}_2$, moves continuously as λ goes continuously from 0 to 2π. Let $r = r(\lambda)$ and $r^* = r^*(\lambda)$ denote the distances of P_λ from the origin and from the point $\theta\mathbf{h}$ respectively. Then r^* is a continuous function of λ, which attains its upper and lower bounds $r^*_{\max}$ and $r^*_{\min}$ in $0 \leqslant \lambda \leqslant 2\pi$. The continuous closed curve Γ, traced out by P_λ, will intersect a circle of radius $|n_3|\pi$ centred at $\theta\mathbf{h}$ for at least one even integer n_3 provided that

$$r^*_{\max} - r^*_{\min} \geqslant 2\pi \ . \tag{7.7}$$

In that event, there will exist a value of λ for which (6.1) is true for this integer n_3, and case 1 will have a 3–move solution. Hence it suffices to verify (7.7). The triangle inequality shows that

$$r - \theta \leqslant r^* \leqslant r + \theta \tag{7.8}$$

for all values of λ. Hence

$$r^*_{\max} \geqslant r(0) - \theta \ , \quad r^*_{\min} \leqslant r(\tfrac{1}{2}\pi) + \theta \ ; \tag{7.9}$$

and so

$$r^*_{\max} - r^*_{\min} \geqslant r(0) - r(\tfrac{1}{2}\pi) - 2\theta \ . \tag{7.10}$$

Now

$$r(\lambda) = \sqrt{(\varphi_1^2 + \varphi_2^2 + 2\varphi_1\varphi_2 \cos\chi_\lambda)} \ , \tag{7.11}$$

where χ_λ is the angle between $\mathbf{h}_1$ and $\mathbf{h}_2$. Also we get from (7.4) and the inequalities $0 < \frac{1}{2}\varphi_p^* < \pi$

$$2(n-1)\pi < \varphi_p < 2(n+1)\pi , \qquad (p=1,2) \tag{7.12}$$

Hence

$$r(0) > 2(n-1)\pi\sqrt{(2+2\cos\chi_0)} , \quad r(\tfrac{1}{2}\pi) < 2(n+1)\pi\sqrt{(2+2\cos\chi_{\frac{1}{2}\pi})}; \tag{7.13}$$

and

$$r(0) - r(\tfrac{1}{2}\pi) > 2n\pi\{\sqrt{(2+2\cos\chi_0)} - \sqrt{(2+2\cos\chi_{\frac{1}{2}\pi})}\} - 8\pi. \tag{7.14}$$

Since $\mathbf{h}_1$ and $\mathbf{h}_2$ have the same directions as $\mathbf{k}\times\mathbf{u}_1$ and $\mathbf{k}\times\mathbf{u}_2$ respectively, χ_λ is the angle between $\mathbf{k}\times\mathbf{u}_1$ and $\mathbf{k}\times\mathbf{u}_2$; and this is equal to the angle between the orthogonal projections of $\mathbf{u}_1$ and $\mathbf{u}_2$ on the horizontal plane Oxy. When $\lambda=0$, these projections of $\mathbf{u}_1$ and $\mathbf{u}_2$ each have the same component $\cos\frac{1}{4}\varphi$ in the direction of $\mathbf{h}_0$, and likewise have equal and opposite components $\pm\gamma\sin\frac{1}{4}\varphi$ perpendicular to $\mathbf{h}_0$. Hence

$$\tan\tfrac{1}{2}\chi_0 = \gamma\tan\tfrac{1}{4}\varphi . \tag{7.15}$$

On the other hand, when $\lambda = \frac{1}{2}\pi$, the projections of $\mathbf{u}_1$ and $\mathbf{u}_2$ each have the same component $\gamma\cos\frac{1}{4}\varphi$ perpendicular to $\mathbf{h}_0$, and have equal and opposite components $\pm\sin\frac{1}{4}\varphi$ in the direction of $\mathbf{h}_0$, giving

$$\tan\tfrac{1}{2}\chi_{\frac{1}{2}\pi} = \gamma^{-1}\tan\tfrac{1}{4}\varphi . \tag{7.16}$$

Thus

$$\sqrt{(2+2\cos\chi_0)} - \sqrt{(2+2\cos\chi_{\frac{1}{2}\pi})} = 2(1+\gamma^2\tan^2\tfrac{1}{4}\varphi)^{-\frac{1}{2}} - 2(1+\gamma^{-2}\tan^2\tfrac{1}{4}\varphi)^{-\frac{1}{2}}. \tag{7.17}$$

Assembly of (7.3), (7.11), (7.14), and (7.17) establishes (7.7) and so completes the 3–move solution of case 1.

8. Even version: case 2.

Since $0 < \varphi < 2\pi$ according to (6.3), we have $e^{\frac{1}{2}\varphi\mathbf{u}} \neq \pm 1$ in (6.2). Whenever $\psi = 0$, as required in cases 2 and 3, $\mathbf{u}$ is perpendicular to $\mathbf{h}_1$ and $\mathbf{h}_2$. Therefore we can invoke not only (2.1) but also its converse (2.2) to show that the *general* solution of (6.2) is

$$\pm(\cos\tfrac{1}{2}\varphi_p + \mathbf{h}_p\sin\tfrac{1}{2}\varphi_p) = \pm e^{\frac{1}{2}\varphi_p\mathbf{h}_p} = \pm\mathbf{h}_p \quad (p=1,2) , \tag{8.1}$$

where $\frac{1}{2}\varphi$ is the angle from $\mathbf{h}_1$ to $\mathbf{h}_2$ along $\mathbf{u} = \mathbf{k}$. Hence

$$\varphi_1 = n_1\pi , \quad \varphi_2 = n_2\pi , \tag{8.2}$$

where n_1 and n_2 are odd integers (not necessarily positive). Hence the closed curve Γ is a circle, centred at the origin O, with radius

$$r = \pi\sqrt{(n_1^2 + n_2^2 + 2n_1n_2c)} . \tag{8.3}$$

in view of (6.3). A sufficient condition that this shall intersect a circle, centred at $\theta\mathbf{h}$, with radius $|n_3|\pi$ for some even integer n_3, is that

$$0 \leqslant r - \theta \leqslant n_3\pi \leqslant r + \theta \ . \tag{8.4}$$

To establish the existence of a 3–move solution in case 2, it therefore suffices to choose integers n_1, n_2 (both odd), and n_3 (even) to satisfy (8.3) and (8.4). We do this as follows.

By hypothesis, $\theta > 0$ in case 2; so (6.3) allows to choose an odd integer

$$n_1 \geqslant \max(16\pi^2/s\theta^2,\ \theta/\pi s) \ . \tag{8.5}$$

Next we choose the even integer n_3 that satisfies

$$0 \leqslant n_3 - n_1 s < 2 \ , \tag{8.6}$$

and then the odd integer n_2 that satisfies

$$-2 < n_2 + n_1 c - \surd\{2n_1 s(n_3 - n_1 s)\} \leqslant 0 \ . \tag{8.7}$$

For brevity we write

$$m = n_2 + n_1 c \ ; \qquad \mu = m^2/n_1^2 s^2 \ . \tag{8.8}$$

Since $c^2 + s^2 = 1$, we can rewrite (8.3) in the form

$$r/\pi = n_1 s\surd(1+\mu) = n_1 s\{1 + \tfrac{1}{2}\mu - \tfrac{1}{8}\mu^2(1 + \tau\mu)^{-\frac{3}{2}}\} \tag{8.9}$$

where $0 < \tau < 1$ according to Taylor's theorem. From (8.7) and (8.8) we get

$$4 - 4\surd\{2n_1 s(n_3 - n_1 s)\} + 2n_1 s(n_3 - n_1 s) \leqslant m^2 \leqslant 2n_1 s(n_3 - n_1 s); \tag{8.10}$$

and hence, by (8.6) and (8.8),

$$(2/n_1 s) - 4(n_1 s)^{-\frac{1}{2}} + n_3 - n_1 s \leqslant \tfrac{1}{2}\mu n_1 s \leqslant n_3 - n_1 s < 2 \ . \tag{8.11}$$

Therefore

$$\tfrac{1}{8}n_1 s\mu^2 = (\tfrac{1}{2}\mu n_1 s)^2/(2n_1 s) < 2/n_1 s \ . \tag{8.12}$$

By (8.5), (8.11), and (8.12)

$$\tfrac{1}{8}n_1 s\mu^2 - (\theta/\pi) \leqslant (2/n_1 s) - 4(n_1 s)^{-\frac{1}{2}} \leqslant -n_3 + n_1 s(1 + \tfrac{1}{2}\mu) \leqslant 0 \ . \tag{8.13}$$

From (8.9) and (8.12)

$$r/\pi \leqslant n_1 s(1 + \tfrac{1}{2}\mu) \leqslant n_3 \leqslant n_1 s(1 + \tfrac{1}{2}\mu - \tfrac{1}{8}\mu^2) + (\theta/\pi) \leqslant (r+\theta)/\pi \ . \tag{8.14}$$

On the other hand, by (8.5) and (8.9)

$$r/\pi \geqslant n_1 s \geqslant \theta/\pi \ ; \tag{8.15}$$

whereupon (8.4) follows from (8.14) and (8.15); and we have established the existence of a 3–move solution in case 2.

9. Even version: case 3.

In (8.1) and (8.2) we derived the general solution of (6.2) for case 3 as well as case 2. The curve Γ remains a circle with centre O and radius r given by (8.3). Because $\theta = 0$ in case 3, r must be an even multiple of π in order to satisfy (6.1) Hence the necessary and sufficient condition for a 3–move solution in case 3 is that

$$\cos\tfrac{1}{2}\varphi = (n_3^2 - n_1^2 - n_2^2)/2n_1n_2 , \tag{9.1}$$

where n_1 and n_2 are odd integers and n_3 is an even integer. This now includes two extra cases $\varphi = 0$, $n_1 = n_2 = 1$, $n_3 = 2$ and $\varphi = 2\pi$, $n_1 = n_2 = 1$, $n_3 = 0$, already known to have 3–move solutions though previously set aside by the temporary assumption $0 < \varphi < 2\pi$ employed for the solution of (6.2) in §8. The right of (9.1) is equal to $2l/2m$, where l and m are both odd; so (9.1) becomes $(m + l)n(n - n_3) = (m - l)(n - n_1)(n - n_2)$ where $n = \tfrac{1}{2}(n_1 + n_2 + n_3)$. Hence $m \pm l$ are both even, and just one of them is divisible by 8 because $n \equiv n - n_3 \not\equiv n - n_1 \equiv n - n_2$ (modulo 2). Thus

$$\cos\tfrac{1}{2}\varphi = l/m; \quad 1 \leqslant l^2 \leqslant m^2; \quad l^2 \equiv m^2 \text{ (modulo 16)}; \quad l,m \text{ odd.} \tag{9.2}$$

Conversely, given any odd integers l and m satisfying (9.2), we can satisfy (9.1) by $n_1 = m$, $n_2 = \tfrac{1}{8}(m^2 - l^2) - l - 2$, $n_3 = \tfrac{1}{8}(m^2 - l^2) + 2$. We write Φ for the set of all angles φ expressible in the form (9.2), and conclude that case 3 has a 3–move solution if and only if $\varphi \in \Phi$. Clearly, Φ is a dense countable subset of the set of all angles. This, together with the results already proved in cases 1 and 2, rigorously justifies the claim, made in §5 on heuristic grounds, that any final state can be approximated arbitrarily closely in 3 moves. On the other hand, it disproves (5.1). In view of (4.1), $N_{\text{even}} = 4$ is the correct result. In particular, $\pi \notin \Phi$ and the case $\psi = \theta = 0$, $\varphi = \pi$ has no 3–move solution in the even version of Kendall's problem.

10. The odd and integral versions of Kendall's problem.

We shall now show that the particular case

$$\psi = \theta = 0 , \quad \varphi = \pi \tag{10.1}$$

has no 3–move solution in the odd version of Kendall's problem. Taken with the remark at the end of §9, and with (4.1), this will prove that

$$N_{\text{even}} = N_{\text{odd}} = N_{\text{integral}} = 4 . \tag{10.2}$$

Suppose, for the sake of a contradiction, that a 3–move solution of the odd version exists for (10.1). Then we can put (3.4) in the form

$$q_3q_2q_1 = \pm\mathbf{k} , \tag{10.3}$$

where

$$q_p = \cos\tfrac{1}{2}\varphi_p + \mathbf{h}_p\sin\tfrac{1}{2}\varphi_p = c_p + s_p\mathbf{h}_p , \quad (p = 1,2,3) \tag{10.4}$$

and, for at least one of these three values of p,

$$q_p = \pm \mathbf{h}_p . \tag{10.5}$$

Also (3.3) becomes

$$\sum_{p=1}^{3} \varphi_p \mathbf{h}_p = \mathbf{0} . \tag{10.6}$$

If (10.3) holds then, by (2.12)

$$\bar{q}_1 \bar{q}_2 \bar{q}_3 = \mp \mathbf{k} . \tag{10.7}$$

Replacing φ_p by $-\varphi_p$ ($p=1,2,,3$) and noting that the addition in (10.6) is commutative, we see that if there is a solution with $p = 1$ in (10.5) there is also a solution with $p = 3$ in (10.5), and vice versa. Hence it is enough to consider only the two cases $p = 1$ and $p = 2$ in (10.5). We may also suppose that

$$s_p \neq 0 \quad (p=1,2,3) , \tag{10.8}$$

for otherwise one of the moves would be an even move, and no solution could exist by virtue of §9.

If $q_1 = \pm \mathbf{h}_1$, then (10.3) gives

$$\begin{aligned} &(c_3 c_2 - s_3 s_2 \mathbf{h}_3.\mathbf{h}_2) + (c_3 s_2 \mathbf{h}_2 + c_2 s_3 \mathbf{h}_3 + s_3 s_2 \mathbf{h}_3 \times \mathbf{h}_2) \\ &= q_3 q_2 = \pm \mathbf{k} \bar{\mathbf{h}}_1 = \mp \mathbf{k} \mathbf{h}_1 = \mp \mathbf{k} \times \mathbf{h}_1 = \mathbf{h}_0 , \end{aligned} \tag{10.9}$$

where $\mathbf{h}_0$ is a unit horizontal vector. Taking the vector component of each side and resolving it in the direction $\mathbf{k}$, we have

$$\mathbf{h}_3 \times \mathbf{h}_2 = \mathbf{0} , \tag{10.10}$$

because of (10.8). Hence $\mathbf{h}_3 = \pm \mathbf{h}_2$ and we may combine the second and third moves into a single move. Therefore there is a two move solution of (10.1), which contradicts the remark in the first paragraph of §4.

If $q_2 = \pm \mathbf{h}_2$, (10.3) gives

$$\begin{aligned} &(- s_3 \mathbf{h}_3.\mathbf{h}_2) + (c_3 \mathbf{h}_2 + s_3 \mathbf{h}_3 \times \mathbf{h}_2) = \pm q_3 q_2 = \pm \mathbf{k} \bar{q}_1 \\ &= \pm (c_1 \mathbf{k} - s_1 \mathbf{k} \times \mathbf{h}_1). \end{aligned} \tag{10.11}$$

The scalar component of (10.11) gives

$$s_3 \mathbf{h}_3.\mathbf{h}_2 = 0 , \tag{10.12}$$

and the vector component of (10.11) resolved horizontally gives

$$c_3 \mathbf{h}_2 = \mp s_1 \mathbf{k} \times \mathbf{h}_1 . \tag{10.13}$$

In view of (10.8), we see from (10.12) and (10.13) that $\mathbf{h}_3$ and $\mathbf{h}_1$ are both perpendicular to $\mathbf{h}_2$. Hence, resolving (10.6) in the direction $\mathbf{h}_2$, we get $\varphi_2 = 0$, which contradicts (10.8). This completes the proof of (10.2).

In §§7, 8, 9 we showed that the only final states requiring 4 moves in the even version were those for which $\psi = \theta = 0$ and $\varphi \notin \Phi$. We leave it as an exercise for the reader to determine completely which final states require 4 moves in the odd (and hence the integral) versions of Kendall's problem. This will give an opportunity to those readers, who wish to practise manipulating quaternions, and it will also save this paper from becoming unduly long.

It is worth recording that in 1950s there were a few professional mathematicians, who so much believed in the possibility of a topological argument along the lines of §5 as to declare Kendall's problem too trivial to be worth their attention. In retrospect, having established (10.2), we realise at once why topology must fail to verify (5.1): the infinite Euclidean plane is not compact. But now we can appeal to topology to provide an answer in the opposite direction. Since every bounded set of the plane is a subset of some closed set, there must be final states which *cannot* be approximated arbitrarily closely in 3 moves, one of which is integral, on a bounded table. In cases 1 and 2 of the even version, the inequalities (7.3) and (8.5) require very long moves if ψ or θ is very small; and, although we do not claim that (7.3) or (8.5) are as economical as possible, arbitrarily long moves must be inescapable in these circumstances. The question of economising the total length of track covered will be examined further in §§12, 13, and 14.

11. The isosceles version of Kendall's problem.

In deriving (3.5) we assumed that the restriction placed on one move was independent of other moves. Without that assumption, (3.5) may be false. For example, under the restriction that all moves be integral, the final state can be attained if and only if the initially vertical diameter of the ball is also finally vertical (up or down), no matter how many moves are allowed. The *isosceles version* of Kendall's problem arises when two moves must be of equal length; and, since this condition on one such move is not independent of other moves, (3.5) might fail. However, in the procedure adopted in §4, the third and fourth moves had equal lengths. So a 4–move isosceles solution is always possible, and

$$3 \leqslant N_{\text{isosceles}} \leqslant 4 . \qquad (11.1)$$

We shall examine case 3 of the isosceles version; so now we assume that $\mathbf{u} = \mathbf{k}$ and $\theta = 0$. We shall roll the ball around the perimeter of an isosceles triangle having sides of length α, α, $2a$, starting and finishing at the vertex where the two equal sides of length α meet and going round the triangle in an anticlockwise sense when looking down on the table. We wish to choose α and a so that the ball undergoes a rotation through a prescribed angle φ along $\mathbf{k}$. We shall write β for the semi–vertical angle of the triangle; so

$$a = \alpha \sin \beta , \qquad (0 \leqslant \beta < \tfrac{1}{2}\pi) . \qquad (11.2)$$

There is no loss of generality in choosing the direction of the axis Ox to be the axis of symmetry of the triangle. The axes of rotation of the three moves are perpendicular to the sides of the triangle, and hence

$$\varphi_1\mathbf{h}_1 = \alpha(\mathbf{i}\sin\beta + \mathbf{j}\cos\beta), \quad \varphi_2\mathbf{h}_2 = -2a\mathbf{i}\,, \quad \varphi_3\mathbf{h}_3 = \alpha(\mathbf{i}\sin\beta - \mathbf{j}\cos\beta)\,, \tag{11.3}$$

where **i** and **j** are unit vectors in the directions Ox and Oy. From (11.2) and (11.3), we see that (3.3) is trivially satisfied. We shall satisfy (3.4) if

$$\begin{aligned}\pm\ (\cos\tfrac{1}{2}\varphi + \mathbf{k}\sin\tfrac{1}{2}\varphi) &= \pm e^{\frac{1}{2}\varphi\mathbf{k}} = e^{\frac{1}{2}\varphi_3\mathbf{h}_3}e^{\frac{1}{2}\varphi_2\mathbf{h}_2}e^{\frac{1}{2}\varphi_1\mathbf{h}_1}\\ &= \{\cos\tfrac{1}{2}\alpha + (\mathbf{i}\sin\beta - \mathbf{j}\cos\beta)\sin\tfrac{1}{2}\alpha\}\{\cos a - \mathbf{i}\sin a\}\\ &\qquad\qquad \{\cos\tfrac{1}{2}\alpha + (\mathbf{i}\sin\beta + \mathbf{j}\cos\beta)\sin\tfrac{1}{2}\alpha\}\\ &= \{(\cos^2\beta + \sin^2\beta\cos\alpha)\cos a + \sin\alpha\sin\beta\sin a\}\\ &\qquad + \mathbf{i}\{\sin\alpha\sin\beta\cos a - \cos\alpha\sin a\}\\ &\qquad + \mathbf{k}\{(1-\cos\alpha)\sin\beta\cos\beta\cos a - \sin\alpha\cos\beta\sin a\}\end{aligned} \tag{11.4}$$

on performing the multiplications. In view of (11.2), the coefficient of **i** on the right of (11.4) will vanish if we take

$$a\cot a = \alpha\cot\alpha. \tag{11.5}$$

From (11.4) and (11.5) we find, after some straightforward trigonometrical manipulation,

$$\pm(\cos\tfrac{1}{2}\varphi + \mathbf{k}\sin\tfrac{1}{2}\varphi) = \tfrac{1}{2}\cos a(1-\sec\alpha)(\cos 2\beta - \cot^2\tfrac{1}{2}\alpha + \mathbf{k}\sin 2\beta). \tag{11.6}$$

Here we must equate the scalar components on each side, and also the vector components on each side. This gives two equations; but they are not independent, because $\cos^2\frac{1}{2}\varphi + \sin^2\frac{1}{2}\varphi = 1$. To ensure equality in (11.6) it is necessary and sufficient to choose the $\pm$ sign appropriately and to have the single equation of the coefficient ratios

$$\cot\tfrac{1}{2}\varphi = \cot 2\beta - \operatorname{cosec} 2\beta\cot^2\tfrac{1}{2}\alpha. \tag{11.7}$$

Therefore we shall have a 3–move isosceles solution of case 3 if and only if we can simultaneously satisfy (11.2), (11.5), and (11.7) for any prescribed φ in the interval $0 \leqslant \varphi < 2\pi$.

To achieve that, we regard a as a primary variable in the range $0 \leqslant a \leqslant \pi$, and $\alpha = \alpha(a)$ as the root of (11.5) that satisfies $\frac{5}{4}\pi < \alpha \leqslant 2\pi$. Since $a\cot a$ decreases steadily from $+1$ to $-\infty$ as a goes from 0 to π, while $\alpha\cot\alpha$ decreases steadily from $\frac{5}{4}\pi$ to $-\infty$ as α goes from $\frac{5}{4}\pi$ to 2π, this defines $\alpha(a)$ uniquely as a strictly increasing continuous function of a. Hence $0 \leqslant \sin\beta = a/\alpha < \frac{4}{5}$, which validates the inequality in (11.2) and shows that $\beta = \beta(a)$ is a continuous function of a such that $\sin 2\beta = 0$ if and only if $a = 0$. Hence the right–hand side of (11.7) is a finite continuous function of a for $0 < a < \pi$. Now $\frac{5}{4}\pi < \alpha(0) < \frac{3}{2}\pi$ by (11.5); so $0 < 1 - \cot^2\frac{1}{2}\alpha(0) < 1$. Hence the right–side of (11.7) tends to $+\infty$ as $a \to 0$, and to $-\infty$ as $a \to \pi$. Consequently we can always find a value of a such that (11.7) is satisfied for any prescribed φ in $0 \leqslant \varphi \leqslant 2\pi$. This completes the existence of a 3–move isosceles solution of case 3; and then, from §7 and §8, we conclude that $N = 3$ answers the question for a demy in §1.

It is easy to calculate α and φ as functions of a from (11.2), (11.5), and (11.7) with the aid of a pocket calculator. We get the results shown in Table 1, in which the lengths $2a$ and α are quoted in radians, and in which φ° denotes φ quoted in degrees.

TABLE 1

$2a$	α	φ°	$2a$	α	φ°	$2a$	α	φ°
0.0	4.4934	0.00	2.2	4.5910	155.47	4.4	5.0211	295.85
0.2	4.4942	14.29	2.4	4.6116	169.21	4.6	5.0957	306.62
0.4	4.4964	28.57	2.6	4.6346	182.82	4.8	5.1806	316.81
0.6	4.5001	42.84	2.8	4.6606	196.30	5.0	5.2775	326.30
0.8	4.5054	57.08	3.0	4.6897	209.62	5.2	5.3884	334.97
1.0	4.5123	71.30	3.2	4.7223	222.76	5.4	5.5153	342.68
1.2	4.5208	85.48	3.4	4.7588	235.68	5.6	5.6600	349.24
1.4	4.5310	99.61	3.6	4.7997	248.37	5.8	5.8237	354.44
1.6	4.5430	113.68	3.8	4.8456	260.77	6.0	6.0052	358.04
1.8	4.5569	127.69	4.0	4.8972	272.86	6.2	6.2001	359.83
2.0	4.5729	141.63	4.2	4.9553	284.57	2π	2π	360.00

By inverse interpolation in this Table, we can determine the lengths $2a$ and α required for any prescribed φ°. Thus, when $\varphi^\circ = 180$, we get

$$2a = 2.5584 \,, \qquad \alpha = 4.6296 \,. \tag{11.8}$$

In fact, we can obtain any φ from just the first half of Table 1 up to the values in (11.8); for, if we trace the triangle clockwise (looking down on the table) instead of anticlockwise, we shall get $-\varphi = 2\pi - \varphi$ (modulo 2π). Thus, we can always spin the ball, initially and finally at the centre of a *finite* circular table of radius 4.6296, through any desired angle about its vertical axis in three moves. The total distance travelled, using the first half of Table 1, never exceeds 11.8176; and it is a reasonable guess that no other 3–move strategy can have a shorter total travel if it has to cater for all possible spins with a return to the starting point on the table. The second half of Table 1 for $2.5584 \leqslant 2a \leqslant 2\pi$ also permits, with clockwise or anticlockwise travel around the triangle, any angle of spin, and serves a different purpose: given changes in a now cause small changes in φ°, especially when φ° is near 360; so we can expect to get a more accurate value of φ, should errors occur in traversing the triangle. This raises a stochastic version of Kendall's problem. Suppose that, in attempting to roll the ball along some desired path, we deviate from that path under the influence of some stochastic process. What then will be the resulting error in the final state, and how should we choose a rolling strategy to minimize this final error?

12. RCB solutions and other problems.

Ruler–and–compasses constructions, traditional in Euclidean plane geometry, have an aesthetic appeal; and there is really no reason why we should not contemplate their extension to three dimensions. For example, given a point P not on a plane Π, we could use the compasses in an obvious way to trace out the circle that is the intersection of Π with a sphere of given radius centred at P. Can one obtain a ruler–and–compass construction for the 4–move solution of §4? The answer is "No" because, if it were possible, we should be able to construct a length π on the table given the length 2 of the diameter of the ball; whereas it is well known that the circle cannot be squared. However, besides the ruler and compasses, we also have a ball; and we might regard the ball as supplementary equipment, thus entertaining the idea of a ruler–and–compasses–and–ball solution (or, for short, an RCB construction).

It is easy to check that the 4–move solution in §4 is RCB–constructible, since essentially it involves nothing more than (i) halving a given angle, (ii) finding the intersection of two circles, and (iii) rolling the ball along a straight line until a prescribed point on its surface comes into contact with the table. Here, the operation (iii) is, by hypothesis, a legitimate operation in an RCB solution,

I do not know, and the reader may care to investigate, whether or not a 3–move RCB solution exists. The treatment of case 2 in §8 is RCB–constructible, since all the multiplications and the trigonometrical ratios and the square roots involved in the specific formulae, like (8.7), can be achieved by ordinary Euclidean constructions (though it would be tedious to spell out the details). On the other hand, the treatment of case 1 in §7 and of case 3 in §11 would appear to be beyond the reach of an RCB method.

Although we did manage to prove in §11 that $N = 3$, we can hardly be satisfied with the method, which is little more than that lowest of all forms of mathematical argument, a mere existence proof. Moreover, the solution has an unpleasant and regrettable feature of an unboundedly long track even when the initial and final states are quite close together. If T_N denotes the total length of the track on the table in an N–move solution, we must of course have the inequality $T_N \geqslant \Delta$, where Δ is the distance between the points of the table where the ball rests initially and finally. However if $N \geqslant 4$, there exist best–possible constants δ_N such that some optimum N–move strategy will always guarantee that

$$\Delta \leqslant T_N \leqslant \Delta + \delta_N \,. \tag{12.1}$$

We can show that $\delta_4 \leqslant 8\pi$ by simple modification of the analysis of §4. In §4, the first move had length not exceeding 2π and served to bring the antipodal point of B onto the table at the end of the move. But we can bring this antipodal point onto the table directly in a first move whose length does not exceed π. The second move has length π. At the end of the second move the distance of the ball from its final position will not exceed $\Delta + 2\pi$, and for the third and fourth moves we can choose two even moves, each of equal length not exceeding $\frac{1}{2}\Delta + 3\pi$.

Hence $\delta_4 \leqslant 8\pi$. Clearly δ_N is, virtually by its definition, a non–increasing function of N. In §14 we shall prove that $\delta_N \geqslant \pi\sqrt{3}$ for every N. Hence

$$8\pi \geqslant \delta_4 \geqslant \delta_5 \geqslant \ldots \geqslant \pi\sqrt{3}. \tag{12.2}$$

This raises the problem of determining the exact values of these constants δ_N, and of finding the associated optimum strategies. It is a likely guess that δ_3 is also finite, though I do not know how to prove this.

To sketch briefly a possible line of attack on the finiteness of δ_3, consider rolling the ball around the perimeter of a triangle whose circumradius is R and whose angles are A_1, A_2, A_3. The total length of track is then

$$T = 2R(\sin A_1 + \sin A_2 + \sin A_3) \; . \tag{12.3}$$

The resulting rotation is represented by the quaternion $\pm q$, given by

$$\begin{aligned} q = {} & (c_1c_2c_3 + c_1s_2s_3\cos A_1 + c_2s_3s_1\cos A_2 + c_3s_1s_2\cos A_3) \\ & + \mathbf{i}(s_1s_2s_3\sin A_1 + c_1c_2s_3\sin A_2 - c_3c_1s_2\sin A_3) \\ & + \mathbf{j}(c_2c_3s_1 + s_1s_2s_3\cos A_1 - c_1c_2s_3\cos A_2 - c_3c_1s_2\cos A_3) \\ & + \mathbf{k}(-c_1s_2s_3\sin A_1 + c_2s_3s_1\sin A_2 - c_3s_1s_2\sin A_3) \end{aligned} \tag{12.4}$$

where

$$c_p = \cos(R\sin A_p), \qquad s_p = \sin(R\sin A_p), \qquad (p=1,2,3). \tag{12.5}$$

In obtaining (12.4), we have supposed that the triangle is described anticlockwise (as seen from above the table) and in the order of the sides opposite A_1, A_2, A_3 respectively; and we have chosen the co–ordinate axes so that the side opposite A_1 lies along the x–axis. Now suppose that we could prove that, for any given unit quaternion q, the equations (12.4) and (12.5) posssess a solution (R, A_1, A_2, A_3) with

$$A_1 + A_2 + A_3 = \pi \quad \text{and} \quad R \geqslant 0 \quad \text{and} \quad A_p \geqslant 0 \quad (p=1,2,3); \tag{12.6}$$

and let T^* be an upper bound for T in (12.3), taken over all such solutions for all q. Then we should be able to conclude that $\delta_3 \leqslant T^*$. For, in going from any initial to any final state, we could first roll the ball from its initial to its final position on the table (that is to say a distance Δ), we could also choose the co–ordinate system so that the x–axis lay along the direction thus far travelled, and we could complete the process by rolling around the perimeter of the aforementioned triangle (this being an additional distance not exceeding T^*). The difficulty in this suggested argument lies in showing that (12.4) and (12.5) actually have solutions satisfying (12.6), let alone in obtaining these solutions in a sufficiently explicit form to permit an estimate of T^*. Indeed, this difficulty is the reason for side–stepping (12.4) and (12.5) by means of the even version of Kendall's problem used in §6, 7, 8.

Another problem is to evaluate the limit of the sequence δ_N as $N \to \infty$. When $N \to \infty$, the path, along which the ball rolls, may become a curve; and we now turn to this question. For simplicity, we shall assume that this path has finite curvature.

13. Rolling along a curve.

I have not yet been able to solve completely the problem (*a*) of finding on the table a curved path Γ, of shortest total length T, that will deliver the ball to a prescribed final state. However, there is a simple solution to the much easier problem (*b*) of delivering the ball just to a prescribed final orientation without requiring it to have a prescribed final position on the table: in fact, Γ must then be either a straight line or an arc of a circle and $0 \leqslant T \leqslant \pi\sqrt{3}$, the upper bound being attained if and only if if the desired re–orientation is a rotation through π about a vertical axis.

It will cast some interesting light on the argument if we replace the spherical ball by a smooth convex body B that rolls geodesically on the table. We shall suppose that B has a finite strictly positive Gaussian curvature K at each point of its surface (for example, B might be a rugger ball). Then, at any instant, there will be a unique point P on the table and a unique point P^* on the surface of B such that P and P^* are in contact; and, as the body rolls along the curved path Γ, P will describe Γ on the table and P^* will describe a curve Γ^* on the surface of B. A little thought shows that (i) the lengths of Γ and Γ^* are equal, and (ii) the geodesic curvature of Γ^* at each P^* equals the ordinary curvature of Γ at the corresponding P. Since B is to be delivered from a given initial orientation to a given final orientation, the initial and final points on Γ^* (call them P_0^* and P_1^*) must be given points on B. Let Γ_0^* be a geodesic on B from P_0^* to P_1^*; let Γ_1^* be any other curve on B from P_0^* to P_1^*; and let Γ_0 and Γ_1 be the corresponding curves on the table. Let Δ be the region on the surface of B enclosed by the circuit from P_0^* to P_1^* along Γ_1^* followed by Γ_0^* traced backwards from P_1^* to P_0^*: here each element of area $\mathrm{d}S$ in Δ is to be counted with due (positive or negative) multiplicity according to the sense and multiplicity that this circuit surrounds it. Now, if the body is rolled from P_0 geodesically along the whole of either Γ_0 or Γ_1, the final orientations of B will differ only by a spin discrepancy equal to $\iint_\Delta K\mathrm{d}S$, by virtue of the Gauss–Bonnet theorem. Hence our problem (call it problem 1) reduces to this: Γ^* must be chosen to have shortest length T subject to prescribed end–points P_0^* and P_1^* and a prescribed value of $\iint_\Delta K\mathrm{d}S$, where Δ is now taken as the region with $\Gamma^* = \Gamma_1^*$.

Now consider a different problem (problem 2): choose Γ^* with shortest length T subject to prescribed end–points P_0^* and P_1^* and a prescribed value of $\iint_\Delta \mathrm{d}S$. Problem 2 has the known solution that the geodesic curvature of Γ^* must be constant [Weatherburn *Differential Geometry* (1927) Vol. 1, p. 122]. When the body B is a unit sphere, $K = 1$ and problems 1 and 2 coincide. Hence, by (i) and (ii) in the previous paragraph, Γ has constant curvature and is a straight line or a circular arc of length T. To show that $0 \leqslant T \leqslant \pi\sqrt{3}$ when B is a unit sphere, one has to calculate the area between a small circle and a great circle on B: this is straightforward and the details are left to the reader.

Can the foregoing results persist if B is not a unit sphere? There is a theorem, which seems to be due to Hilbert, that spheres are the only compact surfaces with constant Gaussian curvature [T. J. Willmore *Introduction to Differential Geometry* (1959) p. 131]; but that theorem is too restrictive here. For problem 1 and 2 to coincide, we only require that $\iint_\Delta K dS = \iint_\Delta dS$ for all regions Δ on the surface; so it is sufficient if $K = 1$ almost everywhere on a closed strictly convex surface. The condition $K = 1$ becomes

$$rt - s^2 = (1 + p^2 + q^2)^2 \tag{13.1}$$

in terms of the usual notation for the partial derivatives $p = \partial z/\partial x$, $q = \partial z/\partial y$, $r = \partial^2 z/\partial x^2$, $s = \partial^2 z/\partial x \partial y$, $t = \partial^2 z/\partial y^2$. Although (13.1) is ostensibly an equation of Monge's type [H. T. H. Piaggio *Elementary Treatise on Differential Equations* (1962) p. 183], I have not succeeded in obtaining its general solution by Monge's method. However, by writing

$$z = f(u), \quad u = x^2 + y^2, \quad g(u) = 4u[f'(u)]^2 \tag{13.2}$$

we can determine those particular solutions of (13.1) that are surfaces of revolution. From (13.1) and (13.2) we find

$$g'(u) = [1 + g(u)]^2; \tag{13.3}$$

whence

$$z = \int_0^\varphi (1 - k^2 \sin^2 \theta)^{\frac{1}{2}} d\theta, \quad x^2 + y^2 = k^2 \cos^2 \varphi, \tag{13.4}$$

where k is a constant satisfying $0 < k \leqslant 1$ and φ is a parameter that ranges over $-\frac{1}{2}\pi \leqslant \varphi \leqslant \frac{1}{2}\pi$. This represents a unit sphere when $k = 1$; for smaller values of k, it is a spindle-shaped surface of revolution with a sharp point at each of the ends $\varphi = \pm\frac{1}{2}\pi$; and, as $k \to 0$, the surface tends to a line segment of length π. The integral in (13.4) is the incomplete elliptic integral of the second kind, usually denoted by $E(\varphi, k)$ and tabulated in the *Smithsonian Elliptic Functions Tables* (1947). It is easy to verify that (13.4) is a strictly convex closed surface; and, of course, $K = 1$ everywhere except at the two end-points $\varphi = \pm\frac{1}{2}\pi$. The Gaussian curvature K is, by definition, the product of the two principal curvatures: as we approach the sharp end-points, one of these principal curvatures tends to zero while the other tends to infinity in such a way that their geometric mean remains equal to 1, so, if we like, we may *define* $K = 1$ by continuity at these two isolated end-points. We can also construct other surfaces of revolution from (13.4). For example, there is a discus-shaped surface obtained by cementing together two equal and opposite spherical caps; on such a discus we have $K = 1$ everywhere, except that K is infinite on the sharp rim between the two halves. Similarly, we can obtain a doubly-pointed discus with a fixed k in $0 < k < 1$, or a multi-rimmed discus by using different constants k in the different bands between successive rims. An amusing idea is to make k equal to a suitable Cantor ternary function of θ, thereby getting a non-spherical convex surface without sharp points or sharp rims. The result of this section (apart from suitable modifications to the value for the upper bound $\pi\sqrt{3}$) will persist for any non-spherical body B of the foregoing kind.

Notice also that, although there is a solution of (13.1) with $|\varphi| \leqslant \sin^{-1}(1/k)$ for a constant $k > 1$, such a solution is not a convex surface.

I am not aware of literature on the convex solutions of partial differential equations satisfied almost everywhere; and this topic merits further investigation.

14. Quaternion calculus of variations.

In this section I shall examine how the calculus of variations may be applied to problems (*a*) and (*b*) of §13. Standard techniques seem to give a nonsensical answer, but an *ad hoc* parametric technique is more fruitful. We shall roll the ball along the curved path Γ on the table, starting from the origin and writing t for the arc length along Γ measured from $t = 0$ at the origin. We write $q = q(t)$ for the quaternion such that $\pm q$ represents the resultant rotation of the ball at t. We take $q(0) = 1$ and suppose that $q(t)$ is a twice–differentiable function of t. We use dots to denote differentiation with respect to t. Let $\mathbf{h} = \mathbf{h}(t)$ be the unit vector along the instantaneous axis of rotation at t; so, in rolling from t to $t + \mathrm{d}t$, the ball rotates through $\mathrm{d}t$ about $\mathbf{h}$. Thus

$$q(t+\mathrm{d}t) = \mathrm{e}^{\frac{1}{2}\mathbf{h}\,\mathrm{d}t}\,q(t) = \{1 + \tfrac{1}{2}\mathbf{h}\,\mathrm{d}t + o(\mathrm{d}t)\}q(t); \tag{14.1}$$

whence

$$\dot{q} = \tfrac{1}{2}\mathbf{h}q. \tag{14.2}$$

Thus

$$\tfrac{1}{2}\mathbf{h} = \dot{q}\bar{q} = f_0 + f_1\mathbf{i} + f_2\mathbf{j} + f_3\mathbf{k}, \tag{14.3}$$

where the f_p denote the quaternion coordinates of $\frac{1}{2}\mathbf{h}$. If the coordinates of q are

$$q = \rho_0 + \rho_1\mathbf{i} + \rho_2\mathbf{j} + \rho_3\mathbf{k}\ , \tag{14.4}$$

then (14.3) can be written explicitly in the form

$$\begin{aligned} f_0 &= \dot{\rho}_0\rho_0 + \dot{\rho}_1\rho_1 + \dot{\rho}_2\rho_2 + \dot{\rho}_3\rho_3 \\ f_1 &= -\dot{\rho}_0\rho_1 + \dot{\rho}_1\rho_0 - \dot{\rho}_2\rho_3 + \dot{\rho}_3\rho_2 \\ f_2 &= -\dot{\rho}_0\rho_2 + \dot{\rho}_1\rho_3 + \dot{\rho}_2\rho_0 - \dot{\rho}_3\rho_1 \\ f_3 &= -\dot{\rho}_0\rho_3 - \dot{\rho}_1\rho_2 + \dot{\rho}_2\rho_1 + \dot{\rho}_3\rho_0. \end{aligned} \tag{14.5}$$

Since $\mathbf{h}$ is a horizontal vector, we have the constraints

$$f_0 = f_3 = 0 \tag{14.6}$$

identically for all t. Also in problem (*a*), though not in problem (*b*), $\int_0^T \mathbf{h}\,\mathrm{d}t$ gives the prescribed final position of the ball on the table; so $\int_0^T f_1\,\mathrm{d}t$ and $\int_0^T f_2\,\mathrm{d}t$ are also prescribed. We need to minimize $T = \int_0^T 1\,\mathrm{d}t$, subject to these constraints and subject to a prescribed final orientation $q(T)$. First variations will not alter a stationary value of T; so, for first variations, we can treat T as a constant. Thus we look for stationary values of $\int_0^T f\,\mathrm{d}t$, where

$$f = 1 + \lambda_0 f_0 + \lambda_1 f_1 + \lambda_2 f_2 + \lambda_3 f_3. \tag{14.7}$$

Here the λ_p are Lagrangian multipliers; λ_0 and λ_3 are functions of t, while

λ_1 and λ_2 are constants. We can cover problem (*b*) as well as problem (*a*) by taking $\lambda_1 = \lambda_2 = 0$ in problem (*b*).

The foregoing procedure is standard practice (covered in Theorem 5.1 of M. R. Hestenes (1966) *Calculus of Variations and Optimal Control Theory*, p. 263); but unfortunately it leads to a contradiction. I do not understand what has gone wrong, despite consulting one or two experts in the calculus of variations. Perhaps the reader can locate the cause of this trouble.

The Euler equations

$$\frac{\partial f}{\partial \rho_p} = \frac{\mathrm{d}}{\mathrm{d}t}\left(\frac{\partial f}{\partial \dot{\rho}_p}\right) \qquad (p=0,1,2,3) \tag{14.8}$$

become in matrix form

$$\begin{bmatrix} \rho_0 & \dot{\rho}_1 & \dot{\rho}_2 & \dot{\rho}_3 & \rho_3 \\ \rho_1 & -\dot{\rho}_0 & -\dot{\rho}_3 & \dot{\rho}_2 & \rho_2 \\ \rho_2 & \dot{\rho}_3 & -\dot{\rho}_0 & -\dot{\rho}_1 & -\rho_1 \\ \rho_3 & -\dot{\rho}_2 & \dot{\rho}_1 & -\dot{\rho}_0 & -\rho_0 \end{bmatrix} \begin{bmatrix} -\dot{\lambda}_0 \\ 2\lambda_1 \\ 2\lambda_2 \\ 2\lambda_3 \\ \dot{\lambda}_3 \end{bmatrix} = \begin{bmatrix} 0 \\ 0 \\ 0 \\ 0 \end{bmatrix}. \tag{14.9}$$

Taking norms in (14.3), we have

$$\tfrac{1}{4} = \dot{q}\bar{q}q\dot{\bar{q}} = \dot{q}\dot{\bar{q}}, \qquad q\bar{q} = 1; \tag{14.10}$$

and hence, on premultiplying (14.9) by

$$\begin{bmatrix} \rho_1 & -\rho_0 & \rho_3 & -\rho_2 \\ \rho_2 & -\rho_3 & -\rho_0 & \rho_1 \\ \dot{\rho}_1 & -\dot{\rho}_0 & \dot{\rho}_3 & -\dot{\rho}_2 \\ \dot{\rho}_2 & -\dot{\rho}_3 & -\dot{\rho}_0 & \dot{\rho}_1 \end{bmatrix}$$

and using (14.6) and (14.10) we get

$$\begin{bmatrix} 0 & 0 & 0 & -f_2 & 0 \\ 0 & 0 & 0 & f_1 & 0 \\ f_1 & \frac{1}{4} & 0 & 0 & f_2 \\ f_2 & 0 & \frac{1}{4} & 0 & -f_1 \end{bmatrix} \begin{bmatrix} -\dot{\lambda}_0 \\ 2\lambda_1 \\ 2\lambda_2 \\ 2\lambda_3 \\ \dot{\lambda}_3 \end{bmatrix} = \begin{bmatrix} 0 \\ 0 \\ 0 \\ 0 \end{bmatrix}. \tag{14.11}$$

The first two equations in (14.11) yield

$$\lambda_3 f_1 = \lambda_3 f_2 = 0. \tag{14.12}$$

But, according to (14.3) and (14.6), **h** is a unit vector; so $f_1^2 + f_2^2 = \frac{1}{4}$, and (14.12) now implies $\lambda_3 = 0$ and thence $\dot{\lambda}_3 = 0$. Thereupon the last two equations in (14.11) give

$$\dot{\lambda}_0 f_1 = \tfrac{1}{2}\lambda_1, \quad \dot{\lambda}_0 f_2 = \tfrac{1}{2}\lambda_2. \tag{14.13}$$

So $\frac{1}{4}\dot{\lambda}_0^2 = \dot{\lambda}_0^2(f_1^2 + f_2^2) = \frac{1}{4}(\lambda_1^2 + \lambda_2^2)$, which is constant. If $\dot{\lambda}_0 = 0$, then (14.13) provides $\lambda_1 = \lambda_2 = 0$, and the Euler equations become vacuous. So $\dot{\lambda}_0$ must be taken as a non–zero constant, and then

$$\mathbf{h} = (\lambda_1\mathbf{i} + \lambda_2\mathbf{j})/\dot{\lambda}_0 \tag{14.14}$$

is a constant unit vector, and Γ can only be a straight line.

To circumvent this stumbling block we pursue a parametric argument as follows. We write $\mathfrak{K}$ for the set of quaternions having the form $q = a + b\mathbf{k}$, where a and b are scalars. Since $\mathbf{k}$ a fixed square root of -1, we can regard $\mathfrak{K}$ as an embedding of the complex field in the quaternion division ring; and we can treat the members of $\mathfrak{K}$ as ordinary complex numbers. Accordingly we rewrite (14.4) as

$$q = q_1 + q_2\mathbf{i}, \quad q_1 = \rho_0 + \rho_3\mathbf{k} \in \mathfrak{K}, \quad q_2 = \rho_1 + \rho_2\mathbf{k} \in \mathfrak{K}. \tag{14.15}$$

Suppose $\mathbf{h}$ subtends an angle $\psi = \psi(t)$ with the x–axis:

$$\mathbf{h} = \mathbf{i}\cos\psi + \mathbf{j}\sin\psi = e^{\psi\mathbf{k}}\mathbf{i} = \mathbf{i}e^{-\psi\mathbf{k}}. \tag{14.16}$$

The curvature of Γ at the point t is

$$\kappa = \dot{\psi}. \tag{14.17}$$

Equation (14.2) becomes

$$\dot{q}_1 + \dot{q}_2\mathbf{i} = \tfrac{1}{2}e^{\psi\mathbf{k}}\mathbf{i}(q_1 + q_2\mathbf{i}) = \tfrac{1}{2}e^{\psi\mathbf{k}}(-\bar{q}_2 + \bar{q}_1\mathbf{i}), \tag{14.18}$$

in which we may equate the pre–coefficients of $\mathbf{i}$:

$$\dot{q}_1 = -\tfrac{1}{2}e^{\psi\mathbf{k}}\bar{q}_2, \quad \dot{q}_2 = \tfrac{1}{2}e^{\psi\mathbf{k}}\bar{q}_1. \tag{14.19}$$

From (14.17) and (14.19) we deduce that q_1 and q_2 are both solutions of the differential equation

$$\ddot{q}_p - \kappa\mathbf{k}\dot{q}_p + \tfrac{1}{4}q_p = 0 \qquad (p = 1,2) \tag{14.20}$$

with respective boundary conditions

$$q_1(0) = 1, \quad \dot{q}_1(0) = 0; \quad q_2(0) = 0, \quad \dot{q}_2(0) = \tfrac{1}{2}e^{\psi(0)\mathbf{k}} \tag{14.21}$$

obtained from inserting $q(0) = 1$ into (14.19). Moreover, because q is a unit quaternion, there exist real functions $a = a(t)$, $b = b(t)$, $c = c(t)$ such that

$$q_1 = e^{\frac{1}{2}b\mathbf{k}}\cos\tfrac{1}{2}a, \quad q_2 = e^{\frac{1}{2}c\mathbf{k}}\sin\tfrac{1}{2}a. \tag{14.22}$$

Substituting q_1 from (14.22) into (14.20), we find

$$(1 + 2\dot{b}\kappa - \dot{a}^2 - \dot{b}^2 + 2\ddot{b}\mathbf{k})\cos\tfrac{1}{2}a + 2(\dot{a}\kappa\mathbf{k} - \ddot{a} - \dot{a}\dot{b}\mathbf{k})\sin\tfrac{1}{2}a = 0. \tag{14.23}$$

The real and imaginary parts of this complex number must both vanish. This gives two equations, between which we eliminate κ to get

$$(1-\dot{a}^2)(\dot{a}\sin\tfrac{1}{2}a\cos\tfrac{1}{2}a) - 2\dot{a}\ddot{a}\sin^2\tfrac{1}{2}a = 2\dot{b}\ddot{b}\cos^2\tfrac{1}{2}a - \dot{b}^2\dot{a}\sin\tfrac{1}{2}a\cos\tfrac{1}{2}a; \tag{14.24}$$

and this integrates to

$$(1-\dot{a}^2)\sin^2\tfrac{1}{2}a = \dot{b}^2\cos^2\tfrac{1}{2}a + \text{constant}. \tag{14.25}$$

From (14.21) and (14.22) we have

$$a(0) = b(0) = \dot{b}(0) = 0; \tag{14.26}$$

so the constant in (14.25) is zero, and hence

$$\dot{b} = \varepsilon_1(1-\dot{a}^2)^{\frac{1}{2}}\tan\tfrac{1}{2}a \tag{14.27}$$

where $\varepsilon_1 = \pm 1$. Similarly, substituting q_2 from (14.22) into (14.20), we get

$$\dot{c} = \varepsilon_2(1-\dot{a}^2)^{\frac{1}{2}}\cot\tfrac{1}{2}a \tag{14.28}$$

with $\varepsilon_2 = \pm 1$. We can now calculate κ from (14.20) in two different ways corresponding to $p = 1,2$:

$$\kappa = \varepsilon_1\{\ddot{a}(1-\dot{a}^2)^{-\frac{1}{2}} - (1-\dot{a}^2)^{\frac{1}{2}}\cot a\} = -\varepsilon_2\{\ddot{a}(1-\dot{a}^2)^{-\frac{1}{2}} - (1-\dot{a}^2)^{\frac{1}{2}}\cot a\}. \tag{14.29}$$

Hence $\varepsilon_1 = -\varepsilon_2$. Next integrate (14.27) and (14.28), using the boundary conditions (14.21) and (14.26), and insert the result into (14.19) to obtain

$$\dot{a}(0)e^{\psi\mathbf{k}} = \{\dot{a} + \mathbf{k}(1-\dot{a}^2)^{\frac{1}{2}}\}\exp\{\mathbf{k}\psi(0) + \mathbf{k}\varepsilon_2\int_0^t (1-\dot{a}^2)^{\frac{1}{2}}\cot a\,dt\}. \tag{14.30}$$

Then differentiate (14.30) logarithmically in a sufficiently small neighbourhood of $t = 0$ to cater for the principal values, and compare the result with (14.17) and (14.29). This gives $\varepsilon_2 = +1$. We can also assume without essential loss of generality that $\dot{a}(0) = +1$; for otherwise we can simply replace $a(t)$ and $\psi(0)$ by $a(-t)$ and $\psi(0) + \pi$ respectively. Hence we obtain the required parametrization

$$q_1 = \cos\tfrac{1}{2}a\,\exp\{-\tfrac{1}{2}\mathbf{k}\int_0^t (1-\dot{a}^2)^{\frac{1}{2}}\tan\tfrac{1}{2}a\,dt\}, \tag{14.31}$$

$$q_2 = \sin\tfrac{1}{2}a\,\exp\{\mathbf{k}\psi(0) + \tfrac{1}{2}\mathbf{k}\int_0^t (1-\dot{a}^2)^{\frac{1}{2}}\cot\tfrac{1}{2}a\,dt\}, \tag{14.32}$$

$$-\mathbf{hi} = e^{\psi\mathbf{k}} = \{\dot{a} + \mathbf{k}(1-\dot{a}^2)^{\frac{1}{2}}\}\exp\{\mathbf{k}\psi(0) + \mathbf{k}\int_0^t (1-\dot{a})^{\frac{1}{2}}\cot a\,dt\}. \tag{14.33}$$

where $\psi(0)$ is an arbitrary parameter and $a(t)$ is an arbitrary parametric function for $t \geqslant 0$, satisfying the boundary conditions

$$a(0) = 0, \qquad \dot{a}(0) = 1, \qquad -1 \leqslant \dot{a}(t) \leqslant 1. \tag{14.34}$$

When the ball rotates through $\mathrm{d}t$ about $\mathbf{h}$, its centre suffers a translation $\mathbf{hk}\mathrm{d}t$. So the final position on the table will be

$$\mathbf{v}(T) = \int_0^T \mathbf{hk}\,\mathrm{d}t = \int_0^T \mathrm{e}^{\psi\mathbf{k}}\mathbf{ik}\,\mathrm{d}t. \tag{14.35}$$

When the final orientation $q(T)$ is prescribed, then $a(T)$, $b(T)$, $c(T)$ are prescribed via (14.15) and (14.22). So we require

$$b(T) = -\int_0^T (1-\dot{a}^2)^{\frac{1}{2}} \tan\tfrac{1}{2}a\,\mathrm{d}t, \tag{14.36}$$

$$c(T) = 2\psi(0) + \int_0^T (1-\dot{a}^2)^{\frac{1}{2}} \cot\tfrac{1}{2}a\,\mathrm{d}t, \tag{14.37}$$

$$\mathbf{v}(T)\mathbf{j} = \int_0^T \mathrm{e}^{\psi\mathbf{k}}\,\mathrm{d}t. \tag{14.38}$$

We can secure (14.37) by choice of the arbitrary $\psi(0)$, whereupon (14.38) becomes

$$\mathbf{v}(T)\mathbf{j}\mathrm{e}^{-\frac{1}{2}c(T)\mathbf{k}} = \int_0^T \mathrm{e}^{g\mathbf{k}}\,\mathrm{d}t, \tag{14.39}$$

where

$$g = \cos^{-1}\dot{a} - \tfrac{1}{2}\int_0^T (1-\dot{a}^2)^{\frac{1}{2}}\cot\tfrac{1}{2}a\,\mathrm{d}t + \int_0^t (1-\dot{a}^2)^{\frac{1}{2}}\cot a\,\mathrm{d}t. \tag{14.40}$$

For problem (a), we have to minimize $T = \int_0^T 1\,\mathrm{d}t$ subject to (14.36) and (14.39). So, introducing constant Lagrangian multipliers μ and $\mu\sigma$, where μ is real and σ is a complex number belonging to $\mathfrak{K}$, we have to minimize the real part of

$$I = \int_0^T 1 - \mu(1-\dot{a}^2)^{\frac{1}{2}}\tan\tfrac{1}{2}a + \mu\sigma\mathrm{e}^{g\mathbf{k}}\,\mathrm{d}t \tag{14.41}$$

for variations in $a(t)$ subject to (14.34) and prescribed $a(T)$, where T is constant. Hence the required Euler equation is

$$\begin{aligned}\frac{1}{\dot{a}}\frac{\mathrm{d}}{\mathrm{d}t}\left\{\frac{\tan\frac{1}{2}a}{(1-\dot{a}^2)^{\frac{1}{2}}}\right\} &= \left(\frac{\partial}{\partial a} - \frac{\mathrm{d}}{\mathrm{d}t}\frac{\partial}{\partial\dot{a}}\right)(1-\dot{a}^2)^{\frac{1}{2}}\tan\tfrac{1}{2}a \\ &= \Re\left(\frac{\partial}{\partial a} - \frac{\mathrm{d}}{\mathrm{d}t}\frac{\partial}{\partial\dot{a}}\right)\sigma\mathrm{e}^{g\mathbf{k}},\end{aligned} \tag{14.42}$$

in which $\Re$ denotes the real part of the subsequent expression.

This is a non–linear integro–differential equation, which looks forbidding; and I do not know how to solve it analytically. Of course, it can be written out explicitly by substituting (14.40) into the right–hand side of (14.42). It could then be solved numerically, though this would be quite a formidable task even on a large computer because σ and T are adjustable parameters to be chosen so that the solution satisfies the constraints (14.36) and (14.39).

On the other hand (14.42) easily yields the solution for problem (*b*), for which there is no constraint (14.39); so we may take $\sigma = 0$. Thus (14.42) guarantees that $(1 - \dot{a}^2)^{-\frac{1}{2}} \tan \frac{1}{2}a$ must be a constant, say $1/\gamma$. If $1/\gamma = 0$, then $a = 0$ for all t. This corresponds to the trivial case in which the ball remains at the origin because its initial and final orientations are the same. Setting aside this trivial case, we have

$$(1 - \dot{a}^2)^{\frac{1}{2}} = \gamma \tan \tfrac{1}{2}a, \tag{14.43}$$

and hence

$$\ddot{a}(1 - \dot{a}^2)^{-\frac{1}{2}} = -\tfrac{1}{2}\gamma \sec^2 \tfrac{1}{2}a. \tag{14.44}$$

Substituting (14.44) into (14.29) and recalling that $\varepsilon_2 = +1$, we obtain

$$\kappa = \tfrac{1}{2}\gamma \sec^2 \tfrac{1}{2}a + \gamma \tan \tfrac{1}{2}a \cot a = \gamma. \tag{14.45}$$

Thus the curvature of Γ is constant, and Γ must be the arc of the some circle or a segment of a straight line. This confirms the result already obtained for problem (*b*) in §13. When κ is constant, (14.20) and the initial conditions (14.21) yield an explicit solution for $q_1(t)$, whose real and imaginary parts provide

$$\begin{aligned} 2\omega \cos \tfrac{1}{2}a \cos \tfrac{1}{2}b &= (\omega - \kappa)\cos \tfrac{1}{2}(\omega + \kappa)t + (\omega + \kappa)\cos \tfrac{1}{2}(\omega - \kappa)t \\ 2\omega \cos \tfrac{1}{2}a \sin \tfrac{1}{2}b &= (\omega - \kappa)\sin \tfrac{1}{2}(\omega + \kappa)t + (\omega + \kappa)\sin \tfrac{1}{2}(\omega - \kappa)t, \end{aligned} \tag{14.46}$$

where $\omega = \sqrt{(1 + \kappa^2)} \geqslant 1$. Squaring and adding these two equations, we get

$$\omega \sin \tfrac{1}{2}a = \sin \tfrac{1}{2}\omega t \tag{14.47}$$

after taking a square root, whose ambiguity of sign is resolved by conformity with $a(0) = +1$ in (14.34). Remembering that $T \geqslant 0$ by hypothesis, we can now define α and β by means of

$$\omega^{-1} = \sin \alpha, \quad \kappa/\omega = \cos \alpha, \quad 0 < \alpha < \pi; \quad \beta = \tfrac{1}{2}\omega T \geqslant 0. \tag{14.48}$$

Hence

$$\kappa = \cot \alpha, \quad T = 2\beta \sin \alpha. \tag{14.49}$$

To complete the solution of problem (*b*), we must express the initial direction, the curvature, and the length of Γ in terms of the prescribed final orientation $q(T) = \cos \frac{1}{2}a \cos \frac{1}{2}b + \mathbf{i} \sin \frac{1}{2}a \cos \frac{1}{2}c + \mathbf{j} \sin \frac{1}{2}a \sin \frac{1}{2}c + \mathbf{k} \cos \frac{1}{2}a \sin \frac{1}{2}b$. Here and hereafter, for brevity, we have written $a = a(T)$, $b = b(T)$, $c = c(T)$ for their prescribed terminal values. If we change (a,b,c) to $(a + 2\pi, b + 2\pi, c + 2\pi)$ or to $(-a, b, c + 2\pi)$ or $(a + 2\pi, b, c)$, we either leave $q(T)$ unchanged or else change its sign. Since $\pm q(T)$ both represent the same final orientation, there is no loss of generality in imposing the terminal conditions

$$0 \leqslant \tfrac{1}{2}a \leqslant \tfrac{1}{2}\pi, \quad 0 \leqslant \tfrac{1}{2}b < \pi, \quad 0 \leqslant \tfrac{1}{2}c < 2\pi. \tag{14.50}$$

The initial direction of Γ at the origin is inclined to the x–axis at an angle

$$\psi(0) - \tfrac{1}{2}\pi = \tfrac{1}{2}c - \tfrac{1}{2}\pi - \tfrac{1}{2}\int_0^T (1 - \dot{a}^2)^{\frac{1}{2}} \cot \tfrac{1}{2}a\, dt = \tfrac{1}{2}(c - \pi - \kappa T), \tag{14.51}$$

by virtue of (14.37), (14.43), and (14.45). So it only remains to determine α and β as functions of a and b, whence κ and T will be available from (14.49). We shall discover that there are infinitely many extremals Γ satisfying the Euler equations and the terminal conditions, namely (14.46) with $t = T$; and we shall need to extract from those the *least extremal*, that is to say the one with the smallest value of T. For example, in the special case $a = \pi$, it is easy to verify that the only solutions of (14.46) are $\kappa = 0$, $\omega = 1$, $t = T = n\pi$ where n is any odd positive integer, amongst which $n = 1$ provides the least extremal. Setting aside this trivial special case, we can assume that $0 \leqslant \frac{1}{2}a < \frac{1}{2}\pi$ and hence we can eliminate $\omega\cos\frac{1}{2}a > 0$ from (14.48) with $t = T$, using (14.49), to get after some manipulation

$$\tan(\tfrac{1}{2}b - \beta\cos\alpha) = -\cos\alpha\tan\beta. \tag{14.53}$$

Also (14.47) with $t = T$ becomes

$$\sin\tfrac{1}{2}a = \sin\alpha\sin\beta. \tag{14.54}$$

It is straightforward to reverse this argument, to show that any solution of (14.53) and (14.54) with $0 < \alpha < \pi$ and $\beta \geqslant 0$ gives a solution of (14.46) with $t = T$. There are two occasions in this reverse argument where we have to use (14.50) to resolve an ambiguity of sign. So now it suffices to solve (14.53) and (14.54) for α and β in terms of a and b. But before doing this, it is convenient to get rid of the special case $a = 0$. We can ignore the trivial subcase $a = b = 0$, in which $q(T) = 1$ and there is no need to roll the ball at all because the initial and final orientations are the same and $T = 0$. So $0 < \frac{1}{2}b < \pi$; and then (14.54), together with $0 < \alpha < \pi$, implies $\beta = m_1\pi$ where m_1 is a positive integer. Hence, by (14.53), $\cos\alpha = (\frac{1}{2}b + m_2\pi)/m_1\pi$, with m_2 an integer and $|\frac{1}{2}b + m_2\pi| < m_1\pi$. Substituting those results into (14.49) and choosing m_1 and m_2 to minimize T, we find $m_1 = 1$ and $m_2 = -1$ or 0 according as $0 < \frac{1}{2}b \leqslant \frac{1}{2}\pi$ or $\frac{1}{2}\pi < \frac{1}{2}b < \pi$. Hence the least extremal in the special case $a = 0$ has

$$T = \sqrt{\{4\pi^2 - (\pi + |b - \pi|)^2\}} \leqslant \pi\sqrt{3} \quad (0 = a \leqslant b < 2\pi) \tag{14.55}$$

with equality if $b = \pi$. Note that (14.55) includes the subcase $a = b = T = 0$.

Having disposed of the special cases $a = 0$ and $a = \pi$, we can hereafter assume

$$0 < \tfrac{1}{2}a < \tfrac{1}{2}\pi, \quad 0 \leqslant \tfrac{1}{2}b < \pi. \tag{14.56}$$

For the fixed prescribed $\frac{1}{2}a$, we define the function $\alpha = \alpha(\beta)$ as the solution of

$$\sin\tfrac{1}{2}a = \sin\alpha\sin\beta, \quad \tfrac{1}{2}\pi \leqslant \alpha < \pi \tag{14.57}$$

in which β is a variable in the interval $J_l = [2l\pi + \frac{1}{2}a,\ 2l\pi + \pi - \frac{1}{2}a]$ for some integer $l \geqslant 0$. This interval ensures that (14.57) has a unique solution $\alpha = \alpha(\beta)$. Then we define the function

$$F(\beta) = \beta\cos\alpha(\beta) + \tan^{-1}\{-\cos\alpha(\beta)\tan\beta\}, \tag{14.58}$$

in which we interpret $0 \leqslant \tan^{-1}\{-\cos\alpha(\beta)\tan\beta\} \leqslant \pi$. We shall then get a solution of (14.53) and (14.54) whenever we have a solution of

$$F(\beta) = \tfrac{1}{2}b \quad (\text{modulo } \pi), \quad \beta \in J_l. \tag{14.59}$$

Elementary calculations show that $F(\beta)$ is a continuous function of β in J_l, with

$$F(2l\pi + \tfrac{1}{2}a) = 0, \quad F(2l\pi + \pi - \tfrac{1}{2}a) = \pi \tag{14.60}$$

$$\mathrm{d}F/\mathrm{d}\beta = (\beta\cot\beta - 1)\tan\alpha(\beta)\sin\alpha(\beta). \tag{14.61}$$

This derivative cannot vanish in the interior of J_l unless $\beta = \beta_l$, where

$$\tan\beta_l = \beta_l, \quad 2l\pi < \beta_l < 2l\pi + \pi. \tag{14.62}$$

Hence, if $\beta_l \notin J_l$, $F(\beta)$ will increase steadily from 0 to π as β traverses J_l, and (14.59) will have a unique solution. On the other hand, if $\beta_l \in J_l$, $F(\beta)$ will decrease from 0 to $F(\beta_l)$ and next increase from $F(\beta_l)$ to π. Since $F(\beta_l) \geqslant -\beta_l \geqslant -(2l + 1)\pi$, (14.59) will have at least one and at most $4l + 3$ solutions when $\beta_l \in J_l$. There is just one other way of getting solutions to (14.53) and (14.54): namely, we could have replaced the inequality in (14.57) by $0 < \alpha \leqslant \tfrac{1}{2}\pi$. In that event, we get

$$F(\beta) = \pi - \tfrac{1}{2}b \quad (\text{modulo } \pi), \quad \beta \in J_l \tag{14.63}$$

instead of (14.59). Once again, we see that (14.63) will have at least one and at most $4l + 3$ solutions. So altogether for each integer $l \geqslant 0$, we get at least one and at most $8l + 6$ solutions to (14.53) and (14.54). From the resulting infinity of extremals, we must now select the least extremal.

To this end, we now consider the particular case $l = 0$. Since $\beta_0 \notin J_0$, (14.62) and (14.63) each have unique solutions, denoted by $\beta = \hat{\beta}$ and $\beta = \tilde{\beta}$ respectively. The corresponding values of T obtained from (14.49) and (14.57) will be

$$\hat{T}_0(a,b) = 2\hat{\beta}\operatorname{cosec}\hat{\beta}\sin\tfrac{1}{2}a, \quad \tilde{T}_0(a,b) = 2\tilde{\beta}\operatorname{cosec}\tilde{\beta}\sin\tfrac{1}{2}a = \hat{T}_0(a, 2\pi - b). \tag{14.64}$$

Let $T_0(a,b)$ denote the smaller of $\hat{T}_0(a,b)$ and $\hat{T}_0(a,2\pi - b)$. But $F(\beta)$ and $\beta\operatorname{cosec}\beta$ are both increasing functions of β in J_0. Hence

$$T_0(a,b) \leqslant \hat{T}_0(a,\pi) \tag{14.65}$$

with equality if and only if $b = \pi$. When $b = \pi$, (14.59) reduces to

$$\cot\{\beta\cos\alpha(\beta)\} = -\cos\alpha(\beta)\tan\beta, \quad \tfrac{1}{2}\pi \leqslant \alpha(\beta) < \pi, \quad \beta \in J_0. \tag{14.66}$$

Now (14.66) implies $\tfrac{1}{2}\pi \leqslant \beta \leqslant \pi$. We now regard $\alpha = \alpha(\beta)$ as a function of β defined by (14.66) instead of (14.57): this is legitimate, provided that we also regard (14.57) with $\alpha = \alpha(\beta)$ as defining $a = a(\beta)$ as a function of β. Then, by elementary calculations, (14.64) yields

$$\frac{\mathrm{d}\hat{T}_0(a,\pi)}{\mathrm{d}\beta} = \frac{2 - 2\beta^{-1}\tan\beta}{1 + \cos^2\alpha(\beta)\tan^2\beta - \beta^{-1}\tan\beta} > 0, \tag{14.67}$$

because $\tan\beta \leqslant 0$. When $\beta = \tfrac{1}{2}\pi + 0$, $\beta\cos\alpha = 0$ from (14.66); so $\alpha = \tfrac{1}{2}\pi$ and (14.57) gives $a = \pi$. Hence $T_0(\pi,\pi) = \pi$. On the other hand, when $\beta = \pi$, (14.66) gives $\alpha = \frac{2}{3}\pi$ and (14.57) gives $a = 0$, and therefore $\hat{T}_0(0,\pi) = \pi\sqrt{3}$ by (14.49). The strict monotonicity implied by (14.67) now shows that $\pi = T_0(\pi,\pi) \leqslant \hat{T}_0(a,\pi) \leqslant \hat{T}_0(0,\pi) = \pi\sqrt{3}$, the upper bound being attained only if $a = 0$. From (14.55) and (14.65), the least extremal must have $T \leqslant T_0(a,b) \leqslant \pi\sqrt{3}$ with equality if and only if $a = 0$, $b = \pi$, that is to say if and only if the required reorientation is a rotation through π about a vertical axis.

By (14.49) and (14.54), $T = 2\beta \operatorname{cosec}\beta \sin\frac{1}{2}a$ which (for fixed a in $0 < a < \frac{1}{2}\pi$) has a unique minimum in $2l\pi \leqslant \beta \leqslant 2l\pi + \pi$ at β_l. Also $2\beta_l \operatorname{cosec}\beta_l = \sec\beta_l \to \infty$ as $l \to \infty$. Choose an integer $L = L(a)$ such that $(\frac{1}{2}\pi\sqrt{3})\operatorname{cosec}\frac{1}{2}a < \sec\beta_l$ for all $l > L$. Then any extremal obtained from the solution of (14.59) or (14.63) with $l > L$ will yield $T > \pi\sqrt{3}$, and so will be inadmissible as a least extremal. Therefore, in seeking the least extremal we may confine attention to the *finite* number of solutions of (14.59) and (14.63) with $0 \leqslant l \leqslant L$, and pick that amongst them that gives the minimal T. Indeed, a reasonable conjecture is that $l = 0$ provides the least extremal; but I cannot yet prove this conjecture.

15. Variants for the twenty-first century.

Suppose the ball is initially at the centre of a circular table of radius r, and $N(r)$ denotes the necessary and sufficient number of moves, each confined to this finite table, to attain any final state. By considering the particular case when the ball is rotated through π about the vertical axis, finishing at the centre of the table, we find that $2r\{N(r) - 1\} > \pi\sqrt{3}$; and hence $\liminf_{r\to 0} rN(r) \geqslant \frac{1}{2}\pi\sqrt{3}$. Is it true that $rN(r)$ tends to a finite limit λ as $r \to 0$; and, if so, what is the numerical value of λ and is $N(r) - \lambda r^{-1}$ bounded for $r > 0$? Calculate the values of r at which $N(r)$ is discontinuous. Does $N(r) \to N$ as $r \to \infty$? What strategy of moves on a table of radius r will achieve a path of shortest total length to a desired terminal state, and what is this minimum length? What happens if we replace the table by a fixed ball of radius R, on which the original unit ball can roll? More generally, consider one smooth convex body rolling geodesically on another such body, or inside a curved shell whose maximum curvature exceeds the minimum curvature of the rolling body. To what extent is geodesic rolling homeomorphically transformable; and can we, for example, infer the answers for an ellipsoid rolling on a sphere from the corresponding answers for the ball on the flat table? What occurs when the paths of contact are restricted to prescribed bounded subsets of the surfaces? Are modifications required for surfaces of higher genus, as with a ball inside a toroidal tube? Discuss the bifurcation theory for rolling on smoothly dimpled surfaces. Solve the multidimensional generalizations, such as an n–dimensional sphere rolling on an $(n - 1)$–dimensional hyperplane or a bounded subset thereof.

16. Typographical acknowledgment: OUCS Lasercomp.

This article has been typeset on a Lasercomp at Oxford University Computing Service. I am indebted to Catherine Griffin, Giuseppe Mazzarino, and Caroline Wise for their help with this typesetting.

INVARIANT MEASURES AND THE Q-MATRIX

F.P. Kelly

Abstract

This paper provides a necessary and sufficient condition for a measure to be invariant for a Markov process. The condition is expressed in terms of the q-matrix assumed to generate the process.

1. Introduction

Let $Q = (q_{ij}, i, j \in S)$ be a stable, conservative, regular and irreducible q-matrix over a countable state space S, and let $P(t) = (p_{ij}(t), i,j \in S)$ be the matrix of transition probabilities of the Markov process determined by Q. If (the Markov process determined by) Q is recurrent then the relations

$$\left.\begin{aligned} \sum_i m_i q_{ij} &= 0 \\ m_j &> 0 \end{aligned}\right\} \quad j \in S \qquad (1)$$

have a solution $m = (m_i, i \in S)$, unique up to constant multiples. Call m an invariant measure for $P(t)$ if

$$\sum_i m_i p_{ij}(t) = m_j \qquad t > 0,\ j \in S.$$

When Q is positive recurrent it is known (Doob [5], Kendall and Reuter [13]) that a solution m to (1) is an invariant measure for $P(t)$. This conclusion also holds when Q is null recurrent, but may not when Q is transient. When Q is transient the set of solutions to (1) may be empty or it may contain linearly independent elements: we obtain a necessary and sufficient condition for a given element of the set to be an invariant measure for $P(t)$.

The basic properties of Markov processes which will be needed are taken from Kendall [11] and are briefly stated in Section 2: they can also be found in [3], [6], [10], [12], [13] and [17]. Section 3 contains the main result of the present paper. Here it is shown that a solution to (1) is an invariant measure for $P(t)$ if and only if a time-reversed q-matrix $\tilde{Q}$, defined in terms of m and Q, is regular. It is convenient to obtain the result assuming only that Q is stable and conservative, with $P(t)$ the minimal (Feller) transition matrix determined by Q. The method of proof is a straightforward generalization of an argument used by Kendall [12] in the case where Q is symmetrically reversible. Section 4 is devoted to a cycle criterion relating Q and $\tilde{Q}$, modelled on the criteria discussed by Kolmogorov [14], Reich [16], Kendall [10] and Whittle [19].

My interest in the topic of this paper arose from an observation in an applied probability context which is perhaps worth mentioning here. A technique useful in the study of certain forms of queueing network involves solving relations (1) for each of the q-matrices corresponding to the individual queues of the network operating in isolation and then combining these solutions appropriately to obtain a solution to relations (1) for the q-matrix corresponding to the network [9]. Now the Markov process representing the network may well be positive recurrent even when some or all of the processes representing individual queues fail to be recurrent or even regular. The solution for the network will then have a straightforward interpretation as an invariant measure and it is of interest to ask when the individual solutions from which it has been constructed have interpretations as invariant measures for the individual queues.

2. Preliminaries

Suppose that we are given a stable, conservative q-matrix, that is a collection of real numbers $Q = (q_{ij}, i,j \in S)$ where S is a countable set and

$$\begin{aligned} q_{ij} &\geq 0 && i \neq j \\ \sum_{j \neq i} q_{ij} &= -q_{ii} && i \in S \qquad (2) \\ -q_{ii} &\triangleq q_i < \infty && i \in S \end{aligned}$$

A Markov process with transition rates Q can be constructed by the standard method, due to Feller and roughly indicated as follows. Starting from state i allow the process to stay there for a period exponentially distributed with parameter q_i, and then move the process to state j with probability q_{ij}/q_i; let the process remain in state j for a period exponentially distributed with parameter q_j, and so on. This construction defines a Markov process $(X(t),\ 0 \le t < T)$ with initial state $X(0) = i$ and with stationary transition rates

$$\begin{aligned} q_{ij} &= \lim_{t\to 0} \frac{1}{t}\, p_{ij}(t) \qquad\qquad i \neq j \\ q_i &= \lim_{t\to 0} \frac{1}{t}\, [1 - p_{ii}(t)] \end{aligned} \tag{3}$$

where

$$p_{ij}(t) = P\{t < T,\ X(t) = j \mid X(0) = i\} \quad .$$

The terminal time T is the sum of the (random) sequence of exponentially distributed holding times, and may well be finite. The process will then have made infinitely many jumps in a finite time, and will have "run out of instructions". A necessary and sufficient condition for T to be infinite with probability one, whatever the initial state i, is that the equations

$$\sum_j q_{ij}\, y_j = y_i \qquad\qquad i \in S$$

have no non-trivial non-negative bounded solution, and in this case Q is said to be *regular*.

Occasionally it requires some effort to show that a matrix Q is regular, but there are a number of sufficient conditions. If $(q_i,\ i \in S)$ is bounded above then Q is regular. Call the sequence of states occupied by the process $(X(t),\ 0 \le t < T)$ the jump chain, and call the sequence $(X(r\delta),\ r = 0,1,\dots,\lfloor T/\delta \rfloor)$ the δ-skeleton. These are both Markov chains and recurrence of either of them implies recurrence of the Markov process, and hence regularity of Q.

From the construction of the process $(X(t),\ 0 \le t < T)$ it follows that $p_{ij}(t)$ is the limit of the non-decreasing sequence $(f_{ij}(t,n),\ n = 0,1,\dots)$ where $f_{ij}(t,n)$ is the probability that the process is in state j at time t after at most n jumps. Clearly

$$f_{ij}(t,0) = \delta_{ij}\, e^{-q_j t} \tag{4}$$

and the collection $(f_{ij}(t,n))$ can be generated using either the backward integral recurrence

$$f_{ij}(t,n+1) = \delta_{ij}\, e^{-q_i t} + \sum_{k \neq i} \int_0^t q_{ik}\, f_{kj}(u,n) e^{-q_i(t-u)}\, du \tag{5}$$

or the forward integral recurrence

$$f_{ij}(t,n+1) = \delta_{ij}\, e^{-q_j t} + \sum_{k \neq j} \int_0^t f_{ik}(u,n) q_{kj}\, e^{-q_j(t-u)}\, du \tag{6}$$

The transition probabilities $P(t) = (p_{ij}(t),\ i,j \in S)$ thus constructed satisfy

$$\sum_j p_{ij}(t) = P\{t < T | X(0) = i\} \le 1 \quad .$$

For each fixed $t > 0$ equality holds in this relation for all $i \in S$ if and only if Q is regular.

The matrix Q is *irreducible* if for each pair (i,j) of distinct states there exists a finite sequence of states $i, k_1, k_2, \ldots, k_r, j$ satisfying

$$q_{ik_1}\, q_{k_1k_2} \cdots q_{k_rj} > 0 \ .$$

This is equivalent to the condition that for each pair (i,j) of distinct states $p_{ij}(t) > 0$ for $t > 0$ ([10], proof of Theorem IV (i); [3], Theorem 18.4).

A collection of positive numbers $m = (m_i,\ i \in S)$ is an invariant measure for the transition probabilities $P(t)$ if

$$\sum_i m_i\, p_{ij}(t) = m_j \qquad t > 0,\ j \in S\ . \tag{7}$$

If $\sum m_j = 1$ and Q is irreducible then the process is positive recurrent (since its δ-skeleton is) and the invariant probability distribution m has an interpretation as either a stationary or a limiting distribution. It also has an ergodic interpretation: the ratio of the time spent in state i to that spent in state j over the interval $[0,t]$ tends to m_i/m_j with probability one as $t \to \infty$. If $\sum m_j = \infty$ and Q is irreducible and the process is recurrent then the measure m retains the ergodic interpretation. In both these cases recurrence implies that Q is regular and the ergodic interpretation shows that m is essentially unique.

The work of Derman [4] and Brown [2] provides an alternative interpretation of an invariant measure m. This interpretation remains available when Q fails to be recurrent or even regular, and can be described informally as follows. At time $t = 0$ place N_i particles at site i, $i \in S$, where $(N_i, i \in S)$ are independent random variables and N_i has a Poisson distribution with mean m_i. From time $t = 0$ onwards allow particles to move independently from site to site, each moving in accordance with a Markov process constructed from the matrix Q. Then at any time $t > 0$ the number of particles at site i, $N_i(t)$, has a Poisson distribution with mean m_i, and $(N_i(t), i \in S)$ are independent. Note that if Q is not regular then it is possible for the Markov process describing a particle's motion to have a finite terminal time - in this case the particle disappears from the set of states S at the terminal time. It is quite possible that a q-matrix might admit an invariant measure and yet not be regular: we shall give an example later.

3. The conditions for invariance

This section explores the connection between the relations (1) and the invariance property (7). We begin with our main result.

Theorem. Let Q be a stable, conservative q-matrix and let $m = (m_i, i \in S)$ be a collection of positive numbers. Then the following statements are equivalent :

(i) m is an invariant measure for the transition probabilities $P(t)$ constructed from Q;

(ii) m satisfies

$$\sum_i m_i q_{ij} = 0 \qquad j \in S \tag{8}$$

and the relations

$$\sum_i z_i q_{ij} = z_j \qquad\qquad j \in S \qquad (9)$$
$$z_j \le m_j$$

have no non-trivial non-negative solution.

Proof. Suppose that m is an invariant measure. Then from equation (7)

$$\sum_{i \neq j} m_i \frac{p_{ij}(t)}{t} = m_j \frac{1-p_{jj}(t)}{t} \qquad j \in S$$

Relations (3) and Fatou's lemma thus imply

$$\sum_{i \neq j} m_i q_{ij} \le m_j \ q_j \qquad j \in S \qquad (10)$$

Define

$$\tilde{q}_{ij} = \frac{m_j}{m_i} q_{ji} \qquad i,j \in S \qquad (11)$$

Then

$$\tilde{q}_i \triangleq - \tilde{q}_{ii} = -q_{ii} < \infty \quad ,$$

and

$$\sum_j \tilde{q}_{ij} \le 0$$

from inequality (10). Define $\tilde{q}_{i\partial}$ by

$$\sum_j \tilde{q}_{ij} + \tilde{q}_{i\partial} = 0$$

and let

$$\tilde{q}_{\partial j} = 0 \qquad j \in S \cup \{\partial\} \quad .$$

Then $\tilde{Q} = (\tilde{q}_{ij},\ i,\ j \in S \cup \{\partial\})$ is a stable, conservative q-matrix on the state space $S \cup \{\partial\}$. Define $(\tilde{f}_{ij}(t,n))$ by the initial condition (4) and the forward integral recurrence (6), and let

$$\tilde{p}_{ij}(t) = \lim_{n\to\infty} \tilde{f}_{ij}(t,n) \qquad i,j \in S \cup \{\partial\} \ .$$

Clearly

$$m_i\, f_{ij}(t,0) = m_j\, \tilde{f}_{ji}(t,0) \qquad i,j \in S \ .$$

Assume for the moment the inductive hypothesis that

$$m_i\, f_{ij}(t,n) = m_j\, \tilde{f}_{ji}(t,n) \qquad i,j \in S \tag{12}$$

for some $n \geq 0$. Then from the backward integral recurrence (5)

$$m_i\, f_{ij}(t,n+1) = m_i\, \delta_{ij}\, e^{-q_i t} + \sum_{k\neq i} \int_0^t m_i q_{ik}\, f_{kj}(u,n) e^{-q_i(t-u)}\, du \ .$$

But

$$m_i\, q_{ik}\, f_{kj}(u,n) = m_k\, \tilde{q}_{ki}\, f_{kj}(u,n)$$

$$= m_j\, \tilde{q}_{ki}\, f_{jk}(u,n)$$

from equation (11) and the inductive hypothesis (12). The definition of $\tilde{f}_{ji}(t,n+1)$ by means of the forward integral recurrence (6) thus shows that

$$m_i\, f_{ij}(t,n+1) = m_j\, \tilde{f}_{ji}(t,n+1) \qquad i,j \in S.$$

The additional state ∂ causes no difficulty, since $q_{\partial i} = 0$. The inductive hypothesis (12) is established, and so, on letting n tend to infinity,

$$m_i\, p_{ij}(t) = m_j\, \tilde{p}_{ji}(t) \qquad i,j \in S \ .$$

The assumed invariance of m thus implies

$$\sum_{i\in S} \tilde{p}_{ji}(t) = 1 \tag{13}$$

and so

$$\tilde{p}_{j\partial}(t) = 0 \qquad j \in S \ .$$

But $(\tilde{f}_{j\partial}(t,n),\ n = 0,1,\ldots)$ is a non-decreasing sequence whose limit is $\tilde{p}_{j\partial}(t)$: thus

$$\tilde{f}_{j\partial}(t,n) = 0 \qquad j \in S$$

and so, from the case $n = 1$ of the recurrence (6),

$$\tilde{q}_{j\partial} = 0 \qquad j \in S \ .$$

Thus

$$\sum_j \tilde{q}_{ij} = 0 \qquad i \in S$$

and so equations (8) follow from the definition (11). The equality in relation (13) implies that the stable conservative q-matrix $(\tilde{q}_{ij},\ i,\ j \in S)$ is regular, and so the equations

$$\sum_{j\in S} \tilde{q}_{ij}\ y_j = y_i \qquad i \in S$$

have no non-trivial non-negative bounded solution. These equations can be rewritten as

$$\sum_j m_j\ y_j\ q_{ji} = m_i\ y_i \qquad i \in S$$

and the substitution $z_j = m_j\ y_j$ then shows that relations (9) can have no non-trivial non-negative solution.

The converse is established similarly. Suppose that statement (ii) holds. Once again define

$$\tilde{q}_{ij} = \frac{m_j}{m_i}\, q_{ji} \qquad i,\ j \in S \tag{14}$$

Then $\tilde{Q} = (\tilde{q}_{ij},\ i, j \in S)$ is stable, and is conservative also, by virtue of the hypothesis (8). The transition probabilities $\tilde{P}(t) = (\tilde{p}_{ij}(t), i, j \in S)$ constructed from $\tilde{Q}$ satisfy

$$m_i\, p_{ij}(t) = m_j\, \tilde{p}_{ji}(t) \qquad i, j \in S \qquad (15)$$

by he inductive argument used earlier. Thus, summing over i, the collection $(m_i,\ i \in S)$ is an invariant measure if

$$\sum_i \tilde{p}_{ji}(t) = 1 \qquad t > 0, \qquad j \in S$$

This is equivalent to the regularity of $\tilde{Q}$, which follows from the assumption that relations (9) have no non-trivial non-negative solution.

□

It is perhaps worth noting that the proof of the theorem makes no use of the assumption that Q is conservative: the conclusions of the theorem remain valid when the conservation condition (2) is relaxed to

$$\sum_{j \neq i} q_{ij} \leq -q_{ii} \qquad i \in S \quad ,$$

with the appropriately extended definition of the transition probabilities $P(t)$ ([6], Section 5.6). Note also that equations (8) above imply that m is subinvariant,

$$\sum_i m_i\, p_{ij}(t) \leq m_j \qquad t > 0,\ j \in S \quad ,$$

from relation (15).

If Q is transient equations (8) may not have a non-trivial solution. A collection of positive numbers $(m_i,\ i \in S)$ satisfies equations (8) if and only if $(m_i\, q_i,\ i \in S)$ is an invariant measure for the jump chain associated with Q, and the work of Harris [7] and Veetch [18] provides a necessary and sufficient condition for the existence of such a measure when Q is irreducible.

Call the matrix $\tilde{Q}$ defined by relation (14) the time-reverse of Q with respect to m. This terminology is suggested by the observation that if $(X(t), -\infty < t < \infty)$ is a stationary Markov process with q-matrix Q and stationary distribution m then $\tilde{Q}$ is the q-matrix of the Markov process $(X(-t), -\infty < t < \infty)$. The particle system interpretation discussed in the previous section provides a further insight. If Q and $\tilde{Q}$ are stable, conservative and regular then m will be an invariant measure for both $P(t)$ and $\tilde{P}(t)$, and a stationary particle system constructed fro· m and Q and then reversed in time will have the same distributional law as a stationary particle system constructed from m and $\tilde{Q}$.

Suppose now that Q is stable and conservative and that the positive collection $m = (m_i, i \in S)$ satisfies equations (8) but is not necessarily invariant. Consider a particle system in which at time $t \geq 0$ the number of particles at site i is $N_i(t)$: as before suppose that $(N_i(0), i \in S)$ are independent random variables, $N_i(0)$ Poisson with mean m_i, and that from time $t = 0$ onwards particles move independently from site to site, each in accordance with a Markov process constructed from the matrix Q. If $\tilde{Q}$ is not regular then $(N_i(t), i \in S)$ will be a collection of independent Poisson random variables but $EN_i(t)$ may well be less than m_i. From the work of Reuter ([17], Section 5.3) it is possible to deduce that relations (9) have a maximal non-negative solution z, and that

$$\int_0^\infty \left[\sum_i m_i p_{ij}(t)\right] e^{-t}dt = m_j - z_j .$$

For the particle system this implies that

$$EN_j(\theta) = m_j - z_j$$

where θ is an exponential random variable with unit mean. The vector z thus indicates the extent to which the measure m is subinvariant.

We now illustrate the theorem with a simple example.

<u>Example 1</u> Take the state space S to be the integers $\mathbb{Z}$ and let

$$\begin{aligned} q_{ij} &= q_i & j &= i + 1 \\ &= -q_i & j &= i \\ &= 0 & &\text{otherwise} \end{aligned}$$

where $q_i > 0$ for all $i \in \mathbb{Z}$. The (essentially unique) solution to equations (8) is

$$m_i = q_i^{-1} \qquad\qquad i \in \mathbb{Z}.$$

A solution to the equations $Qy = y$ (using the usual matrix abbreviation) must have the form

$$\begin{aligned} y_j &= y_0 \prod_{i=0}^{j-1} (1 + q_i^{-1}) \qquad & j > 0 \\ &= y_0 \prod_{i=1}^{-j} (1 + q_{-i}^{-1})^{-1} \qquad & j < 0 \end{aligned},$$

and a solution to the equations $zQ = z$ must have the form

$$\begin{aligned} z_j &= z_0\, q_0\, q_j^{-1} \prod_{i=1}^{j} (1 + q_i^{-1})^{-1} \qquad & j > 0 \\ &= z_0\, q_0\, q_j^{-1} \prod_{i=0}^{-j-1} (1 + q_{-i}^{-1}) \qquad & j < 0 \end{aligned}$$

Thus if (and only if)

$$\sum_{i=0}^{\infty} q_i^{-1} < \infty \tag{16}$$

there exists a non-trivial non-negative bounded solution to the equations $Qy = y$. On the other hand if (and only if)

$$\sum_{i=0}^{\infty} q_{-i}^{-1} < \infty \tag{17}$$

there exists a non-trivial non-negative solution to the equations $zQ = z$ which is bounded above by m. When condition (16) fails and condition (17) holds we have an example of a q-matrix which is not regular and yet admits an invariant measure m. When condition (16) holds and condition (17) fails we have an example of a regular q-matrix which does not admit an invariant measure, even though equations (8) admit a positive solution. Note that if conditions (16) and (17) both hold the solution m to equations (8) is summable, although the q-matrix is not regular. This possibility has been pointed out by Miller [15].

The second condition for invariance, that there exist no non-trivial non-negative vector z satisfying (9), should be compared with the condition arising in the investigation by Reuter ([17], Section 6) of the uniqueness of the solution to the forward equations associated with Q. Reuter's condition is that there exist no non-trivial non-negative vector z satisfying

$$\sum_i z_i q_{ij} = z_j \qquad j \in S$$

$$\sum_j z_j < \infty \quad .$$

In Example 1 the two conditions are equivalent, but in general the relationship does not appear to be this straightforward. Note that Reuter's condition is expressed solely in terms of the matrix Q, whereas relations (9) involve both Q and m. When Q is transient there may exist linearly independent positive solutions to equations (8): some of these may be invariant measures while others may not, as the next example illustrates.

<u>Example 2</u> Let $S = \mathbb{Z}$ and set

$$\begin{aligned} q_{ij} &= \lambda q_i & j &= i+1 \\ &= -q_i & j &= i \\ &= \mu q_i & j &= i-1 \\ &= 0 & &\text{otherwise} \end{aligned} \tag{18}$$

where $\lambda + \mu = 1$, $\lambda > \mu$ and $q_i > 0$ for $i \in \mathbb{Z}$. Consider the Markov process constructed from this q-matrix. From the form of the jump chain it is apparent that with probability one the process will remain in the set $\{-1,-2,\ldots\}$ for just a finite time. If these sections of the sample path are deleted the result is equivalent to reflecting the Markov process at the origin, and so the q-matrix (18) is regular if and only if the q-matrix corresponding to the reflected process is regular. Thus, from the analysis of Reuter ([17], Section 8.4), the q-matrix (18) is regular if and only if

$$\sum_{i=0}^{\infty} q_i^{-1} = \infty \qquad (19).$$

Now a solution to equations (8) is

$$m_i = \left(\frac{\lambda}{\mu}\right)^i q_i^{-1} \qquad i \in \mathbf{Z} \qquad (20)$$

and with this choice of m the matrix $\tilde{Q}$ defined by (14) is identical to Q. Thus m, given by expression (20), is an invariant measure for Q if and only if condition (19) holds. Another solution to equations (8) is

$$m_i = q_i^{-1} \qquad i \in \mathbf{Z} \qquad (21)$$

and with this choice of m the matrix $\tilde{Q}$ is given by

$$\begin{aligned} \tilde{q}_{ij} &= \mu q_i && j = i + 1 \\ &= -q_i && j = i \\ &= \lambda q_i && j = i - 1 \\ &= 0 && \text{otherwise} . \end{aligned}$$

This $\tilde{Q}$ is regular if and only if

$$\sum_{i=0}^{\infty} q_{-i}^{-1} = \infty \qquad (22)$$

and so condition (22) is necessary and sufficient for the measure (21) to be invariant.

The methods of this section provide an alternative proof of results, stated in the next Corollary, which are of some use in applied probability contexts. Part (i) is contained in, and part (ii) can be readily deduced from, Theorem 8 of Kendall and Reuter [13]; these statements also follow from the work of Miller [15]. They differ from a number of more widely available results in that the recurrence of Q is not required as a premise. Part (iii) is given for completeness but is hardly surprising: when Q is null recurrent relations (1) have an essentially unique solution, since the jump chain is recurrent, and it is known that $P(t)$ has an essentially unique invariant measure [1].

<u>Corollary</u> Let Q be stable, conservative, regular and irreducible, and suppose that $m = (m_i, i \in S)$ is a collection of positive numbers satisfying

$$\sum_i m_i q_{ij} = 0 \qquad\qquad j \in S$$

(i) If $\sum_i m_i = 1$ then Q is positive recurrent and m is the invariant probability distribution for $P(t)$.

(ii) If $\sum_i m_i = \infty$ then Q is null recurrent or transient.

(iii) If Q is null recurrent then m is the essentially unique invariant measure for $P(t)$.

Proof. Since Q is regular m is an invariant measure for the stable, conservative and irreducible matrix $\tilde{Q}$ defined by relation (14). Thus m is an invariant measure for the δ-skeleton of the Markov process constructed from $\tilde{Q}$.

If $\sum m_i = 1$ then this δ-skeleton is positive recurrent. Thus $\tilde{Q}$ is regular and so m is an invariant measure for Q also. Since $\sum m_i = 1$ the δ-skeleton of the Markov process constructed from Q is positive recurrent, and hence so is $\tilde{Q}$.

If $\sum m_i = \infty$ then the δ-skeleton of the Markov process constructed from $\tilde{Q}$ is null recurrent or transient. In either case

$$\lim_{t\to\infty} \tilde{p}_{ji}(t) = 0 \qquad\qquad i, j \in S$$

where $(\tilde{p}_{ji}(t))$ are the transition probabilities of the Markov process constructed from $\tilde{Q}$. But these probabilities satisfy relation (15), and so

$$\lim_{t\to\infty} p_{ij}(t) = 0 \qquad\qquad i, j \in S \ .$$

Thus Q must be either null recurrent or transient.

The null recurrence of Q is equivalent to

$$\int_0^\infty p_{ii}(t)dt = \infty \ .$$

But, from relation (15),

$$\int_0^\infty p_{ii}(t)dt = \int_0^\infty \tilde{p}_{ii}(t)dt \ .$$

Thus if Q is null recurrent then so is $\tilde{Q}$, and this in turn implies that $\tilde{Q}$ is regular and m is invariant for $P(t)$.

4. A cycle criterion

Recall that a matrix $\tilde{Q}$ is the time-reverse of the matrix Q with respect to a collection of positive numbers $(m_i, i \in S)$ if

$$m_j q_{jk} = m_k \tilde{q}_{kj} \qquad\qquad j, k \in S \qquad\qquad (23).$$

Say that $\tilde{Q}$ is a time-reverse of Q if there exists a collection of positive numbers $(m_i, i \in S)$ satisfying (23).

Theorem Let Q and $\tilde{Q}$ be stable and conservative, and suppose that Q is irreducible. Then $\tilde{Q}$ is a time-reverse of Q if and only if

$$q_{jk_1} q_{k_1k_2} \cdots q_{k_rj} = \tilde{q}_{jk_r} \tilde{q}_{k_rk_{r-1}} \cdots \tilde{q}_{k_1j} \qquad\qquad (24)$$

for each positive integer r and for all $j, k_1, k_2, \ldots, k_r \in S$. When this condition holds the measure m determined up to scalar multiples by

$$m_j q_{jk} = m_k \tilde{q}_{kj} \qquad\qquad j, k \in S$$

is an invariant measure for the transition probabilities $P(t)$ constructed from Q if and only if $\tilde{Q}$ is regular.

Proof. If $\tilde{Q}$ is a time-reverse of Q then (24) follows immediately from (23). Conversely suppose that (24) is satisfied. This condition and the assumed irreducibility of Q imply that $\tilde{Q}$ is irreducible. Choose a base state, labelled 0 say, from the state space S. Set $m_0 = 1$ and define

$$m_j = \frac{\tilde{q}_{0k_1}\,\tilde{q}_{k_1k_2}\cdots\tilde{q}_{k_wj}}{q_{jk_w}\,q_{k_wk_{w-1}}\cdots q_{k_10}} \tag{25}$$

where the path $(j, k_w, k_{w-1},\ldots,k_1,0)$ is chosen so that the denominator in expression (25) is positive. By virtue of the cycle condition (24) m_j also satisfies

$$m_j = \frac{q_{0h_1}\,q_{h_1h_2}\cdots q_{h_vj}}{\tilde{q}_{jh_v}\,\tilde{q}_{h_vh_{v-1}}\cdots\tilde{q}_{h_10}} \tag{26}$$

for any path $(j,h_v,h_{v-1},\ldots,h_1,0)$ chosen so that the denominator in expression (26) is positive. Since $\tilde{Q}$ is irreducible there is at least one such path. The representation (26) together with the cycle condition (24) then imply that every path $(j,k_w,k_{w-1},\ldots,k_1,0)$ which produces a positive denominator in expression (25) defines the same value for m_j.

We now show that m_j is positive. Since Q is irreducible there exists a cycle $(j,k_w,k_{w-1},\ldots,k_1,0,h_1,h_2,\ldots,h_v)$ such that both the denominator of expression (25) and the numerator of expression (26) are positive: the cycle condition (24) then implies that both the numerator of expression (25) and the denominator of expression (26) are positive.

Now consider two states $i,j \in S$, and suppose $q_{ij} > 0$. Then, using the path $(i,j,k_w,k_{w-1},\ldots,k_1,0)$ to define m_i, we have that

$$m_i = \frac{\tilde{q}_{ji}}{q_{ij}}\, m_j .$$

Similarly when $\tilde{q}_{ij} > 0$ expression (26) shows that

$$m_i = \frac{q_{ji}}{\tilde{q}_{ij}}\, m_j .$$

If $i = j$ then the choice $k_1 = j$ in relation (24) implies $q_{jj} = \tilde{q}_{jj}$. Thus

$$m_j \, q_{jk} = m_k \, \tilde{q}_{kj} \qquad j, k \in S \quad , \tag{27}$$

and so $\tilde{Q}$ is the time-reverse of Q with respect to m.

Since $\tilde{Q}$ is assumed conservative equations (27) imply that m satisfies $mQ = 0$, and so the final part of the theorem follows from the results of the previous section. □

The above method can be used to establish a similar cycle criterion for discrete time Markov chains (cf. [8], p.32).

References

[1] AZEMA, J., KAPLAN-DUFLO, M. and REVUZ, D. (1967) Mesure invariante sur les classes récurrentes des processus de Markov. *Z. Wahrscheinlichkeitstheorie verw. Geb. 8*, 157-181.

[2] BROWN, M. (1970) A property of Poisson processes and its application to macroscopic equilibrium of particle systems. *Ann. Math. Statist. 41*, 1935-1941.

[3] CHUNG, K.L. (1967) *Markov Chains with Stationary Transition Probabilities*. Second edition. Springer-Verlag, Berlin.

[4] DERMAN, C. (1955) Some contributions to the theory of denumerable Markov chains. *Trans. Amer. Math. Soc. 79*, 541-555.

[5] DOOB, J.L. (1942) Topics in the theory of Markoff chains. *Trans. Amer. Math. Soc. 52*, 37-64.

[6] FREEDMAN, D. (1971) *Markov Chains*. Holden-Day, San Francisco.

[7] HARRIS, T.E. (1951) Transient Markov chains with stationary measures. *Proc. Amer. Math. Soc. 8*, 937-942.

[8] KELLY, F.P. (1979) *Reversibility and Stochastic Networks*. Wiley, Chichester.

[9] KELLY, F.P. (1982) Networks of quasi-reversible nodes. In R.Disney (Ed.), *Applied Probability - Computer Science, the Interface: Proceedings of the ORSA-TIMS Boca Raton Symposium, Birkhäuser Boston, Cambridge, Mass.*

[10] KENDALL, D.G. (1959) Unitary dilations of one-parameter semigroups of Markov transition operators, and the corresponding integral representations for Markov processes with a countable infinity of states. *Proc. London Math. Soc. (3) 9*, 417-431.

[11] KENDALL, D.G. (1974) *Markov Methods*.(University of Cambridge lecture notes).

[12] KENDALL, D.G. (1975) Some problems in mathematical genealogy. In J. Gani (Ed.), *Perspectives in Probability and Statistics: Papers in Honour of M.S. Bartlett*, Applied Probability Trust, Sheffield. Distributed by Academic Press, London, pp 325-345.

[13] KENDALL, D.G. and REUTER, G.E.H. (1957) The calculation of the ergodic projection for Markov chains and processes with a countable infinity of states. *Acta Math. 97,* 103-144.

[14] KOLMOGOROV, A.N. (1936) Zur Theorie der Markoffschen Ketten. *Math. Ann. 112,* 155-160.

[15] MILLER, R.G., JR. (1963) Stationarity equations in continuous time Markov chains. *Trans. Amer. Math. Soc. 109,* 35-44.

[16] REICH, E. (1957) Waiting times when queues are in tandem. *Ann. Math. Statist. 28,* 768-773.

[17] REUTER, G.E.H. (1957) Denumerable Markov processes and the associated contraction semigroups on ℓ. *Acta Math. 97,* 1-46.

[18] VEECH, W. (1963) The necessity of Harris' condition for the existence of a stationary measure. *Proc. Amer. Math. Soc. 14,* 856-860.

[19] WHITTLE, P. (1975) Reversibility and acyclicity. In J. Gani (Ed.) *Perspectives in Probability and Statistics : Papers in Honour of M.S. Bartlett,* Applied Probability Trust, Sheffield. Distributed Academic Press, London, pp 217-224.

THE APPEARANCE OF A MULTIVARIATE EXPONENTIAL DISTRIBUTION IN SOJOURN TIMES FOR BIRTH-DEATH AND DIFFUSION PROCESSES

John T. Kent

Abstract

One way to obtain a multivariate exponential distribution is by adding together two independent identically distributed random vectors, each of which is obtained by squaring the components of a multivariate normal random vector with mean 0 and Markovian covariance matrix. The purpose of this paper is to explore the relationship between this multivariate exponential distribution and the distribution of the sojourn times for a birth-death process up to a first passage time. The continuous time analogue of this relationship is the Ray-Knight theorem on local times of Brownian motion. Some infinite divisibility properties of sojourn times are also discussed.

1. Introduction

The usual *exponential distribution* on $(0,\infty)$ has two important derivations in probability theory. In the first place, it can be derived from the "lack of memory" property, and hence represents the *waiting time* spent in a given state of a continuous-time Markov process before jumping to a new state. Secondly, it can be derived from the *normal distribution* as the law of $X_1^2 + X_2^2$, where X_1 and X_2 are independent normal random variables with mean zero and the same variance.

This "double derivation" feature of the exponential distribution can be used to help define a *multivariate exponential distribution*. For the first derivation we look at the *sojourn time vector* of a birth-death process up to a first passage time. For the second derivation we use a multivariate normal random vector with mean zero and Markovian covariance matrix.

Further, it is possible to extend our analysis to continuous state-space to define an *exponential stochastic process*. The "double derivation" feature of this process is included in the well-known theorem of Ray (1963) and Knight (1963). The sojourn time process of Brownian motion up to a first passage time can be identified with the sum of two independent squared Brownian motions.

The multivariate exponential distribution used in this paper was first proposed using the normal-based derivation. Using this approach the multivariate exponential distribution is just a particular case (index 1) of a *multivariate gamma distribution* defined for general half-integer index. See Wicksell (1933) and Kibble (1941) for the two-dimensional case, Krishnamoorthy and Parthasarathy (1951,1960) for the general case, and Griffiths (1970) for infinite divisibility properties in the Markovian case. However, the use of this distribution in practical modelling situations has been rather limited.

Hopefully the sojourn time derivation given here will lead to some interesting practical applications. Some other attempts to motivate this distribution have been given by Downton (1970), Hawkes (1972) and Paulson (1973).

For a survey of other possible definitions of a multivariate exponential distribution, see Block (1975,1977), Kotz (1975), and Galambos and Kotz (1978), but these alternatives will not be considered here at all.

We start the paper by deriving the sojourn time distributions for a birth-death process started at m and stopped at n, $m < n$. When $m = o$, the sojourn time vector defines a multivariate exponential distribution. In Section 3 we show that this class of exponential distributions coincides with a class of exponential distributions derived from the multivariate normal distribution.

The sojourn time vector for Brownian motion is studied in Sections 4-5 and the analysis is very similar to the birth-death case. An examination of the "double derivation" feature of the exponential stochastic process arising in this situation leads to a simple direct proof of the Ray-Knight theorem.

Lastly, some properties related to the multivariate infinite divisibility of sojourn times and other stochastic processes are considered in Sections 6-7.

2. Sojourn times for birth-death processes

Let a birth-death process $X(t)$ on $\{0,1,2,\ldots\}$ be defined in terms of a Q-matrix given by

$$q_{i,i+1} = \lambda_i > 0, \; q_{i+1,i} = \mu_{i+1} \geq 0 \quad \text{for} \quad i \geq 0 \tag{2.1}$$

and suppose

$$q_{ii} = -(\lambda_i + \mu_i), \quad i \geq 0 \tag{2.2}$$

(taking $\mu_o = 0$) so that the Q-matrix is *conservative*. As usual write $q_i = -q_{ii}$.

Let $\tau_i^{mn} = \tau_i$ denote the amount of time the process spends in state i if the process is started at m and stopped when it first reaches n, with $m < n$, and let $\boldsymbol{\tau}^{mn} = (\tau_o,\ldots,\tau_{n-1})$ denote the sojourn time vector. When convenient we can also consider $\boldsymbol{\tau}^{mn}$ as a vector of infinite length, taking $\tau_i = 0$ for $i \geq n$.

For a vector $\mathbf{s} = (s_o, s_1 \ldots)$ let $\mathbf{s} \leq \mathbf{0}$ ($\mathbf{s} < \mathbf{0}$) mean that $s_i \leq 0$ ($s_i < 0$) for all i. In order to determine the distribution of $\boldsymbol{\tau}^{mn}$ on $[0,\infty)^n$, we wish to determine its moment generating function (mgf)

$$\begin{aligned} \phi_{mn}(\mathbf{s}) &= E\{\exp(\mathbf{s}'\boldsymbol{\tau}^{mn})\} \\ &= E\{\exp(\sum_{i=o}^{n-1} s_i \, \tau_i^{mn})\} \end{aligned}$$

which is defined for $\mathbf{s} \leq \mathbf{0}$. Note that $\phi_{mn}(\mathbf{s})$ depends only on $s_o,\ldots,s_{n-1}$ Also, we use the moment generating function instead of the more usual Laplace transform because it is more convenient in Sections 6-7.

By the usual conditioning argument on what happens to the process in the first instant of time, we can derive the backwards equation for the mgf (as the starting point m varies):

$$\begin{aligned} &\lambda_o \, \phi_{1n}(\mathbf{s}) = (q_o - s_o) \, \phi_{on}(\mathbf{s}) \\ &\lambda_i \, \phi_{i+1,n}(\mathbf{s}) = (q_i - s_i) \, \phi_{in}(\mathbf{s}) - \mu_i \phi_{i-1,n}(\mathbf{s}), \; i = 1,\ldots,n-1 \\ &\phi_{nn}(\mathbf{s}) = 1. \end{aligned} \tag{2.3}$$

If we define a sequence of polynomials $P_i(\mathbf{s})$ by

$$P_o(\mathbf{s}) = 1$$

$$P_1(\mathbf{s}) = \lambda_o^{-1}(q_o - s_o) \tag{2.4}$$

$$P_{i+1}(\mathbf{s}) = \lambda_1^{-1}\{(q_i - s_i)\, P_i(s) - \mu_i\, P_{i-1}(\mathbf{s})\}\ ,\ i \geq 1,$$

then it is clear that

$$\phi_{mn}(\mathbf{s}) = P_m(\mathbf{s})/P_n(\mathbf{s})\ .$$

Let Q_i denote the upper $(i\times i)$ submatrix of Q and let S_i denote the diagonal matrix $\text{diag}(s_o,\ldots,s_{i-1})$. Then an inductive argument and expansion by minors along the bottom row shows that

$$P_i(s) = \{\prod_{j=o}^{i-1} \lambda_j^{-1}\}\ \det(-Q_i - S_i). \tag{2.5}$$

For convenience define $\det(-Q_o - S_o) = 1$ so that (2.5) holds for all $i \geq o$. Hence we can also write

$$\phi_{mn}(\mathbf{s}) = \{\prod_{j=m}^{n-1} \lambda_j\}\ \det(-Q_m - S_m)/\det(-Q_n - S_n)\ ,\ o \leq m < n. \tag{2.6}$$

Remarks. (1) If $\boldsymbol{\alpha} \geq o$ is a fixed vector then $\boldsymbol{\alpha}'\boldsymbol{\tau}^{mn}$ has mgf $\psi(t) = \phi(t\boldsymbol{\alpha})$, $t \leq o$. Such distributions were studied, for example, by Wang (1980). If $\boldsymbol{\alpha} > \mathbf{0}$ then $\boldsymbol{\alpha}'\tau^{mn}$ can also be viewed as a first-passage time for the time-changed birth-death process obtained by dividing the ith row of Q by α_i, for all i.

(2) If $\boldsymbol{\alpha}$ is concentrated on one point $(\alpha_i = o$ for $i \neq j$ and $\alpha_j = 1$ for some j, $o \leq j \leq n-1)$ then

$$\psi(t) = \frac{1 - t/\gamma_{m,j}}{1 - t/\gamma_{n,j}} \quad \text{if} \quad o \leq j < m, \tag{2.7}$$

and

$$\psi(t) = 1/(1-t/\gamma_{n,j}) \quad \text{if} \quad m \leq j \leq n-1, \tag{2.8}$$

where $\gamma_{i,j}^{-1}$ is the (j,j)th entry of $(-Q_i)^{-1}$.

Note that (2.8) is the mgf of an exponential distribution and that (2.7) is the mgf of an elementary mixture of exponentials (that is the mixture of a point mass at 0 with an exponential distribution).

(3) Thus, if $m = 0$ then all of the marginal distributions of τ^{on} will be exponentially distributed and we can describe the distribution of τ^{on} as a *multivariate exponential distribution*. An alternative derivation of this distribution based on the multivariate normal distribution will be given in Section 3.

(4) Since $\phi_{on}(\mathbf{0}) = 1$ for all $n \geq 1$, we see that

$$\det(-Q_n) = \prod_{j=0}^{n-1} \lambda_j \quad \text{for all} \quad n \geq 1. \tag{2.9}$$

In particular we can also write

$$\phi_{on}(\mathbf{s}) = \det(-Q_n)/\det(-Q_n - S_n). \tag{2.10}$$

(5) Although Q is not in general symmetric, it can be replaced by a symmetric matrix in (2.10). Define a tri-diagonal symmetric matrix Q^* by

$$q_{ii}^* = q_{ii}, \quad q_{i+1,i}^* = q_{i,i+1}^* = (\lambda_i \mu_{i+1})^{\frac{1}{2}} . \tag{2.11}$$

When all of the μ_i are non-zero, Q^* is also given by $Q^* = C\,Q\,C^{-1}$ where $C = \mathrm{diag}(c_i)$, with $c_0 = 1$ and $c_{i+1}/c_i = (\lambda_i/\mu_{i+1})^{\frac{1}{2}}$, $i \geq 0$. It is easily checked that $\det(-Q_n - S_n) = \det(-Q_n^* - S_n)$ for all s, so that (2.10) can also be written

$$\phi_{on}(\mathbf{s}) = \det(-Q_n^*)/\det(-Q_n^* - S_n). \tag{2.12}$$

Further, from (2.9) it follows that $-Q_n^*$ is positive-definite for all $n \geq 1$.

(6) If we replace condition (2.2) by

$$q_i \geq \lambda_i + \mu_i$$

then the birth-death process now has possible killing. Letting $\phi_{mn}(\mathbf{s})$ now denote

$$\phi_{mn}(\mathbf{s}) = E\{I_H \exp(\mathbf{s}'\pi^{mn})\}$$

where I_H is an indicator function on the set H of all sample paths which reach n before they die, we find that the (defective) mgf $\phi_{mn}(\mathbf{s})$ is still given by (2.10).

3. The normal-based derivation of the multivariate exponential distribution

Let $\mathbf{x} = (x_0,\ldots,x_{n-1})$ be a n-dimensional multivariate normal random vector with mean $\mathbf{0}$ and positive definite covariance matrix Σ. Define $\mathbf{y}$ by $y_i = x_i^2$, $i = 0,\ldots,n-1$. Straightforward integration shows that the mgf of $\mathbf{y}$ is given by

$$\{\det(\Sigma^{-1})/\det(\Sigma^{-1} - 2\,S_n)\}^{\frac{1}{2}}. \tag{3.1}$$

This multivariate gamma distribution of index $\frac{1}{2}$ was considered in two dimensions by Kibble (1941) and in higher dimensions by Krishnamoorthy and Parthasarathy (1951, 1960). Taking convolutions of independent copies of distributions with mgf (3.1) leads to a multivariate gamma distribution of general half-integer index.

In particular, if two independent copies of $\mathbf{y}$ are added together, the mgf becomes

$$\det(\Sigma^{-1})/\det(\Sigma^{-1} - 2\,S_n), \tag{3.2}$$

which can be identified with (2.12) if $\frac{1}{2}\Sigma^{-1} = -Q_n^*$.

Recall that a positive definite matrix Σ is called *Markov* if a multivariate normal vector with covariance matrix Σ forms a Markov process (Feller, 1966, p.66). Necessary and sufficient conditions for a positive definite matrix Σ to be Markov are either

(a) $\sigma_{ik} = \sigma_{ij}\sigma_{jk}/\sigma_{jj}$, for all $i \le j \le k$, or

(b) Σ^{-1} is tri-diagonal.

Further if Σ is Markov, direct calculation shows that the non-zero entries of Σ^{-1} are given by

$$\sigma^{ii} = \sigma_{ii}^{-1} \{\sigma_{i-1,i-1}\sigma_{ii}^{2}\sigma_{i+1,i+1} - \sigma_{i-1,i}^{2}\sigma_{i,i+1}^{2}\}/\{D_{i-1,i}D_{i,i+1}\}$$

$$\sigma^{i,i+1} = \sigma^{i+1,i} = -\sigma_{i,i+1}/D_{i,i+1} \tag{3.3}$$

where $D_{i,i+1} = \sigma_{ii}\sigma_{i+1,i+1} - \sigma_{i,i+1}^{2}$, and we interpret $\sigma_{-i,-i} = \sigma_{nn} = 1$ and $\sigma_{-1,o} = \sigma_{n-1,n} = 0$.

The study of mgfs of the form (3.1) - (3.2) for which Σ is Markov was started by Griffiths (1970) and we shall restrict our attention to Markov Σ in this paper.

Property (b) suggests the following theorem which enables us to refer unambiguously (within the context of this paper) to *the class of multivariate exponential distributions*.

Theorem 3.1. The class of sojourn time distributions (2.10) with any Q-matrix Q coincides with the class of distributions (3.2) with any Markov positive definite matrix Σ.

Proof. Given a Q-matrix Q define a symmetric matrix Q_n^* by (2.11) and set $\Sigma^{-1} = -2Q_n^*$. Since $-Q_n^*$ is tri-diagonal and positive definite, Σ will be Markov and we can identify (2.10) with (3.2).

Conversely, suppose we are given a Markov $n \times n$ matrix Σ. By replacing Σ with $C\Sigma C$ if necessary, where $C = \mathrm{diag}(c_i)$ and $c_i = \pm 1$, we may without loss of generality suppose that all of the elements of Σ are nonnegative (since the mgf (3.2) remains unaltered by such a transformation).

Define $Q_n^* = -(2\Sigma)^{-1}$. We wish to find a matrix Q_n such that Q_n is part of a Q-matrix (that is, the sum of the elements of each row equals 0, except for the nth row which sums to something negative), and such that $\det(-Q_n - S_n) = \det(-Q_n^* - S_n)$ for all $\mathbf{s}$.

To begin with suppose that all of the elements of Σ are non-zero, $\sigma_{ij} > 0$, so that all of the elements $q_{i,i+1}^*$ are non-zero. We shall define Q_n by the formula $Q_n = D_n Q_n^* D_n^{-1}$ where D_n is a diagonal matrix constructed inductively as follows. Set $d_o = 1$ and suppose $d_o, \ldots, d_i$ have been defined so that Q_{i+1}, the upper $(i+1)\times(i+1)$ sub-matrix of Q_n, is part of a Q-matrix. In particular this implies that $q_{i,i-1} + q_{ii} < 0$ so we can define $d_{i+1} > 0$ by

$$d_i\, q^*_{i,i+1}/d_{i+1} = -(q_{i,i-1} + q_{i,i}) \ .$$

Then the ith row (and also rows 0,...,i-1) of Q_{i+2} sum to 0. To show Q_{i+2} is part of a Q-matrix we must show that the (i+1)th row sums to something negative. Let **1** denote a vector of ones. Since $-Q^*_{i+2}$ is positive definite we have

$$0 > \mathbf{1}' D^{-1}_{i+2} Q^*_{i+2} D^{-1}_{i+2}\, \mathbf{1} = \mathbf{1}' \mathbf{D}^{-2}_{i+2} \mathbf{Q}_{i+2}\, \mathbf{1}$$

$$= d^{-2}_{i+1}(q_{i+1,i} + q_{i+1,i+1}) \ ,$$

so that this condition is satisfied.

If Σ contains some zero entries (but is still positive definite), then Q^* can be split into disjoint blocks over which the above argument can be applied, and the result is still true.

We have constructed our birth-death process to start at 0 and stop at n. Alternatively, we could have constructed the birth-death process to start at n and stop at 0.

4. Sojourn times for Brownian motion

The analogue of a birth-death process in continuous state space is a diffusion and since all diffusions can be obtained from Brownian motion by a time change (Ito and McKean, 1965, Chapter 5), it is sufficient to study the case of Brownian motion. Let $X(t)$ denote a standard Brownian motion started at a and let T_{ab} denote the first passage time to b, $a < b$. Let $L(t,x)$ be a jointly continuous version of the local time of the process, normalized so that

$$\int_\alpha^\beta L(t,x)\,dx = \int_0^t I[X(t')\in(\alpha,\beta)]\,dt' \ .$$

where $I[\cdot]$ is an indicator function.

We wish to determine the distribution of the sojourn time process $L(T_{ab},x)$, $x<b$. For this it is sufficient to calculate the finite dimensional distributions. Let $x_0<x_1<\ldots<x_{n-1}<b$ denote fixed points and let $\phi_{ab}(\mathbf{s})$ be the mgf of $L(T_{ab},x_0),\ldots,L(T_{ab},x_{n-1})$. The dependence of $\phi_{ab}(\mathbf{s})$ on the choice of $x_0,\ldots,x_{n-1}$ is suppressed in the notation.

Let $\delta_i(x)$ denote the Dirac delta function concentrated at the point x_i. If we solve the equation

$$\tfrac{1}{2} f''(x) + \sum_{i=0}^{n-1} s_i \, \delta_i(x) f(x) = 0 \tag{4.1}$$

for a non-zero function $f(x)$, $x \in R$, which is bounded as $x \to -\infty$, then $\phi_{ab}(\mathbf{s})$ will be given by

$$\phi_{ab}(\mathbf{s}) = f(a)/f(b) \ , \ \mathbf{s} < 0 \ . \tag{4.2}$$

See Ito and McKean (1965), p.74.

For convenience let $f(x) = 1$ for $x \le x_0$ (it must equal a non-zero constant by the boundary condition). Then (4.1) implies that $f(x)$ is a piecewise linear continuous function whose derivative is a piecewise constant function with jump equal to $-2 \ s_i \ f(x_i)$ at x_i.

The function f can be built up using some recurrence formulae. Set $f_i = f(x_i)$, $f_i^+ = f'(x_i+)$ and let $d_i = x_{i+1} - x_i$ (taking $x_n = b$). Then (4.1) implies

$$\begin{aligned} &f_0 = 1 \ , \ f_0^+ = -2s_0 \\ &f_i = 1 - 2s_0 d_0 \ , \ f_1^+ = f_0^+ - 2 \ s_1 f_1 \end{aligned} \tag{4.3}$$

and in general

$$f_{i+1} = f_i + d_i f_i^+ \ , \ f_{i+1}^+ = f_i^+ - 2s_{i+1} f_{i+1}, \ i \ge 1 \ . \tag{4.4}$$

Eliminating the derivatives $\{f_i^+\}$, we can get a recurrence formula in terms of the function values $\{f_i\}$ alone,

$$f_{i+1} = (1 + d_i/d_{i-1} - 2 \ d_i s_i) \ f_i - (d_i/d_{i-1}) \ f_{i-1}, \ i \ge 1. \tag{4.5}$$

Comparing (4.3) and (4.5) with (2.4) we see that the $\{f_i\}$ are the same as the polynomials $P_i(s)$ if the birth-death process has birth and death rates defined by

$$\lambda_i = \mu_{i+1} = (2 \ d_i)^{-1} \quad , \ i \ge 0. \tag{4.6}$$

Thus, if the starting point a equals one of the points x_i, then the Brownian motion has the same finite dimensional sojourn time distribution as the corresponding birth-death process. In particular, if $a \le x_0$ the mgf

$\phi_{ab}(\mathbf{s})$ can be written in the form $\phi_{ab}(\mathbf{s}) = \det(\frac{1}{2}\Sigma^{-1})/\det(\frac{1}{2}\Sigma^{-1} + S_n)$ using Theorem 3.1. Since the Q-matrix defined by (4.6) already symmetric, the tri-diagonal matrix Σ^{-1} is given by

$$\sigma^{oo} = d_o^{-1}, \quad \sigma^{ii} = d_i^{-1} + d_{i-1}^{-1}, \quad 1 \le i \le n-1$$
$$\sigma^{i,i+1} = \sigma^{i+1,i} = d_i^{-1}, \qquad 0 \le i \le n-2 . \tag{4.7}$$

Remarks (1) The finite dimensional mgf (4.2) for the sojourn times also occurs if the Brownian motion has an instantaneously reflecting left-hand boundary point at r_o where $r_o \le x_o$ (and we restrict the starting point so that $a \ge r_o$). But we shall not consider the reflecting Brownian motion further here.

(2) In the next section it will be convenient to consider the sojourn time $L(T_{ab}, b+x)$ for $a>b$ (instead of $a<b$) evaluated at the points $0<x_o<\ldots<x_{n-1} \le a-b$. Letting $d_i = x_{i+1} - x_i$ as before (taking $d_{-1} = x_o$), we find that Σ^{-1} is now given by

$$\sigma^{n-1,n-1} = d_{n-2}^{-1}, \quad \sigma^{i,i} = d_i^{-1} + d_{i-1}^{-1}, \quad 0 \le i \le n-2$$
$$\sigma^{i,i+1} = d_i^{-1} \qquad 0 \le i \le n-2 . \tag{4.8}$$

Note that in (4.7) it is the *last* row of Σ^{-1} which has row sum less than 0, whereas in (4.8) it is the 0th row.

5. The Ray-Knight Theorem

Let $Q_x^d(t)$, $t \ge 0$ denote the law of a d-dimensional squared Bessel process started at x, with differential generator

$$A_d = 2xD^2 + dD, \quad x \in (0,\infty),$$

where $D = d/dx$. For $0<d<2$, 0 is a regular boundary which we make reflecting, and for $d = 0$, 0 is an exit boundary which we make absorbing. For convenience suppose that for $t \le 0$ $Q_x^d(t)$ assigns probability one to the path identically equal to x so that $Q_x^d(t)$ becomes a continuous process for $t \in R$.

Shiga and Watanabe (1973) showed that $Q_x^d(\cdot)$ has the additive property

$$\{Q_x^d(t) \oplus Q_{x'}^{d'}(t) \ , \ t\epsilon R\} \stackrel{D}{=} \{Q_{x+x'}^{d+d'}(t), \ t\epsilon R\} \tag{5.1}$$

for $d,d' \geq 0$ and $x,x' \geq 0$. Here $\oplus$ denote the convolution of independent processes and $\stackrel{D}{=}$ denotes equality in distribution for the two processes. This additive property has been used to yield deep results by Pitman and Yor (1981a).

It will also be convenient to consider a time-inhomogeneous squared Bessel process, $Q_{x,u}^{d_1,d_2}$, which is a diffusion started at x for $t = 0$ (and equal to x for $t \leq 0$ with probability one), with differential generator A_{d_1} for $0 \leq t < u$, and with differential generator A_{d_2} for $t \geq u$. Thus, if $Y(t)$ is a process with law $Q_{x,u}^{d_1,d_2}(\cdot)$, then

$$\{Y(t), \ t \leq u\} \sim \{Q_x^{d_1}(t) \ , \ t \leq u\} \quad ,$$

and conditional on $\{Y(t'), \ t' \leq u\}$, we have

$$\{Y(u + t), \ t \geq 0\} \sim \{Q_{Y(u)}^{d_2}(t), \ t \geq 0\} \ .$$

It is straightforward to show that this process also satisfies the additive property,

$$\{Q_{x,u}^{d_1,d_2}(t) \oplus Q_{x',u}^{d_1',d_2'}(t), \ t\epsilon R\} \stackrel{D}{=} \{Q_{x+x',u}^{d_1+d_1',d_2+d_2'}(t), \ t\epsilon R\}. \tag{5.2}$$

<u>Theorem</u> 5.1. (Ray 1963, and Knight, 1963). Let $L(T_{ab},x)$, $x\epsilon R$, denote the local time process of Brownian motion started at a, up to the first passage time to b, $a>b$. Then

$$\{L(T_{ab}, b+t), \ t\epsilon R\} \sim \{Q_{o,b-a}^{2,o}(t), \ t\epsilon R\} \ . \tag{5.3}$$

<u>Remark</u>. Note that $Q_o^2(t)$ can be obtained as the sum of two independent squared Brownian motions started at 0, $X_1^2(t) + X_2^2(t)$. Hence the statement of this theorem for $0 \leq t \leq a-b$ describes a "double derivation" feature of an *exponential stochastic process*, analogous to Theorem 3.1 for the multivariate exponential distribution. Further, note the way in which the roles of time and space are interchanged in this theorem. For a very different approach to this theorem see, for example, Walsh (1978).

Proof. With the work done in earlier sections we are now able to provide a concise and straightforward proof of this result. The argument here is essentially the same as that given in Ray (1963).

First, note that the theorem is trivially true for $t \le 0$ because both processes equal 0 with probability 1.

Next, let $0 < x_0 < \ldots < x_{n-1} < a-b$ be a collection of "space" points for the left-hand side and "time" points for the right-hand side. Then the right-hand side of (5.3) has mgf given by (3.2) where Σ is the covariance matrix for Brownian motion, namely

$$\sigma_{ii} = x_i \ , \ 0 \le i \le n-1$$

$$\sigma_{i,i+1} = x_i \ , \ 0 \le i \le n-2 \ ,$$

with the other σ_{ij} determined by the Markov condition.

Let $d_i = x_{i+1} - x_i$, $i=0,\ldots,n-2$, taking $d_{n-1}^{-1} = 0$ and $d_{-1} = x_0$. Then using (3.3) we find that Σ^{-1} is tri-diagonal with

$$\sigma^{ii} = d_i^{-1} + d_{i-1}^{-1} \ , \qquad 0 \le i \le n-1$$

$$\sigma^{i,i+1} = \sigma^{i+1,i} = d_i^{-1} \ , \qquad 0 \le i \le n-2 \ .$$

Then using (4.8) and Theorem 3.1, we see that the two sides of (5.3) have the same joint mgf for $t = x_0,\ldots,x_{n-1}$. Since this holds for all the finite dimensional distributions, the two processes must have the same joint law for all $0 \le t \le b-a$.

The full theorem now follows easily from the above result and the strong Markov property, which implies that for $c>a>b$, the local time processes $L(T_{ca},\cdot)$ and $L(T_{ab},\cdot)$ are independent. Since

$$L(T_{cb},\cdot) = L(T_{ca},\cdot) + L(T_{ab},\cdot) \ ,$$

we see from the part of the theorem already proved that

$$\{Q_0^2(t),\ 0 \le t \le c-b\} \overset{D}{=} \{Q_0^2(b-a+t) + P(t), \qquad 0 \le t \le c-b\}$$

where $P(t)$ denotes the law of $L(T_{ab}, b+t)$, $t>0$, which we wish to determine. Conditioning on

$$\{L(T_{cb}, b+t),\ 0 \le t \le a-b\} = \{L(T_{ab}, b+t),\ 0 \le t \le a-b\},$$

and using the additive property (5.2), we find that (5.3) also holds for $a-b \le t \le c-b$. Since c is arbitrary, (5.3) holds for all $t \in R$. □

This theorem enables us to write down immediately the probability density corresponding to the γth convolution power of the multivariate exponential distribution with mgf

$$\phi(\mathbf{s}) = \{|\Sigma^{-1}|/|\Sigma^{-1} + 2S|\}^{\gamma} \quad ,$$

for any real $\gamma > 0$, where Σ is Markov. For simplicity suppose all of the off-diagonal elements $\sigma^{i,i+1}$ are non-zero. By rescaling the components of this distribution if necessary (that is replacing Σ by $C\Sigma C$ where C is diagonal), we may identify Σ with the covariance matrix of Brownian motion at a finite set of points. Hence $\phi(\mathbf{s})$ is the mgf of a finite-dimensional distribution from a $Q_0^{2\gamma}(\cdot)$ process. Since the transition density of a Bessel process is well-known (Molchanov, 1967, or Kent, 1978), the probability density corresponding to $\phi(\mathbf{s})$ can be written down. The two-dimensional case was given in Vere-Jones (1967).

6. Multivariate infinite divisibility

Let $\underline{\underline{A}}$ denote some class of mgfs of infinitely divisible distributions on $[0,\infty)$. It will usually be convenient to suppose that $\underline{\underline{A}}$ has at least the first three of the following properties.

(i) The class $\underline{\underline{A}}$ is closed under weak convergence.

((ii) If X is a random variable with mgf in $\underline{\underline{A}}$, so is αX, for $\alpha > 0$. (That is, if $\phi(s) \in A$, then $\phi(\alpha s) \in \underline{\underline{A}}$.)

(iii) If $F(dx)$ is a probability distribution with mgf $\phi(s)$ in $\underline{\underline{A}}$ so is $e^{-\beta x} F(dx)/\phi(-\beta)$ for $\beta \ge 0$. (That is, if $\phi(s) \in \underline{\underline{A}}$ then $\psi(s) = \phi(s-\beta)/\phi(-\beta) \in \underline{\underline{A}}$.)

(iv) The class $\underline{\underline{A}}$ is closed under convolutions.

(v) The class $\underline{\underline{A}}$ is closed under convolution roots (that is, if $\phi(s) \in \underline{\underline{A}}$, then $\phi(s)^{\gamma} \in \underline{\underline{A}}$ for all $0 \le \gamma \le 1$).

We can now define a multivariate class of infinitely divisible mgfs on $[0,\infty)^n$ as follows. Here $\mathbf{s} \le 0$ is an n-dimensional vector.

Definition 6.1. Say that a function $\phi(\mathbf{s})$, $\mathbf{s} \le 0$ belongs to the *multivariate class* $\underline{\underline{A}}_n$ if and only if

a) $\phi(\mathbf{s})$ is an infinitely divisible mgf on $[0,\infty)^n$, and

b) $\psi(t) = \phi(t\boldsymbol{\alpha}-\boldsymbol{\beta})/\phi(-\boldsymbol{\beta})$ lies in the univariate class $\underline{\underline{A}}$ for all $\boldsymbol{\beta}\geq\mathbf{0}$ and $\boldsymbol{\alpha}\geq\mathbf{0}$ $(\boldsymbol{\alpha}\neq\mathbf{0})$.

Remarks. (1) Since $\underline{\underline{A}}$ is assumed to satisfy conditions (i)-(iii), we see that $\underline{\underline{A}}_1 = \underline{\underline{A}}$.

(2) If $\mathbf{X}$ is an n-dimensional random vector with mgf in $\underline{\underline{A}}_n$, then $\boldsymbol{\alpha}'\mathbf{X}$ has its mgf in $\underline{\underline{A}}$ for all $\boldsymbol{\alpha}\geq\mathbf{0}$, $\boldsymbol{\alpha}\neq\mathbf{0}$.

(3) A more relaxed definition of the multivariate class $\underline{\underline{A}}_n$ would be obtained if property (b) was only required to hold for $\boldsymbol{\beta}=\mathbf{0}$. However, the more restrictive definition seems better suited to our purposes.

(4) If condition (iv) holds then $\underline{\underline{A}}_n$ includes distributions whose one-dimensional components are all independent and lie in $\underline{\underline{A}}$.

(5) If condition (v) holds and $\phi(\mathbf{s})\in\underline{\underline{A}}_n$, then $\phi(\mathbf{s})^\gamma\in\underline{\underline{A}}_n$ for all $0\leq\gamma\leq 1$.

An important tool in the study of infinite divisibility is the concept of an absolutely monotone function. Say that a real-valued function $f(\mathbf{s})$, $\mathbf{s}\leq\mathbf{0}$, is *absolute monotone* if f and all of its partial derivatives of all orders are nonnegative for $\mathbf{s}<\mathbf{0}$. Similarly, say that $f(\mathbf{s})$ has an *absolutely monotone derivative* if each of the first partial derivatives is absolutely monotone.

As for the one-dimensional case (Feller, 1966, p.417), it is easy to show that

(i) if $f(\mathbf{s})$ and $g(\mathbf{s})$ are absolutely monotone, so is $f(\mathbf{s})g(\mathbf{s})$,

(ii) if $f(t)$ is an absolutely monotone function of a univariate variable $t<0$, and if $g(\mathbf{s})<0$ for $\mathbf{s}<0$ has an absolutely monotone derivative, then $f(g(\mathbf{s}))$ is absolutely monotone.

Then we can state necessary and sufficient conditions for multivariate infinite divisibility which are analogous to those in the one-dimensional case.

Theorem 6.1. If $\phi(\mathbf{s})$ a real-valued function with $\phi(\mathbf{0}-) = 1$ and $\log\phi(\mathbf{s})<0$ for $\mathbf{s}<\mathbf{0}$, and if $\log\phi(\mathbf{s})$ has an absolutely monotone derivative, then $\phi(\mathbf{s})$ is an infinitely divisible mgf.

Another useful tool in the study of infinite divisibility is the concept of a *Pick function*. Say that a real-valued function $f(\mathbf{s})$, $\mathbf{s}<\mathbf{0}$ is a *multivariate Pick function* if it can be analytically continued to the region $\operatorname{Im}\mathbf{s}>\mathbf{0}$ and satisfies $\operatorname{Im} f(\mathbf{s})\geq 0$ there. Clearly, if $\phi(\mathbf{s})$ is a multivariate Pick function, then $\psi(t) = \phi(t\boldsymbol{\alpha}-\boldsymbol{\beta})$, as a function of t, is a univariate Pick function for all $\boldsymbol{\beta}\geq\mathbf{0}$, $\boldsymbol{\alpha}\geq\mathbf{0}$ $(\boldsymbol{\alpha}\neq\mathbf{0})$.

Three useful classes of univariate infinitely divisible mgfs are the following classes, listed together with necessary and sufficient conditions for a function $\phi(s)$, satisfying $\phi(0-) = 1$, to lie in the appropriate class.

(a) $\underline{\underline{B}}$, the Bondesson (1981) class, or generalized convolutions of mixtures of exponential distributions; $\log \phi(s)$ is a Pick function.

(b) $\underline{\underline{T}}$, the Thorin (1977) class, or generalized gamma convolutions; $\phi'(s)/\phi(s)$ is a Pick function.

(c) $\underline{\underline{GS}}$, the Goldie (1967) - Steutel (1967) class, or mixtures of exponential distributions; $\phi(s)$ is a Pick function.

These three classes satisfy conditions (i) - (v) at the beginning of the section (except $\underline{\underline{GS}}$ is not closed under convolution), so we can define the corresponding multivariate classes of mgfs. In particular we can obtain the following result.

Theorem 6.2. Let $\mathbf{s}$ be n-dimensional. A function $\phi(\mathbf{s})$ lies in the multivariate Bondesson class $\underline{\underline{B}}_n$ if and only if

(a) $\phi(\mathbf{s})$ is an infinitely divisible mgf, and

(b) $\log\phi(\mathbf{s})$ is a multivariate Pick function.

Similar descriptions hold for the class $\underline{\underline{T}}_n$ and $\underline{\underline{GS}}_n$, with $\log\phi(\mathbf{s})$ replaced by $\phi'(\mathbf{s})/\phi(\mathbf{s})$ and $\phi(\mathbf{s})$, respectively.

□

This study of infinite divisibility for a *finite* parameter process $X(t)$, $t = 0,\ldots,n-1$, can be extended to a *countable* and even *continuous* parameter process. We say that a stochastic process is infinitely divisible if all of its finite dimensional distributions are. Similarly, we say that a stochastic process is an $\underline{\underline{A}}$-*process* if all of its finite dimensional distributions lie in the multivariate $\underline{\underline{A}}$ class.

If $X(\cdot)$ is a stochastic process with *continuous sample paths*, then it is possible to choose all of the convolution powers $X(\cdot)^{*\gamma}$, for real $\gamma \geq 0$, to also have continuous sample paths. Further the mapping from γ to $X(\cdot)^{*\gamma}$, from $[0,\infty)$ to the space of probability measures on continuous sample paths, is a *continuous mapping* in the sense of weak convergence of probability measures. For further details, see Araujo and Giné (1980).

7. Infinite divisibility of sojourn times

We can now describe the infinite divisibility properties of the sojourn times.

Theorem 7.1. (i) The law of the sojourn time $(\tau_x^{mn}, x = 0,\ldots,n-1)$ for the birth-death process lies in the class $\underline{\underline{B}}_n$. (ii) Further the restricted sojourn time $(\tau_x^{mn}, x = m,\dots,n-1)$ lies in the class $\underline{\underline{T}}_{n-m}$. (iii) When $n = m + 1$, the whole sojourn time $(\tau_x^{m,m+1}, x = 0,\dots,m)$ lies in the class $\underline{\underline{GS}}_n$.

<u>Proof</u>. The proof is similar to arguments in Bondesson (1981, pp.64-65), and Paranjape (1978). Let $\phi_i^+(\mathbf{s}) = \phi_{i,i+1}(\mathbf{s})$. Then by (2.3) we have

$$\phi_0^+(\mathbf{s}) = \lambda_0/(q_0-s_0) ,$$

$$\log \phi_i^+(\mathbf{s}) = - \log \{q_i-s_i-\mu_i\phi_{i-1}^+(s)\}+\log \lambda_i, \quad i\geq 1.$$

Clearly $\phi_0^+(\mathbf{s})$ is absolutely monotone. Also since $-\log(1-x)$ has an absolutely monotone derivative for $x<0$ we see by induction that $\log\phi_i^+(\mathbf{s})$ has an absolutely monotone derivative and hence $\phi_i^+(\mathbf{s})$ is absolutely monotone for all $i\geq 0$.

Therefore $\phi_i^+(\mathbf{s})$ is infinitely divisible for each i. Since $\phi_{mn}(\mathbf{s}) = \prod_{i=m}^{n-1} \phi_i^+(\mathbf{s})$ we see that $\phi_{mn}(\mathbf{s})$ is infinitely divisible.

Similarly we can write

$$\phi_i^+(\mathbf{s}) = \lambda_i/\{q_i-s_i-\mu_i\phi_{i-1}^+(\mathbf{s})\} , \quad i\geq 1$$

Clearly $\phi_0^+(\mathbf{s})$ is a Pick function and by induction it is easily proved that $\phi_i^+(\mathbf{s})$ is a Pick function. Hence, part (iii) of the theorem is proved. To prove (i) note that the log of a Pick function is a Pick function and the sum of Pick functions is a Pick function. Hence

$\log \phi_{mn}(\mathbf{s}) = \sum_{i=m}^{n-1} \log \phi_i^+(\mathbf{s})$ is a Pick function, so $\phi_{mn}(\mathbf{s})\in\underline{\underline{B}}_n$.

To prove part (ii) let $\boldsymbol{\alpha}\geq 0$ be a non-zero n-vector with $\alpha_i = 0$, $i = 0,\dots m-1$, and let $\boldsymbol{\beta}\geq 0$. Since $\det(-Q_m - S_m)$ depends only on $s_0,\dots,s_{m-1}$ we see that

$$\psi(t) = \phi_{mn}(t\boldsymbol{\alpha}-\boldsymbol{\beta})/\phi(-\boldsymbol{\beta}) = U(t)^{-1} = \Pi(1-t/u_i)^{-1}$$

where $U(t)$ is a polynomial in t of degree $n-m$, with roots $u_1,\dots,u_{n-m}$, say. Since $\psi(t)$ is an mgf, all of the u_i must be positive. Hence $\psi(t)$ represents a convolution of exponential distributions, so $\psi(t)\in\underline{\underline{T}}$. Hence the restricted sojourn time vector lies in $\underline{\underline{T}}_{n-m}$.

Using the results of Sections 4-5, we can obtain corresponding results for Brownian motion.

Corollary 7.1. Let $\{L(T_{ab},x)\ x\in R\}$ denote the sojourn time process for Brownian motion up to a first passage time, $a<b$. Then this process is a Bondesson process. Further the restricted process $\{L(t_{ab},x),\ a\leq x\leq b\}$ is a Thorin process.

Remarks. (1) Some of these results can also be proved directly from the Ray-Knight Theorem. The additivity property (5.2) shows that the sojourn time process is infinitely divisible. Further a finite linear combination $\Sigma\alpha_i\ L(T_{ab},x_i)$, where $a\leq x_i\leq b$ for all i, is distributed as a convolution of two independent quadratic forms from the multivariate normal distribution with mean **0**. Since such forms can be expressed as convolutions of multiples of χ_1^2 variates (Johnson and Kotz, 1970, Chapter 29), we have part of the proof that the restricted process is a $\underline{\underline{T}}$ process.

(2) Since any diffusion hitting time which is finite with probability 1 can be written as an additive functional, $\frac{1}{2}\int L(T_{ab},\ x)\ m(dx)$, where $m(dx)$ is a *speed measure* (Ito and McKean, 1965, Chapter 5), we have here another proof that *all diffusion hitting times lie in the Bondesson class* (see Kent, 1980, 1982, and Bondesson, 1981).

(3) The joint distribution of two additive functionals lies in the bivariate Bondesson class. An example of such a distribution was considered by Pitman and Yor (1981b, p.296). Another example can be constructed from the joint distribution of the hitting angle and hitting time arising in the two-dimensional Brownian motion with polar drift of Kendall (1974).

(4) The multivariate gamma distribution of Section 3 can be viewed as the diagonal part of a Wishart distribution on the set of positive-definite matrices. The whole Wishart distribution is *not* infinitely divisible. In fact for index $\nu = \frac{1}{2}$, it is *indecomposable*; see Levy (1948), Davidson (1973) and Kendall (1973). However, Shanbhag (1976) has shown that all of the elements of a given row of a Wishart matrix do have an infinitely divisible distribution.

ACKNOWLEDGEMENT

Part of this paper was written during a visit to the University of Umeå, Sweden. Also, I am grateful to Marc Yor for drawing my attention to the Ray-Knight theorem.

REFERENCES

1. Araujo, A., and Giné, E. (1980) *The Central Limit Theorem for Real and Banach Valued Random Variables*. Wiley, New York.

2. Block, H.W. (1975) Continuous Multivariate exponential extensions. in *Reliability and Fault Tree Analysis*, edited by R.E. Barlow et al. SIAM, Philadelphia.

3. Block, H.W. (1977) Multivariate Reliability Classes. in *Applications of Statistics*, edited by P.R. Krishnaiah. North-Holland, Amsterdam, 79-88.

4. Bondesson, L. (1981) Classes of infinitely divisible distributions and densities. *Z. Wahrscheinlichkeitstheorie verw. Gebiete 57*, 39-71.

5. Davidson, R. (1973) Amplification of some remarks of Levy concerning the Wishart distribution. in *Stochastic Analysis*, edited by D.G. Kendall and E.F. Harding. Wiley, New York, 212-214.

6. Downton, F. (1970) Bivariate exponential distributions in reliability theory. *J. Royal Statist. Soc. B32*, 408-417.

7. Feller, W. (1966) *An Introduction to Probability Theory and its Applications*, Vol. 2. Wiley, New York.

8. Galambos, J., and Kotz, S. (1978) *Characterizations of Probability Distributions*. Lecture Notes in Maths. No. 675, Springer-Verlag, Berlin.

9. Goldie, C.M. (1967) A class of infinitely divisible distributions. *Math. Proc. Cambridge Philos. Soc. 63*, 1141-1143.

10. Griffiths, R.C. (1970) Infinitely divisible multivariate gamma distributions. *Sankhya A32*, 393-404.

11. Hawkes, A.G. (1972) A bivariate exponential distribution with applications to reliability theory. *J. Royal Statist. Soc. B34*, 129-131.

12. Itô, K., and McKean, H.P., Jr. (1965) *Diffusion Processes and Their Sample Paths*. Springer-Verlag, Berlin.

13. Johnson, N.L., and Kotz, S. (1970) *Continuous Univariate Distributions-2*. Houghton Mifflin, Boston.

14. Kendall, D.G. (1973) Appendix to Davidson (1973). in *Stochastic Analysis*, edited by D.G. Kendall and E.F. Harding. Wiley, New York, 214-219.

15. Kendall, D.G. (1974) Pole-seeking Brownian motion and bird navigation (with discussion). *J. Royal Statist. Soc. B36*, 365-417.

16. Kent, J.T. (1978) Some probabilistic properties of Bessel functions. *Ann. Probab. 6*, 760-770.

17. Kent, J.T. (1980) Eigenvalue expansions for diffusion hitting times. *Z. Wahrscheinlichkeitstheorie verw. Gebiete. 52*, 309-319.

18. Kent, J.T. (1982) The spectral decomposition of a diffusion hitting time. *Ann. Probab. 10*, 207-219.

19. Kibble, W.F. (1941) A two-variate gamma type distribution. *Sankhya 5*, 137-150.

20. Knight, F.B. (1963) Random walks and the sojourn density process of Brownian motion. *Trans. Amer. Math. Soc. 109*, 56-86.

21. Kotz, S. (1975) Multivariate distributions at a cross road. in *Statistical Distributions in Scientific Work, Vol. I.* D. Riedel Publishing Co., Dordrecht, Holland, 247-270.

22. Krishnamoorthy, A.S., and Parthasarathy, M. (1951) A multivariate gamma type distribution. *Ann. Math. Statist. 22*, 549-557.

23. Krishnamoorthy, A.S., and Parthasarathy, M. (1960) A correction to Krishnamoorthy and Parthasarathy (1951). *Ann. Math. Statist. 31*, 229.

24. Levy, P. (1948) The arithmetical character of the Wishart distribution. *Proc. Cambridge Philos. Soc. 44*, 295-297.

25. Molchanov, S. (1967) Martin boundaries for invariant Markov processes on a solvable group. *Theory of Probab. and its Appl. 12*, 310-314.

26. Paranjape, S.R. (1978) Simpler proofs for the infinite divisibility of multivariate gamma distributions. *Sankhya A40*, 393-398.

27. Paulson, A.S. (1973) A characterization of the exponential distribution and a bivariate exponential distribution. *Sankhya A35*, 69-78.

28. Pitman, J., and Yor, M. (1981a) A decomposition of Bessel bridges. manuscript.

29. Pitman, J., and Yor, M. (1981b) Bessel processes and infinitely divisible laws. in *Stochastic Integrals*, edited by D. Williams, Lecture Notes in Maths. No. 851, Springer-Verlag, Berlin.

30. Ray, D.B. (1963) Sojourn times of diffusion processes. *Ill. J. of Maths. 7*, 615-630.

31. Shanbhag, D.N. (1976) On the structure of the Wishart distribution. *J. Multivariate Analysis*, 347-355.

32. Shiga, T., and Watanabe, S. (1973) Bessel diffusions as a one-parameter family of diffusion processes. *Z. Wahrscheinlichkeitstheorie verw. Gebiete 27*, 37-46.

33. Steutel, F.W. (1967) Note on the infinite divisibility of exponential mixtures. *Ann. Math. Statist. 38*, 1303-1305.

34. Thorin, O. (1977) On the infinite divisibility of the Pareto distribution. *Scand. Actuarial J.*, 31-40.

35. Vere-Jones, D. (1967) The infinite divisibility of a bivariate gamma distribution. *Sankhya A31*, 421-422.

36. Walsh, J.B. (1978) Excursions and local time. *Astérisque 52-53*, 159-192.

37. Wang, Z.K. (1980) Sojourn times and first passage times for birth and death processes. *Scientia Sinica 23*, 269-279.

38. Wicksell, S.D. (1933) On correlation functions of Type III. *Biometrika 25*, 121-133.

THREE UNSOLVED PROBLEMS IN DISCRETE MARKOV THEORY

J.F.C. Kingman,

The theory of Markov processes with a discrete time parameter and at most a countable number of states is often considered a rather well understood branch of probability theory, not least as a result of the work of David Kendall. But it does harbour some unsolved problems, three of which form the subject of this paper.

All three have been in the literature for some years, and appear to be moderately difficult. Their interest is purely mathematical, and it is unlikely that any practical result would attend their solution. In each case there is partial progress to report, and the problem has various equivalent formulations.

They are presented here in affectionate homage to David Kendall, whose own knack of proposing interesting problems has been the stimulus of much of the research of his students and friends.

1. Powers of renewal sequences

A *renewal sequence* is a sequence $(u_n ; n = 0,1,2,\ldots)$ of real numbers which is generated by the recurrence relation

$$u_0 = 1 , \quad u_n = \sum_{r=1}^{n} f_r u_{n-r} \quad (n \geq 1) \tag{1.1}$$

for some sequence $(f_n ; n = 1,2,\ldots)$ satisfying

$$f_n \geq 0 , \quad \sum_{n=1}^{\infty} f_n \leq 1 . \tag{1.2}$$

Such sequences play an important part in Markov chain theory because of an observation of Chung (recorded by Feller [4]) that (u_n) is a renewal sequence if and only if there is a Markov process $(X_n ; n = 0,1,2,\ldots)$ on a countable state space S, such that

$$u_n = p_{ii}^{(n)} = \text{Prob}\,(X_n = i \mid X_0 = i) \tag{1.3}$$

for some $i \in S$.

Chung's theorem leads at once to the result that, if (u'_n) and (u''_n) are renewal sequences, then so is (u_n), where

$$u_n = u'_n u''_n \tag{1.4}$$

This motivates the Kendall-Davidson theory of the arithmetic of renewal sequences [6], an account of which in the discussion to [7] led me to conjecture that, *if* (u_n) *is a renewal sequence, then so is* (u_n^{α}) for any $\alpha > 1$. That this is so for integer values of α is clear, and it is sufficient to consider values of α in the range $1 < \alpha < 2$.

A sequence (u_n) is a *generalised renewal sequence* [9] if it satisfies (1.1) for $f_n \geq 0$ not necessarily limited by the second inequality of (1.2). It is easy to see that a generalised renewal sequence (u_n) is a renewal sequence if and only if $u_n \leq 1$ for all n, and therefore the conjecture is equivalent if formulated in terms of generalised renewal sequences.

The advantage of doing so is in focussing attention on the inequalities $f_n \geq 0$. If (1.1) is written in the form

$$f_n = u_n - \sum_{r=1}^{n-1} f_r\, u_{n-r}\,,$$

it can be solved recursively to give

$$f_n = \phi_n\,(u_1, u_2, \ldots, u_n)\,, \tag{1.5}$$

where the ϕ_n are polynomials determined by the recurrence relation

$$\phi_n(x_1, x_2, \ldots, x_n) = x_n - \sum_{r=1}^{n-1} \phi_r(x_1, x_2, \ldots, x_r)\, x_{n-r}\,. \tag{1.6}$$

Thus we can express the conjecture as the conjunction of a sequence of conjectures $\underset{\sim}{C}_n$ $(n = 1,2,\ldots)$:

$\underset{\sim}{C}_n$: *for any generalised renewal sequence* (u_n) *and any* $\alpha > 1$,

$$\phi_n\,(u_1^\alpha, u_2^\alpha, \ldots, u_n^\alpha) \geq 0\ . \tag{1.7}$$

Consider for example the conjecture $\underset{\sim}{C}_4$. From (1.6),

$$\phi(x_1,x_2,x_3,x_4) = x_4 - 2x_1x_3 - x_2^2 + 3x_1^2x_2 - x_1^4\ ,$$

so that $\underset{\sim}{C}_4$ asserts that

$$u_4^\alpha - 2u_1^\alpha\, u_3^\alpha - u_2^{2\alpha} + 3u_1^{2\alpha}u_2^\alpha - u_1^{4\alpha} \geq 0 \tag{1.8}$$

in $\alpha > 1$ for every generalised renewal sequence (u_n). Substituting for u_n from (1.1), this means that

$$(f_4 + 2f_1f_3 + f_2^2 + 3f_1^2f_2 + f_1^4)^\alpha - 2f_1^\alpha\,(f_3 + 2f_1f_2 + f_1^3)^\alpha$$
$$- (f_2 + f_1^2)^{2\alpha} + 3f_1^{2\alpha}\,(f_2 + f_1^2)^\alpha - f_1^{4\alpha} \geq 0$$

for all $f_1, f_2, f_3, f_4 \geq 0$, $\alpha > 1$. By homogeneity, there is no loss of generality in taking $f_1 = 1$. The left hand side is a non-decreasing function of f_3 and f_4, and hence it suffices to prove the inequality when $f_3 = f_4 = 0$. Hence $\underset{\sim}{C}_4$ is equivalent to the assertion that

$$\begin{aligned}(f^2 + 3f + 1)^\alpha - 2(2f + 1)^\alpha - (f + 1)^{2\alpha} &+ \\ + 3(f + 1)^\alpha - 1 &\geq 0\end{aligned} \tag{1.9}$$

whenever $f \geq 0$, $\alpha > 1$. I am indebted to John Hammersley for showing me how to prove (1.9), and will not spoil the reader's enjoyment by revealing his proof.

Thus $\underset{\sim}{C}_4$ is true, and so likewise (but more easily) are $\underset{\sim}{C}_1$, $\underset{\sim}{C}_2$ and $\underset{\sim}{C}_3$. No proof is known of any $\underset{\sim}{C}_n$ for $n \geq 5$. A certain amount of numerical experiment suggests that they are true, but become more delicate as n increases.

For a particular renewal sequence and a particular value of n with $f_n > 0$, (1.9) holds by continuity for $\alpha - 1$ sufficiently small. The same will be true even if $f_n = 0$ so long as $d\phi_n/d\alpha > 0$ at $\alpha = 1$. Expressing this derivative in terms of the partial derivatives

$$\phi_{nr} = \partial\phi_n/\partial x_r \,, \tag{1.10}$$

this means that it is sufficient that

$$\sum_{r=1}^{n} u_r \log u_r \, \phi_{nr} \, (u_1, u_2, \ldots, u_n) > 0 \,. \tag{1.11}$$

The weak version of (1.11) is necessary for $\underset{\sim}{C}_n$ to hold, but it seems likely that the strict inequality (1.11) is in fact true except for geometric sequences (that is, sequences with $u_n = u_1^n$, for which the conjecture is trivially true). Hence a companion sequence of conjectures $\underset{\sim}{C}_n^*$ might be considered, which has the advantage of being free of the parameter α.

$\underset{\sim}{C}_n^*$: *The inequality (1.11) holds for any non-geometric generalised renewal sequence with* $f_r = 0$ $(r \geq n)$.

What is not perhaps obvious is that the original conjecture, that $\underset{\sim}{C}_n$ holds for all $n \geq 1$, is implied by the conjunction of the $\underset{\sim}{C}_n^*$ for all $n \geq 3$ ($\underset{\sim}{C}_1^*$ and $\underset{\sim}{C}_2^*$ are vacuous). To establish this implication, it is sufficient (by induction on n) to prove $\underset{\sim}{C}_n$ on the assumption that $\underset{\sim}{C}_r$ is true for all $r < n$, and that $\underset{\sim}{C}_n^*$ is true.

To do this by contradiction, assume $\underset{\sim}{C}_r$ $(r < n)$ and $\underset{\sim}{C}_n^*$, together with the negation of $\underset{\sim}{C}_n$. Let (u_m) be a generalised renewal sequence negating $\underset{\sim}{C}_n$, and let β be the infimum of those $\alpha > 1$ for which (1.7) fails. Then, by $\underset{\sim}{C}_r$,

$$f_r^* = \phi_r \, (u_1^\beta, u_2^\beta, \ldots, u_r^\beta) \geq 0 \quad (r \leq n-1) \,,$$

and, by continuity,

$$f_n^* = \phi_n \, (u_1^\beta, u_2^\beta, \ldots, u_n^\beta) = 0 \,.$$

Define $f_m^* = 0$ $(m > n)$ and let (u_m^*) be the generalised renewal sequence generated by (1.1) with these f_m^*. Then $u_r^* = u_r^\beta$ for $r \leq n$, and $\underset{\sim}{C}_n^*$ applies to (u_m^*) to show that $d\phi_n/d\alpha > 0$ at $\alpha = \beta$. Hence for $\alpha > \beta$ with $\alpha - \beta$ sufficiently small, contradicting the definition of β.

In other words, the conjecture of [7] will be established if we can prove $\underset{\sim}{C}_n^*$ for all $n \geq 3$. It is known that $\underset{\sim}{C}_3^*$ and $\underset{\sim}{C}_4^*$ are true, but $\underset{\sim}{C}_5^*$ remains undecided.

2. The Markov group problem

This problem was posed by Kendall himself in [5], and concerns a Markov semigroup on the countable set S, an array of continuous functions $p_{ij} : [0,\infty) \to [0,1]$ indexed by $i,j \in S$, which satisfies

$$\begin{aligned} p_{ij}(0) &= \delta_{ij}, \quad \sum_{j\in S} p_{ij}(t) = 1, \\ p_{ij}(s+t) &= \sum_{k\in S} p_{ik}(s)\, p_{kj}(t) . \end{aligned} \tag{2.1}$$

The definitive account of such systems is of course the book by Chung [1].

The semigroup is so called because it determines a collection of contractions on the space $\ell = \ell_1(S)$ of sequences $x = (x_i\ ;\ i \in S)$ with

$$||x|| = \sum_{i\in S} |x_i| < \infty ,$$

by the formula

$$(P_t x)_j = \sum_{i\in S} x_i\, p_{ij}(t) , \tag{2.2}$$

and these form a strongly continuous semigroup of (bounded linear) operators on ℓ in the sense that

$$P_{s+t} = P_s P_t \ (s,t \geq 0), \ \lim_{t\downarrow 0} ||P_t x - x|| = 0 . \tag{2.3}$$

Kendall noted that, if the semigroup is uniformly continuous in the sense that

$$\lim_{t\downarrow 0} ||P_t - I|| = 0 , \tag{2.4}$$

I being the identity on ℓ, then P_t can be written as

$$P_t = \exp(t\Omega) \tag{2.5}$$

for a bounded operator Ω, that (2.5) makes sense even for negative t, and that (2.5) then extends the semigroup to a strongly continuous group of operators on ℓ. He conjectured that, conversely, P_t can be so extended only when (2.4) holds.

This *Markov group conjecture* is equivalent [5] to the assertion that, *if* P_1 *is invertible as a bounded linear operator on* ℓ, *then (2.4) holds*. It has been the subject of research by Kendall [5], Speakman [11], Davidson [3] and Cuthbert [2], but the best result is due to Mountford [10], who proves by a most ingenious argument that (2.4) is implied by the condition that

$$\|P_1 - I\| < 1 . \tag{2.6}$$

The inequality (2.6) implies, but is not implied by, the invertibility of P_1.

At this stage it may reasonably be objected that this is not a discrete problem at all, since (2.4) depends crucially on the continuous nature of the parameter t. In fact, it can be argued that the conjecture is really about finite stochastic matrices, the simplest of all the tools of the discrete Markov enthusiast.

I was led to this conclusion by trying to construct a counter-example to the conjecture. Take as S a countable union

$$S = \bigcup_{n=1}^{\infty} S_n \tag{2.7}$$

of disjoint finite sets, and for P_t the direct product of finite Markov semigroups. Thus for each n let Q_n be a Q-matrix on S_n (so that Q_n has non-negative off-diagonal elements and zero row sums), and let

$$\exp(tQ_n) = (p_{ij}^n(t) \; ; \; i,j \in S_n) . \tag{2.8}$$

Define P_t on a typical element of ℓ, written as

$$x = (x_i^n \; ; \; n \geq 1, \, i \in S_n)$$

by

$$(P_t x)_j^n = \sum_{i \in S_n} x_i^n \, p_{ij}^n \, (t) \; . \tag{2.9}$$

If P_1 has an inverse, it must be given by setting $t = -1$ in (2.8) and (2.9), and the result is a bounded operator if and only if

$$\sup_n \|\exp(-Q_n)\| < \infty \; , \tag{2.10}$$

where $\|\cdot\|$ is the matrix norm

$$\|A\| = \max_i \sum_j |a_{ij}| \; .$$

On the other hand the condition for uniform continuity of (2.8) is that

$$\sup_n \|Q_n\| < \infty \; . \tag{2.11}$$

Hence we shall have a counter-example to the Markov group conjecture if we can find a sequence of Q-matrices for which (2.10) holds but $\|Q_n\| \to \infty$.

Efforts to construct such a sequence having so far failed, one is led to the following conjecture.

$\underset{\sim}{C}$: *for any* $M > 0$, *there is a constant* $K = K_1(M)$ *such that every finite Q-matrix (of whatever order) with*

$$\|\exp(-Q)\| \leq M \tag{2.12}$$

satisfies

$$\|Q\| \leq K \; . \tag{2.13}$$

If $\underset{\sim}{C}$ is false, then so is the Markov group conjecture. On the other hand, if $\underset{\sim}{C}$ is true, an approximation argument leads to a stronger form of the conjecture, that if

$$\|P_{-1}\| \le M ,$$

then (2.5) holds with

$$\|\Omega\| \le K_1(M) .$$

Thus the essence of Kendall's conjecture is a property of finite Q-matrices.

The condition (2.12) is difficult to handle, and is conveniently replaced by a weaker condition implied by (2.12), that the eigenvalues of the matrix $\exp(-Q)$ should not exceed M in absolute value. These eigenvalues are of the form $e^{-\mu}$, where μ runs over the eigenvalues of Q, so that $\underset{\sim}{C}$ is implied by the conjecture

$\underset{\sim}{C}'$: *for any* $m > 0$, *there exists* $K = K_2(m)$ *such that every finite Q-matrix* Q, *all of whose eigenvalues* μ *satisfy*

$$\mathrm{Re}\ \mu \ge -m , \tag{2.14}$$

satisfies (2.13).

It is important of course that the constants in $\underset{\sim}{C}$ and $\underset{\sim}{C}'$ be independent of the order of the matrix. For matrices of given order, these assertions are true by trace arguments, and Speakman [11] was the first to observe that no counter-example to Kendall's conjecture could be constructed from direct products of finite matrices of bounded order.

If $\underset{\sim}{C}'$ could be proved, $\underset{\sim}{C}$ would follow. The converse is not true, but it seems likely that a class of counter-examples to $\underset{\sim}{C}'$ would contain some which would negate $\underset{\sim}{C}$, and thus the Markov group conjecture.

The norm of a Q-matrix is given by

$$\|Q\| = \max_i (-q_{ii}) ,$$

and therefore a Q-matrix with norm q can be written

$$Q = q(P - I) , \tag{2.15}$$

where P is a stochastic matrix with at least one zero on its diagonal. The eigenvalues of Q are given in terms of those of P by

$$\mu = q(\lambda - 1) \quad ,$$

so that (2.14) is equivalent to

$$\text{Re } \lambda \geq 1 - mq^{-1} \; .$$

It follows easily that $\underset{\sim}{C}'$ is equivalent to the following conjecture about finite stochastic matrices.

$\underset{\sim}{C}''$: *there exists* $\delta > 0$ *such that every finite stochastic matrix (of whatever order), with at least one zero diagonal element, has at least one eigenvalue* λ *with*

$$\text{Re } \lambda < 1 - \delta. \tag{2.16}$$

In this form, the conjecture appears rather improbable, but I know of no disproof.

3. The non-diagonal Markov characterisation problem

One way of looking at the theorem of Chung which was cited in §1 is as a characterisation of those sequences which can arise as diagonal transition probabilities (1.3) of discrete time, countable state space Markov processes. It is an effective characterisation in the sense that it allows one to test any given sequence (u_n), by computing the polynomials ϕ_n and checking the inequalities

$$\phi_n \, (u_1, u_2, \ldots, u_n) \geq 0, \quad u_n \leq 1 \quad ,$$

for all $n \geq 1$. The continuous time analogue of this problem is much more difficult, but also has an effective solution in this sense (see Chapter 6 of [8], or §III.89 of [12]).

It is natural to ask for a similar characterisation in the non-diagonal case. Given a sequence $(a_n \; ; \; n = 1,2,\ldots)$, how can one decide whether there exists a Markov process $(X_n \; ; \; n = 0,1,2,\ldots)$ such that

$$a_n = p_{ij}^{(n)} = \text{Prob } (X_n = j | X_0 = i) \tag{3.1}$$

for all $n \geq 1$ and some pair $i \neq j$ of states?

The obvious way to attack this problem is through the *first passage decomposition*

$$p_{ij}^{(n)} = \sum_{r=1}^{n} {}_jp_{ij}^{(r)} \, p_{jj}^{(n-r)} , \tag{3.2}$$

using Chung's "taboo" notation

$${}_jp_{ij}^{(r)} = \text{Prob}\,(X_1, X_2, \ldots, X_{r-1} \neq j, X_r = j \,|\, X_0 = i) \; . \tag{3.3}$$

This shows that a necessary condition for (a_n) to be of the form (3.1) is that it should be expressible in the form

$$a_n = \sum_{r=1}^{n} b_r \, u_{n-r} \quad (n = 1,2,\ldots), \tag{3.4}$$

where $(b_r \; ; \; r = 1,2,\ldots)$ satisfies

$$b_r \geq 0, \; \sum_{r=1}^{\infty} b_r \leq 1 , \tag{3.5}$$

and (u_n) is a renewal sequence.

It is moreover easy to construct a Markov chain which will realise the sequence (a_n) for any (b_r) satisfying (3.5) and any renewal sequence (u_n). (Consider a process with $p_{ji}^{(n)} = 0$ for all n.) Hence the necessary condition is also sufficient, and we have a characterisation as required. But it is not effective, for how can a particular sequence be tested to decide whether it is of the form (3.4) for some suitable (b_r) and (u_n)?

One step in the right direction is to eliminate (b_r). Using generating functions

$$A(z) = \sum_{n=1}^{\infty} a_n z^n, \; B(z) = \sum_{n=1}^{\infty} b_n z^n, \; U(z) = \sum_{n=0}^{\infty} u_n z^n ,$$

which converge in $|z| < 1$, (3.4) gives

$$A(z) = B(z)\ U(z) = B(z)\ \{1 - F(z)\}^{-1} ,$$

where $F(z)$ is the generating function of the numbers f_n which generate (u_n) by (1.1). Hence

$$B(z) = A(z) - F(z)A(z) ,$$

or

$$b_n = a_n - \sum_{r=1}^{n-1} f_r\, a_{n-r} . \tag{3.6}$$

The condition $b_n \geq 0$ is therefore the inequality

$$a_n \geq \sum_{r=1}^{n-1} f_r\, a_{n-r} \quad (n \geq 1) . \tag{3.7}$$

The second inequality in (3.5) is somewhat more tricky, and is perhaps best expressed as

$$\lim_{z\uparrow 1} A(z) \quad \{1 - F(z)\} \leq 1 . \tag{3.8}$$

When $\sum a_n$ converges, this is equivalent to

$$\sum_{n=1}^{\infty} a_n \left\{1 - \sum_{n=1}^{\infty} f_n\right\} \leq 1 , \tag{3.9}$$

but when $\sum a_n$ diverges, (3.8) is stronger than

$$\sum_{n=1}^{\infty} f_n = 1 . \tag{3.10}$$

In particular, if a_n converges to a non-zero limit a_∞, (3.8) is equivalent to (3.10) together with

$$\sum_{n=1}^{\infty} nf_n \leq a_\infty^{-1} . \tag{3.11}$$

Thus a given sequence (a_n) is of the form (3.1) if and only if there exists a sequence (f_r) satisfying (1.2), (3.7) and (3.8). This necessary and sufficient condition still does not provide an effective characterisation, but converts the question into a linear programming problem of a somewhat odd type. Notice that (3.7) is a weak increasing property for the sequence (a_n).

If a fully effective test could be found, the next challenge would be to extend the analysis to continuous time. The first step in this direction, corresponding to (3.4), may be found on page 149 of [8].

References

[1] K.L. Chung, *Markov chains with stationary transition probabilities*, Springer, Berlin, 1967.

[2] J.R. Cuthbert, An inequality with relevance to the Markov group problem, *J. London Math. Soc. 11* (1975) 104-106.

[3] R. Davidson, Smith's phenomenon, and 'jump' p-functions, *Stochastic Analysis* (ed. D.G. Kendall and E.F. Harding), Wiley, London, 1973, pp. 234-247.

[4] W. Feller, *An introduction to probability theory and its applications*, Wiley, New York, 1950.

[5] D.G. Kendall, On Markov groups, *Proc. Fifth Berkeley Symp. Math. Statistics and Probability*, University of California Press, 1973, pp. 47-72.

[6] D.G. Kendall, Renewal sequences and their arithmetic, *Stochastic Analysis* (ed. D.G. Kendall and E.F. Harding), Wiley, London, 1973, pp. 47-72.

[7] J.F.C. Kingman, An approach to the study of Markov processes, *J. Roy. Statist. Soc. B 28*(1966) 417-447.

[8] J.F.C. Kingman, *Regenerative phenomena*, Wiley, London, 1972.

[9] J.F.C. Kingman, Semi-p-functions, *Trans. Amer. Math. Soc. 174* (1972) 257-273.

[10] D. Mountford, On the Markov group problem, *Bull. London Math. Soc. 9* (1977) 57-60.

[11] J.M.O. Speakman, Some problems relating to Markov groups, *Proc. Fifth Berkeley Symp. Math. Statistics and Probability*, University of California Press, Vol. II, pp. 175-186.

[12] D. Williams, *Diffusions, Markov processes and martingales*, Wiley, Chichester, 1979.

THE ELECTROSTATIC CAPACITY OF AN ELLIPSOID

P.A.P. Moran

Consider an ellipsoid of semi-axes A, B, and C, where $A \geq B \geq C$. Write $K(A,B,C)$ and $S(A,B,C)$ for the electrostatic capacity and the surface area. Then it is known (Pólya and Szegö (1945)) that

$$K(A,B,C)^{-1} = \text{Average } (A^2\ell_1^2 + B^2\ell_2^2 + C^2\ell_3^2)^{-\frac{1}{2}}, \tag{1}$$

where the average is taken over a uniform distribution of direction cosines (ℓ_1, ℓ_2, ℓ_3) in all directions in three-dimensional space. It is also known (Jeans, (1925), p.247) that

$$K(A,B,C)^{-1} = \tfrac{1}{2} \int_0^\infty \Delta^{-\frac{1}{2}}\, dx , \tag{2}$$

where

$$\Delta = (A^2 + x)(B^2 + x)(C^2 + x) . \tag{3}$$

By rather complicated methods Pólya and Szegö (1945) have proved that

$$K(A,B,C) \leq \frac{1}{3}(A+B+C) . \tag{4}$$

$$\geq \frac{1}{3}(\sqrt{(AB)} + \sqrt{(BC)} + \sqrt{(AC)}) . \tag{5}$$

From now on we put $A=1$. Then putting $B=C=D$, say, we have a prolate spheroid whose capacity for $0 < D < 1$ is known to be given by

$$K(A,D,D)^{-1} = \log\left(\frac{1+e}{1+e}\right)/2e , \tag{6}$$

where $e^2 = 1-D^2$. For the oblate spheroid we have

$$K(1,1,C)^{-1} = (\sin^{-1}e)/e \quad , \tag{7}$$

where $e^2 = 1-C^2$.

In the present paper we discuss the problem of calculating $K(A,B,C)$, and establish some other inequalities. Write $2D = B+C$. Then we shall show that keeping D fixed, $K(1,B,C)$ is greatest when $B=C$, and smallest when B is as large as possible consistent with $B \le 1$, i.e. when $B = 1$, $C= 2D-1$ if $1 \le 2D$, and when $B = 2D$, $C = 0$, if $2D \le 1$. The value of the capacity in the latter case is easily seen to be given in terms of a complete elliptic integral of the first kind. Thus all the bounds can be expressed in terms of known functions. We therefore have

$$K(1,B,C) \le K(1,\tfrac{1}{2}(B+C),\ \tfrac{1}{2}(B+C)) \ , \tag{8}$$

and

$$K(1,B,C) \ge K(1,B+C,0) \ , \tag{9}$$

$$\text{when} \quad B+C \le 1 \ ,$$

$$\ge K(1,1,B+C-1) \ , \tag{10}$$

$$\text{when} \quad B+C \ge 1 \ .$$

We first show that (8) is a stronger inequality than (4), i.e. that if $D < 1$

$$K(1,D,D) \le \frac{1}{3}(1 + 2D) \ . \tag{11}$$

Putting $e^2 = 1 - D^2$, this is equivalent to showing that

$$2e/\log(\frac{1+e}{1+e}) \le \frac{1}{3}(1+2\surd(1-e^2)) \ ,$$

or

$$e \le \frac{1}{3}\,(1 + 2\surd(1 - e^2))\,\tanh^{-1} e\ .$$

Putting $e = \tanh\theta$, this becomes

$$\tanh\theta \le \frac{1}{3}\theta(1 + 2\ \mathrm{sech}\ \theta)$$

or
$$\sinh\theta \le \frac{1}{3}\theta(2 + \cosh\theta)\ .$$

Using power series this becomes

$$\theta \sum_{n=0} \theta^{2n}/(2n+1)! \le \theta + \frac{1}{3}\,\theta \sum_{n=1} \theta^{2n}/(2n)!$$

This is true because $3(2n)! \le (2n+1)!$ for $n \ge 1$.

To prove (8), (9), and (10) we put $A = 1$ and suppose $B+C=2D$ is a constant. Then differentiating (2) with respect to B and supposing $B \ge C$ we have

$$\frac{dK(1,B,C)^{-1}}{dB} = \tfrac{1}{2}\int_0^\infty \Delta^{-\frac{1}{2}}\left\{\frac{C}{C^2+x} - \frac{B}{B^2+x}\right\} dx$$

$$= \tfrac{1}{2}(B-C)\int_0^\infty \frac{BC-x}{(1+x)^{\frac{1}{2}}(B^2+x)^{3/2}(C^2+x)^{3/2}}\,dx\ . \qquad (12)$$

When $x < BC$ the integrand is positive and is reduced if we replace the term $(1+x)$ by $(1+BC)$. When $x > BC$ the integrand is negative and is increased in absolute value on replacing $(1+x)$ by $(1+BC)$. Thus (12) is not less than

$$\frac{B - C}{2(1+BC)^{\frac{1}{2}}}\int_0^\infty \frac{BC-x}{(B^2+x)^{3/2}(C^2+x)^{3/2}}\,dx\ . \qquad (13)$$

Using standard integrals (e.g. Dwight (1971) p.214) we find that (13) is zero. Thus

$$\frac{dK^{-1}(1,B,C)}{dB} \ge 0\ ,$$

when $B > C$ and $C = D - B$.

The ratios of (9) and (10) to (8) were tabulated for $D = 0.1(0.1)0.9$. As D increased from zero to 0.5 the ratio of (9) to (8) decreased from unity to 0.968103, and from $D = 0.5$ to 1.0 the ratio of (10) to (8) increased again to unity. Thus the lower and upper bounds are never more than about 3.2% in error. On the other hand (5) is sometimes better than (9) and (10) and sometimes worse. To illustrate this we have $K(1,\cdot 5,\cdot 5) = 0.657\ 595\ 361$, and $K(1,1,0) = 0.636\ 619\ 772$ whilst for $(1,\cdot 5,\cdot 5)$, (5) is 0.638 071 187, and for $(1,1,0)$, (5) is 0.333 333 333 which is much worse than (9) and (10) which are here equal.

We now return to (1) and consider what can be done by probabilistic arguments. Let X_1, X_2, X_3 be independent random variables distributed in normal distributions with zero means and unit standard deviations. Then (1) can be written as

$$K(A,B,C)^{-1} = E\left[\left\{\frac{A^2X_1^2 + B^2X_2^2 + C^2X_3^2}{X_1^2 + X_2^2 + X_3^2}\right\}^{-\frac{1}{2}}\right] \tag{14}$$

where E denotes expectation. It is difficult to obtain any inequalities from this by simple probability arguments. However put $A = 1$, $\alpha^2 = 1 - B^2$, $\beta^2 = 1 - C^2$. Then (14) can be written

$$\begin{aligned} K(1,B,C)^{-1} &= E\left[\left\{1 - \frac{\alpha^2X_2^2 + \beta^2X_3^2}{X_1^2 + X_2^2 + X_3^2}\right\}^{-\frac{1}{2}}\right] \\ &= 1 + \sum_{n=1}^{\infty} \frac{(2n)!}{2^{2n}(n!)^2}\, E\left[\left\{\frac{\alpha^2X_2^2 + \beta^2X_3^2}{X_1^2 + X_2^2 + X_3^2}\right\}^{n}\right] \end{aligned} \tag{15}$$

Consider the distribution of the random variable

$$Z = \alpha^2X_2^2 + \beta^2X_3^2 \quad .$$

It is a standard result that the characteristic function of Z is

$$(1 - 2\alpha^2 i\,\theta)^{-\frac{1}{2}} (1 - 2\beta^2 i\,\theta)^{-\frac{1}{2}}$$

$$= (1 - 2(\alpha^2 + \beta^2)\, i\,\theta - 4\alpha^2\beta^2\theta^2)^{-\frac{1}{2}}$$

$$= \sum_{n=0}^{\infty} (2i\alpha\beta\theta)^n \, P_n\left(\frac{\alpha^2+\beta^2}{2\alpha\beta}\right), \tag{16}$$

where $P_n(z)$ is the Legendre polynomial of order n. Notice that its argument is greater than unity.

It follows that

$$E[Z^n] = (2\alpha\beta)^n \, n! \, P_n\left(\frac{\alpha^2+\beta^2}{2\alpha\beta}\right). \tag{17}$$

Then from a theorem of Pitman (1937) on the moments of the ratios of quadratic forms in normal variates we have

$$E\left[\left\{\frac{\alpha^2 X_2^2 + \beta^2 X_3^2}{X_1^2 + X_2^2 + X_3^2}\right\}\right]^n = \frac{E[(\alpha^2 X_2^2 + \beta^2 X_3^2)^n]}{E[(X_1^2 + X_2^2 + X_3^2)^n]}$$

$$= \frac{2^n (n!)^2 (2\alpha\beta)^n}{(2n+1)!} \, P_n\left(\frac{\alpha^2+\beta^2}{2\alpha\beta}\right), \tag{18}$$

and inserting this in (15) we obtain

$$K^{-1}(1,B,C) = 1 + \sum_{n=1}^{\infty} \frac{(\alpha\beta)^n}{2n+1} \, P_n\left(\frac{\alpha^2+\beta^2}{2\alpha\beta}\right), \tag{19}$$

a formula obtained by Pólya and Szegö (1945) in a somewhat different way. They used (19) in rather involved arguments to prove (4) and (5).

In a previous paper (Moran (1981)) inequalities were obtained for the surface area, $S(A,B,C)$, of the ellipsoid. With X_1, X_2, X_3, defined as above it was shown that

$$S(A,B,C) = 4\pi E\left[\left\{\frac{A^2B^2X_1^2 + B^2C^2X_2^2 + C^2A^2X_3^2}{X_1^2 + X_2^2 + X_3^2}\right\}^{\frac{1}{2}}\right]. \tag{20}$$

and that

$$\frac{4}{3}\pi\ (AB+BC+CA)\ \le S(A,B,C)\ \le 4\pi\surd(\tfrac{1}{3}(A^2B^2 + B^2C^2 + C^2A^2)). \qquad (21)$$

It is possible to express $S(A,B,C)$ in a form analogous to (19). Suppose that $A=1$ and now define α, β differently by putting $\alpha^2 = 1 - C^2$, $\beta^2 = 1 - C^2B^{-2}$. Then

$$S(1,B,C) = 4\pi BE\left[\left\{1 - \frac{\alpha^2X_2^2 + \beta^2X_3^2}{X_1^2 + X_2^2 + X_3^2}\right\}^{\frac{1}{2}}\right]$$

$$= 4\pi B\left[1 - \sum_{n=1}^{\infty}\frac{E\{(\alpha^2X_2^2 + \beta^2X_3^2)^n\}}{2^n\ n!\ (4n^2 - 1)}\right]$$

$$= 4\pi B\left\{1 - \sum_{n=1}^{\infty}\frac{(\alpha\beta)^n}{2^n(4n^2-1)}\ P_n\left(\frac{\alpha^2+\beta^2}{2\alpha\beta}\right)\right\}. \qquad (22)$$

Notice that α and β are defined differently in (19) and (22). These formulae can be usefully employed to calculate $K(A,B,C)$ and $S(A,B,C)$ when α and β are not near unity and experiment shows that (22) is quite satisfactory when $\frac{1}{2} \le CA^{-1}$. In using (19) and (22) it is best to calculate the Legendre polynomials by adapting the usual recurrence formula to apply to

$$(\alpha\beta)^n\ P_n\ \left(\frac{\alpha^2+\beta^2}{2\alpha\beta}\right).$$

However a better method is obtained as follows. Returning to (1) we replace the direction cosines by the polar coordinates ($A \sin\theta \sin\phi$, $B \sin\theta \cos\phi$, $C \cos\theta$), so that we take the polar axis along the C axis. We then get

$$K(A,B,C)^{-1} = (4\pi)^{-1}\int_0^{2\pi}\int_0^{\pi}\{C^2 \cos^2\theta + \sin^2\theta(A^2\cos^2\phi+B^2\sin^2\phi)\}^{-\frac{1}{2}}\sin\theta\ d\theta\ d\phi \qquad (23)$$

$$= (4\pi)^{-1}\int_0^{2\pi}\frac{\sin^{-1}\surd\{1-C^2(A^2\cos^2\phi+B^2\sin^2\phi)^{-1}\}}{\surd\{A^2\cos^2\phi + B^2\sin^2\phi - C^2\}}d\phi \qquad (24)$$

So long as $C < B$, the integrand in (24) is an analytic periodic function of ϕ and numerical integration without any end corrections (since there are no ends) is likely to give very accurate results. This was tried out on a desk computer for $A=1$, $B=0.1(0.1)0.9$ and $C=0.1(0.1)B$. Clearly when $B=1$, the integrand is a constant.

Numerical integration was carried out by taking the average of the integrand at $\phi = \pi/4N, 3\pi/4N, \ldots, (2N-1)\pi/4N$. N was taken successively as 5, 10, 20, 40, and 60 until the ninth significant digit did not alter. $N=5$ gave this degree of accuracy for $B=0.9$ and for $B=0.8, 0.5 \le C$. $N=10$ the same accuracy for $0.5 \le B$, $N=20$ for $0.3 \le B$, $N=40$ for $B=0.2$, and $N=60$ for $B=C=0.1$. It was, in fact, by calculating this table that the inequalities (8), (9) and (10) were discovered. It also appears, as suggested by the proof of (8), (9), (10) that if $B+C$ is held constant, $K(A,B,C)$ is nearly a function of $(B-C)^2$. Quadratic interpolation between the extreme values is thus quite accurate.

It might be suspected that other inequalities similar to (8), (9), (10) could be found. Thus for surface area $S(A,B,C)$ we might vary B and C whilst keeping A and $B+C$ constant. Numerical investigation then shows that $S(A,B,C)$ sometimes increases and sometimes decreases as $(B-C)^2$ increases. A better idea is to study the variation of $S(A,B,C)$ when the two bounds in (21) are held fixed, or to study the variation of $K(A,B,C)$ when $\sqrt{(AB)} + \sqrt{(BC)} + \sqrt{(AC)}$ is held fixed. I have been unable to obtain any results in this direction. For other inequalities see Lehmer (1950).

It is a great pleasure to be able to join in a tribute to Professor D.G. Kendall in view of our long association and friendship.

References

Dwight, H.B. (1971). *Tables of Integrals and other Mathematical Data.* Macmillan, N.Y.

Jeans, H.J. (1925). *The Mathematical Theory of Electricity and Magnetism.* Cambridge University Press.

Lehmer, D.H. (1950). Approximations for the area of an ellipsoid. *Canadian J. Math. 2.* 267-282.

Moran, P.A.P. (1981). The surface area of an ellipsoid in *Statistics and Probability*; Essays in honor of C.R. Rao. North Holland.

Pitman, E.J.G. (1937). The 'closest' estimates of statistical parameters. *Proc. Cam. Phil. Soc. 33.* 212-222.

Pólya, G., and Szegö, G. (1945). Inequalities for the capacity of an ellipsoid. *American J. Math. 67.* 1-32.

STATIONARY ONE-DIMENSIONAL MARKOV RANDOM FIELDS WITH A CONTINUOUS STATE SPACE

F. Papangelou

Abstract

It is well known that a one-dimensional Markov random field need not be a Markov process. In the present paper we prove that if such a random field is stationary, then it is a Markov process, provided it satisfies an appropriate irreducibility requirement. This establishes in a general context a result known for the case of a countable state space ([5]) and implies, as in the latter case, that phase transition is impossible under the constraint of stationarity.

1. Introduction

By a one-dimensional Markov random field we shall mean a stochastic process $\ldots, X_{-1}, X_o, X_1, \ldots,$ with a general state space $(S,\mathcal{B})$ (see §2), satisfying the following two-sided Markov condition: for any integers $k \geq 1$ and n, and any $D \in \mathcal{B}^k = \mathcal{B} \otimes \ldots \otimes \mathcal{B}$ (k factors)

$$P((X_{n+1}, X_{n+2}, \ldots, X_{n+k}) \in D | X_i, \ i \leq n \ \text{ or } \ i > n+k+1)$$

$$= \phi^k(D|X_n, X_{n+k+1}) \quad \text{P-a.s.,} \tag{1}$$

where the two sided conditional probability kernels $\phi^k(D|x,z)$ $(k = 1,2,\ldots)$ are subject to the natural regularity and consistency assumptions stated explicitly in §2 below. The reader should note that this definition involves the requirement that the two-sided conditional probabilities are translation invariant; no other type of Markov random field will be considered below.

If the state space S is finite and $\phi^1(\{y\}|x,z) > 0$ for all $x,y,z \in S$, then ([4], [7, Theorem 3.22]) the associated Markov random field $\{X_i, -\infty < i < \infty\}$ is necessarily stationary (in the sense that its distribution is translation invariant) and also a Markov chain in the standard sense, i.e. it satisfies the one-sided Markov condition. Furthermore, there is exactly one, up to equivalence in distribution, Markov random field having the given system of two-sided conditional probability kernels $\{\phi^k, k \geq 1\}$. If S is countably infinite the situation is very different: a Markov random field satisfying (1) need be neither stationary nor a Markov chain, nor indeed the unique Markov random field associated with the given system $\{\phi^k, k \geq 1\}$ ([8], [2]). In the countable case there are also consistent systems of kernels $\{\phi^k, k \geq 1\}$ admitting no Markov random field.

It was, however, conjectured by Spitzer ([8]) that, if S is countable, there is at most one *stationary* Markov random field satisfying (1) for a given system $\{\phi^k, k \geq 1\}$. This was proved by Kesten ([5]), who showed that $\{\phi^k, k \geq 1\}$ admits a stationary Markov random field if and (more significantly) only if it "arises" from some positive-recurrent transition matrix, and that in this case the stationary Markov chain associated with the latter is the unique stationary Markov random field admitted by $\{\phi^k, k \geq 1\}$.

Our purpose here is to establish an equivalent result for Markov random fields with a general state space S, by proving Theorems 1 and 2 stated below. The condition of irreducibility appearing in these theorems will be defined precisely in §2; at this point it is sufficient to say that, in the countable case, it is analogous to "irreducibility and aperiodicity" for Markov chains, and weakens the condition of strict positivity $\phi^1(\{y\}|x,z) > 0$ mentioned above.

THEOREM 1. All stationary and irreducible one-dimensional Markov random fields are Markov processes, i.e. satify the standard one-sided Markov condition.

THEOREM 2. There is at most one, up to equivalence in distribution, stationary one-dimensional Markov random field associated with a given irreducible system of two-sided conditional probability kernels $\{\phi^k, k \geq 1\}$.

Theorem 2 is in fact essentially a consequence of Theorem 1. The latter is proved in §3 and the former derived from it in §4. In the language of certain physical applications of Markov random fields (nearest neighbour systems), Theorem 2 can be paraphrased as follows: in dimension one, phase transition is impossible under the constraint of stationarity.

In the countable case Theorems 1 and 2 imply Kesten's result. However, the matrix methods employed by Spitzer and Kesten in the countable case do not lend themselves to an immediate extension to the general case and we have instead based our proof on the convergence theorem for backward martingales.

The author wishes to thank H. Kesten and H. Föllmer, who have recently drawn his attention to [1] and [3]. In [3] a uniqueness result is given for a special case, in which Kesten's argument for the positive-recurrent case carries over: namely, when it is already assumed that the two-sided conditional probabilities $\phi^k (k = 1,2,\ldots)$ are those of a stationary Markov process of a rather special type. See the end of §4 for more details. For the Gaussian case see also [1].

2. Preliminaries

Some technical assumptions have to be made about the state space $(S,\mathcal{B})$: that it carries regular versions of all conditional probabilities and that $\mathcal{B}$ is countably generated. Both of these statements are true if, for example, S is a complete separable metric space and $\mathcal{B}$ the σ-field of its Borel sets, and we shall assume throughout that this is the case, although readers not interested in generality may think of S as the real line or a Euclidean space.

For each $k = 1,2,\ldots$ let $\phi^k(D|x,z)$ be a function of $D \in \mathcal{B}^k$, $x \in S$, $z \in S$, which is a probability kernel, i.e. a probability measure in D for fixed (x,z), and $\mathcal{B} \otimes \mathcal{B}$-measurable in (x,z) for fixed D. The sequence $\{\phi^k, k \geq 1\}$ will be said to be a *consistent system of two-sided conditional probability kernels* if it satisfies the obvious consistency conditions implied by (1), in the following strict sense: for any $k \geq 1$, $A \in \mathcal{B}$, $C \in \mathcal{B}$, $D \in \mathcal{B}^k$, $x \in S$, $z \in S$

$$\phi^{k+1}(A \times D|x,z) = \int_{u\in A} \phi^k(D|u,z)\, \phi^{k+1}(du \times S^k|x,z), \tag{2}$$

$$\phi^{k+1}(D \times C|x,z) = \int_{v\in C} \phi^k(D|x,v)\, \phi^{k+1}(S^k \times dv|x,z). \tag{3}$$

It is easy to see that these conditions imply

$$\phi^{k+2}(A \times D \times C|x,z) = \int_{u\in A}\int_{v\in C} \phi^k(D|u,v)\, \phi^{k+2}(du \times S_k \times dv|x,z)$$

and a host of other similar equalities.

2.1 Definition. A stochastic process $\ldots, X_{-1}, X_o, X_1, \ldots$ defined on a probability space (Ω, F, P) and with state space $(S, \mathcal{B})$ will be called a (one-dimensional) *Markov random field* if there is a consistent system of two-sided conditional probability kernels $\{\phi^k, k \geq 1\}$ such that (1) holds for any $k \geq 1$, any n and any $D \in \mathcal{B}^k$. If this is the case, we will say that the Markov random field $\{X_i, -\infty < i < \infty\}$ is associated with, or admitted by, the system $\{\phi^k, k \geq 1\}$.

Conditions (2) and (3) imply that the system $\{\phi^k, k \geq 1\}$ is uniquely determined by the system of simpler probability kernels $\{Q_n^m, m \geq 1, n \geq 1\}$ defined as follows:

$$Q_n^m(x,B,z) = \phi^{m+n-1}(S^{m-1} \times B \times S^{n-1} | x,z) = P(X_o \in B | X_{-m} = x,$$

$$X_n = z) \quad \text{P-a.s.},$$

where $B \in \mathcal{B}$, $x, z \in S$.

For any $x_1, x_2, z_1, z_2 \in S$, $n \geq 1$, $m \geq 1$ we define a measure $r_n^m(x_1, x_2, B, z_1, z_2)$ $(B \in \mathcal{B})$ by stipulating that $r_n^m(x_1, x_2, ., z_1, z_2)$ is the greatest lower bound (in the lattice of probability measures on $\mathcal{B}$) of $Q_n^m(x_1, ., z_1)$ and $Q_n^m(x_2, ., z_2)$. Recall that the greatest lower bound of two probability measures μ and ν is

$$(\mu \wedge \nu)(B) = \inf_{D \in \mathcal{B}} [\mu(B \cap D) + \nu(B \cap D^c)]. \tag{4}$$

Using the tightness of μ and ν one can easily prove that if $\mathcal{R}_o$ is a countable base of open sets in S and $\mathcal{R}$ the class of finite unions of members of $\mathcal{R}_o$ then $\mathcal{B}$ can be replaced by $\mathcal{R}$ in (4). It follows from this that r_n^m is a probability kernel. For the sake of simplicity we shall write $r_n^m(x, B, z_1, z_2)$ for $r_n^m(x, x, B, z_1, z_2)$.

2.2 Definition. A system of two-sided conditional probability kernels $\{\phi^k, k \geq 1\}$ (and, by extension, any Markov random field associated with it) will be called *irreducible* if for every $N = 1,2,3 \ldots$ there are two positive integers i_N, j_N and a set $A_N \in \mathcal{B}$ such that

(i) $A_N \subset A_{N+1}$ $(N = 1,2,\dots)$ and $\bigcup_{N=1}^{\infty} A_N = S$;

(ii) for each N

$$\inf \{r_{j_N}^{i_N}(x_1, x_2, S, z_1, z_2) : x_1, x_2, z_1, z_2 \in A_N\} > 0 .$$

The infimum in condition (ii) will be denoted throughout by δ_N.

Note that condition (ii) is equivalent to

$$\sup \{\|Q_{j_N}^{i_N}(x_1, \cdot, z_1) - Q_{j_N}^{i_N}(x_2, \cdot, z_2)\| : x_1, x_2, z_1, z_2 \in A_N\} < 2$$

(where $\|\mu - \nu\|$ stands for the total variation of $\mu - \nu$), since $2r_n^m(x_1, x_2, S, z_1, z_2) = 2 - \|Q_n^m(x_1, \cdot, z_1) - Q_n^m(x_2, \cdot, z_2)\|$. If the space S is σ-compact, a simple sufficient condition for irreducibility is the following: for any $x_1, x_2, z_1, z_2 \in S$, $\|Q_1^1(x_1, \cdot, z_1) - Q_1^1(x_2, \cdot, z_2\| < 2$ (i.e. $Q_1^1(x_1, \cdot, z_1)$ and $Q_1^1(x_2, \cdot, z_2)$ are not mutually singular), and $\lim_{\substack{x \to x_o \\ z \to z_o}} \|Q_1^1(x, \cdot, z) - Q_1^1(x_o, \cdot, z_o\| = 0$ for arbitrary $x_o, z_o \in S$. In fact, this easily implies that for any compact set $K \subset S$,

$$\sup \{\|Q_1^1(x_1, \cdot, z_1) - Q_1^1(x_2, \cdot, z_2)\| : x_1, x_2, z_1, z_2 \in K\} < 2.$$

This sufficient condition is automatically true if, for example, there is a fixed measure μ on $\mathcal{B}$ and a continuous function $\rho(x,y,z)$ on S^3 such that $\rho(x,y,z) > 0$ for all $x,y,z \in S$ and $Q_1^1(x,B,z) = \int_{y \in B} \rho(x,y,z)\mu(dy)$ for $B \in \mathcal{B}$, $x \in S$, $z \in S$. In the case where S is countable, it is sufficient that $Q_1^1(x,\{y\},z) > 0$ for all $x,y,z \in S$. In fact the condition of irreducibility introduced in Definition 2.2 is a relaxation of this assumption of strict positivity, in the same way that the condition of "irreducibility and aperiodicity" is a relaxation of the condition of strict positivity for the transition matrix of a Markov chain.

We conclude this section by mentioning that in the sequel we shall frequently refer to a measure, say μ, by writing its differential $\mu(dx)$.

§3. The one-sided Markov property

As usual, a stochastic process $\ldots, X_{-1}, X_o, X_1, \ldots$ will be called *stationary* if for any n and any $k \geq 0$ the random vectors $(X_n, X_{n+1}, \ldots, X_{n+k})$ and $(X_{n+1}, X_{n+2}, \ldots, X_{n+k+1})$ have the same distribution. In the present section we shall prove Theorem 1, i.e. that every stationary and irreducible one-dimensional Markov random field $\ldots, X_{-1}, X_o, X_1, \ldots$ is a Markov process. To say that $\ldots, X_{-1}, X_o, X_1 \ldots$ is a Markov process means of course that, for any $B \in \mathcal{B}$ and any integer k, $P(X_{k+1} \in B | X_i, i \leq k) = P(X_{k+1} \in B | X_k)$ P-a.s..

Suppose $\ldots, X_{-1}, X_o, X_1, \ldots$ is a stationary, irreducible one-dimensional Markov random field, for which (1) holds. Define $\pi(B) = P(X_i \in B)$ $(B \in \mathcal{B})$ and let $p^n(x,B)$ $(n = 1,2,\ldots)$ be a version of $P(X_{i+n} \in B | X_i = x)$ $(B \in \mathcal{B}, x \in S)$ which is a probability kernel. It is important to note that the definitions of π and $p^n(\cdot,\cdot)$ involve the distribution of the process $\{X_i, -\infty < i < \infty\}$ and not just the two-sided conditional probability kernels $\{\phi^k, k \geq 1\}$. The stationarity of the process implies that π and $p^n(\cdot,\cdot)$ do not depend on i and that $\pi(B) = \int_S \pi(dx)\, p^n(x,B)$ $(B \in \mathcal{B}, n \geq 1)$. Note also that

$$p^m(x,B) = \int_S p^{m+n}(x,dz)\, Q_n^m(x, B, z)$$

for π-almost all $x \in S$ and all $B \in \mathcal{B}$, $n \geq 1$, $m \geq 1$. However, until we prove Theorem 1, we cannot assert that the "transition functions" $p^n(\cdot,\cdot)$ form a semigroup, i.e. that $p^{m+n}(x, B) = \int_S p^m(x,dy)\, p^n(y, B)$.

Theorem 1 will be deduced from the following.

<u>THEOREM 3</u>. If the one-dimensional Markov random field $\ldots, X_{-1}, X_o, X_1 \ldots$ is stationary and irreducible then, for every $B \in \mathcal{B}$,

$$\lim_{n\to\infty} \int_S \int_S \int_S | Q_n^1(x,B,z_1) - Q_n^1(x,B,z_2) | \pi(dx)$$

$$p^{n+1}(x,dz_1)\, p^{n+1}(x,dz_2) = 0.$$

<u>COROLLARY</u>. Under the hypotheses of Theorem 3, for every $B \in \mathcal{B}$

$$\lim_{n\to\infty} \int_S \int_S | Q_n^1(x,B,z) - p^1(x,B) | \; P(X_{-1} \in dx, X_n \in dz) =$$

$$= \lim_{n\to\infty} \int_S \int_S | Q_n^1(x,B,z) - p^1(x,B) | \pi(dx)\ p^{n+1}(x,dz) = 0.$$

Before proving these results we establish two lemmas. All symbols used have, throughout, the meaning given to them in §2. We begin with the observation that, if $n > i_N$, then

$$r_{j_N}^n (x, \cdot, z_1, z_2) \geq$$

$$\int_{x_1 \in S} \int_{x_2 \in S} Q_{i_N+j_N}^{n-i_N} (x,dx_1,z_1)\ Q_{i_N+j_N}^{n-i_N} (x,dx_2,z_2)$$

$$r_{j_N}^{i_N} (x_1,x_2,\cdot,z_1,z_2)\ . \tag{5}$$

In fact, by (2)

$$Q_{j_N}^n (x,\cdot,z_1) = \int_{x_1 \in S} Q_{i_N+j_N}^{n-i_N} (x,dx_1,z_1)\ Q_{j_N}^{i_N} (x_1,\cdot,z_1) =$$

$$= \int_{x_1 \in S} \int_{x_2 \in S} Q_{i_N+j_N}^{n-i_N} (x,dx_1,z_1)\ Q_{i_N+j_N}^{n-i_N} (x,dx_2,z_2)\ Q_{j_N}^{i_N} (x_1,\cdot,z_1)$$

where $Q_{j_N}^{i_N} (x_1,\cdot,z_1) \geq r_{j_N}^{i_N} (x_1,x_2,\cdot,z_1,z_2)$, and a similar argument applies to $Q_{j_N}^n (x,\cdot,z_2)$.

Let $D_N = \{(x_1,x_2,z_1,z_2) \in S^4 : r_{j_N}^{i_N} (x_1,x_2,S,z_1,z_2) < \delta_N\}$.

<u>3.1 Lemma.</u> If $n > i_N$ then

$$\int_{x\in S} \pi(dx) \int\int\int\int_{(x_1,x_2,z_1,z_2)\in D_N} p^{n+j_N}(x,dz_1)\ p^{n+j_N}(x,dz_2)$$

$$Q_{i_N+j_N}^{n-i_N} (x,dx_1,z_1)\ Q_{i_N+j_N}^{n-i_N}(x,dx_2,z_2) \leq 4\pi(A_N)(1-\pi(A_N))\ . \tag{6}$$

<u>Proof.</u> If we replace D_N by A_N^4, the left-hand side of (6) becomes

$$\int_{x\in S} \pi(dx) \left[\int_{x'\in A_N} \int_{z\in A_N} p^{n+j_N}(x,dz)\, Q_{i_N+j_N}^{n-i_N}(x,dx',z) \right]^2 \geq$$

$$\geq \left[\int_{x\in S} \int_{x'\in A_N} \int_{z\in A_N} \pi(dx)\, p^{n+j_N}(x,dz)\, Q_{i_N+j_N}^{n-i_N}(x,dx',z) \right]^2 \geq$$

$$\geq \left[\left\{ \left(\int_S \int_{A_N} \int_S + \int_S \int_S \int_{A_N} \right) \pi(dx)\, p^{n+j_N}(x,dz) \right. \right.$$

$$\left. \left. Q_{i_N+j_N}^{n-i_N}(x,dx',z) \right\} - 1 \right]^2$$

$$= (2\pi(A_N) - 1)^2 \quad ,$$

since, for any probability measure ρ on $\mathcal{B}^3$, $\rho(S\times A_N\times A_N) = \rho((S\times A_N\times S) \cap (S\times S\times A_N)) \geq \rho(S\times A_N\times S) + \rho(S\times S\times A_N) - 1$. The result now follows if we take complements, since $D_N \subset (A_N^4)^c$ by the definition of δ_N in Definition 2.2.

Next, denote by λ_n the measure $\lambda_n(E) = \int\int\int_{(x,z_1,z_2)\in E} \pi(dx)$ $p^n(x,dz_1)\, p^n(x,dz_2)$ $(E \in \mathcal{B}^3)$.

<u>3.2 Lemma</u>. For every $\varepsilon > 0$, there is $N \geq 1$ such that, for all $n > i_N$, the set $E_{n,N} = \{(x,z_1,z_2) \in S^3 : r_{j_N}^n(x,S,z_1,z_2) < \frac{\delta_N}{2}\}$ satisfies $\lambda_{n+j_N}(E_{n,N}) < \varepsilon$.

<u>Proof</u>. Since $A_N \uparrow S$ as $N \to \infty$, given $\varepsilon > 0$ there is an N such that

$$1 - \pi(A_N) < \frac{\varepsilon}{8} \quad . \tag{7}$$

Fix this N, let $n > i_N$ and suppose that $(x,z_1,z_2) \in E_{n,N}$. By (5)

$$\int_{x_1\in S} \int_{x_2\in S} Q_{i_N+j_N}^{n-i_N}(x,dx_1,z_1)\, Q_{i_N+j_N}^{n-i_N}(x,dx_2,z_2)\, r_{j_N}^{i_N}(x_1,x_2,S,z_1,z_2) < \frac{\delta_N}{2}$$

and hence, by a Chebyshev type inequality, if $D_N^{z_1,z_2} = \{(x_1,x_2) \in S^2 : r_{j_N}^{i_N}(x_1, x_2, S, z_1, z_2) < \delta_N\}$, then

$$\int\int_{(x_1,x_2)\in D_N^{z_1,z_2}} Q_{i_N+j_N}^{n-i_N}(x,dx_1,z_1)\; Q_{i_N+j_N}^{n-i_N}(x,dx_2,z_2)$$

$$\geq 1 - \frac{\frac{\delta_N}{2}}{\delta_N} = \frac{1}{2} \tag{8}$$

The integral, with respect to λ_{n+j_N}, of the left-hand side of (8) over all $(x,z_1,z_2) \in E_{n,N}$ is therefore greater than or equal to $\frac{1}{2}\lambda_{n+j_N}(E_{n,N})$, and hence so is the left-hand side of (6), since the latter is the integral, with respect to λ_{n+j_N}, of the left-hand side of (8) over all $(x,z_1,z_2) \in S^3$. It follows from this fact and Lemma 3.1 that

$$\frac{1}{2}\lambda_{n+j_N}(E_{n,N}) \leq 4(1 - \pi(A_N))$$

and (7) now completes the proof.

We can now turn to the proofs of the theorems.

<u>Proof of Theorem 3</u>. First note that, for fixed $B \in \mathcal{B}$, the process $Q_n^1(X_{-1},B,X_n)$ $(n = 1,2,\ldots)$ is a backward martingale. This follows from (3), which implies $Q_{n+1}^1(x,B,z) = \int Q_n^1(x,B,v)\, Q_1^{n+1}(x,dv,z)$. Hence there is a random variable Y such that

$$\lim_{n\to\infty} E\,|\,Q_n^1(X_{-1},B,X_n) - Y| = 0\,. \tag{9}$$

Given $\varepsilon > 0$, fix an N having the property described in Lemma 3.2 and suppose $n > i_N + j_N$. For any $x,\ y,\ z_1,\ z_2 \in S$

$$|Q_n^1(x,\ B,\ z_1) - Q_n^1(x,\ B,\ z_2)| \leq$$

$$\leq |Q_n^1(x,\ B,\ z_1) - Q_{n-j_N}^1(x,\ B,\ y)| + |Q_n^1(x,B,z_2) - Q_{n-j_N}^1(x,B,y|\ .$$

Integrating with respect to $r_{j_N}^{n-j_N+1}(x,\ dy,\ z_1,\ z_2)$ and recalling the definition of $r_{j_N}^{n-j_N+1}$ we obtain

$$|Q_n^1(x, B, z_1) - Q_n^1(x, B, z_2)| \; r_{j_N}^{n-j_N+1}(x, S, z_1, z_2)| \le$$

$$\le \sum_{i=1}^{2} \int_{y \in S} |Q_n^1(x, B, z_i) - Q_{n-j_N}^1(x, B, y)|$$

$$r_{j_N}^{n-j_N+1}(x, dy, z_1, z_2)$$

$$\le \sum_{i=1}^{2} \int_{y \in S} |Q_n^1(x, B, z_i) - Q_{n-j_N}^1(x, B, y)| \; Q_{j_N}^{n-j_N+1}(x, dy, z_i) .$$

A further integration with respect to

$\lambda_{n+1}(dx, dz_1, dz_2) = \pi(dx) p^{n+1}(x, dz_1) p^{n+1}(x, dz_2)$ leads to

$$\iiint_{(x, z_1, z_2) \in S^3} |Q_n^1(x,B,z_1) - Q_n^1(x,B,z_2)| \; r_{j_N}^{n-j_N+1}(x,S,z_1,z_2)$$

$$\lambda_{n+1}(dx, dz_1, dz_2)$$

$$\le \sum_{i=1}^{2} \iiint_{(x,y,z_i) \in S^3} |Q_n^1(x,B,z_i) - Q_{n-j_N}^1(x,B,y)|$$

$$Q_{j_N}^{n-j_N+1}(x, dy, z_i) \pi(dx) p^{n+1}(x, dz_i)$$

$$= 2 \iiint_{(x,y,z) \in S^3} Q_n^1(x,B,z) - Q_{n-j_N}^1(x,B,y)| \; P(X_{-1} \in dx, X_{n-j_N} \in dy, X_n \in dz)$$

$$= 2 \, E \, |Q_n^1(X_{-1}, B, X_n) - Q_{n-j_N}^1(X_{-1}, B, X_{n-j_N})| .$$

Letting $n \to \infty$

$$\lim_{n \to \infty} \iiint_{(x, z_1, z_2) \in S^3} |Q_N^1(x,B,z_1) - Q_n^1(x,B,z_2)|$$

$$r_{j_N}^{n-j_N+1}(x,S,z_1,z_2) \lambda_{n+1}(dx, dz_1, dz_2) = 0 \qquad (10)$$

by (9). Now from Lemma 3.2

$$\iiint_{(x,z_1,z_2)\in S^3} |Q_n^1(x,B,z_1) - Q_n^1(x,B,z_2)|\ \lambda_{n+1}(dx,dz_1,dz_2) =$$

$$= \iiint_{(x,z_1,z_2)\in E_{n-j_N+1,N}} |Q_n^1(x,B,z_1) - Q_n^1(x,B,z_2)|\ \lambda_{n+1}(dx,dz_1,dz_2) +$$

$$+ \frac{2}{\delta_N} \iiint_{(x,z_1,z_2)\in E^c_{n-j_N+1,N}} |Q_n^1(x,B,z_1) - Q_n^1(x,B,z_2)|\ \frac{\delta_N}{2}\ \lambda_{n+1}(dx,dz_1,dz_2)$$

$$< \varepsilon + \frac{2}{\delta_N} \iiint_{(x,z_1,z_2)\in E^c_{n-j_N+1,N}} |Q_n^1(x,B,z_1) - Q_n^1(x,B,z_2)|\ r_{j_N}^{n-j_N+1}(x,S,z_1,z_2)$$

$$\lambda_{n+1}(dx,dz_1,dz_2)\ .$$

Letting $n \to \infty$, we obtain from this and (10)

$$\lim_{n\to\infty} \int_S \int_S \int_S |Q_n^1(x,B,z_1) - Q_n^1(x,B,z_2)|\ \lambda_{n+1}(dx,dz_1,dz_2) \le \varepsilon\ ,$$

which implies the theorem.

<u>Proof of the Corollary to Theorem 3.</u> This follows from the fact that

$$\int_S \int_S \int_S |Q_n^1(x,B,z_1) - Q_n^1(x,B,z_2)|\ \pi(dx)\ p^{n+1}(x,dz_1)\ p^{n+1}(x,dz_2)$$

$$\ge \int_S \int_S \left| \int_{z_2\in S} Q_n^1(x,B,z_1)p^{n+1}(x,dz_2) - \int_{z_2\in S} Q_n^1(x,B,z_2)p^{n+1}(x,dz_2) \right|$$

$$\pi(dx)p^{n+1}(x,dz_1)$$

$$= \int_S \int_S |Q_n^1(x,B,z_1) - p^1(x,B)|\pi(dx)\ p^{n+1}(x,dz_1)\ .$$

<u>Proof of Theorem 1</u>. Let $B, A_1, A_2, \ldots, A_k \in \mathcal{B}$ and consider the integral

$$\int_{x_k \in A_k} \cdots \int_{x_1 \in A_1} P(X_o \in B | X_{-k} = x_k, \ldots, X_{-1} = x_1)$$

$$P(X_{-k} \in dx_k, \ldots, X_{-1} \in dx_1) \ . \tag{11}$$

This is equal (for arbitrary $n \geq 1$) to

$$P(X_{-k} \in A_k, \ldots, X_{-1} \in A_1, X_o \in B) =$$

$$= \int_{x_k \in A_k} \cdots \int_{x_1 \in A_1} \int_{z \in S} P(X_o \in B | X_{-k} = x_k, \ldots, X_{-1} = x_1, X_n = z) \ .$$

$$.P(X_{-k} \in dx_k, \ldots, X_{-1} \in dx_1, X_n \in dz)$$

$$= \int_{x_k \in A_k} \cdots \int_{x_1 \in A_1} \int_{z \in S} Q_n^1 (x_1, B, z) \ P(X_{-k} \in dx_k, \ldots, X_{-1} \in dx_1, X_n \in dz)$$

$$= \int_{x_k \in A_k} \cdots \int_{x_1 \in A_1} \int_{z \in S} p^1(x_1, B) \ P(X_{-k} \in dx_k, \ldots, X_{-1} \in dx_1, X_n \in dz) +$$

$$+ \int_{x_k \in A_k} \cdots \int_{x_1 \in A_1} \int_{z \in S} (Q_n^1(x_1, B, z) - p^1(x_1, B)) \ P(X_{-k} \in dx_k, \ldots, X_{-1} \in dx_1,$$

$$X_n \in dz) \ .$$

The first term on the right is equal to

$$\int_{x_k \in A_k} \cdots \int_{x_1 \in A_1} P(X_o \in B | X_{-1} = x_1) \ P(X_{-k} \in dx_k, \ldots, X_{-1} \in dx_1) \ , \tag{12}$$

while the second term converges to zero as $n \to \infty$, by the Corollary to Theorem 3. This shows that (11) and (12) are equal for any $A_k, \ldots, A_1$ and hence

$$P(X_o \in B | X_{-k} = x_k, \ldots, X_{-1} = x_1) = P(X_o \in B | X_{-1} = x_1)$$

for $P(X_{-k} \in dx_k, \ldots, X_{-1} \in dx_1)$-almost all $(x_k, \ldots, x_1)$. This is the one-sided Markov property and Theorem 1 is established.

The one-sided Markov property now implies that the transition functions $p^n(\cdot,\cdot)$ $(n = 1,2,\ldots)$ form a semigroup, i.e. there are versions of these transition functions such that

$$p^{m+n}(x,B) = \int_S p^m(x,dy)\, p^n(y,B)$$

for all $x \in S$, $B \in \mathcal{B}$, $m \geq 1$, $n \geq 1$. The following corollary is a trivial consequence of this.

<u>COROLLARY</u>. If a consistent system of two-sided conditional probability kernels $\{\phi^k, k \geq 1\}$ is irreducible and admits a stationary Markov random field, then there is a semigroup of transition functions $p^n(x,B)$ $(x \in S, B \in \mathcal{B}, n \geq 1)$ and a probability measure π on $\mathcal{B}$, invariant under the semigroup, such that for π-almost all $x \in S$ and all $B \in \mathcal{B}$, $m \geq 1$, $n \geq 1$

$$Q_n^m(x, B, z) = \frac{\int_{y \in B} p^m(x, dy)\, p^n(y, dz)}{p^{m+n}(x, dz)},$$

in the sense that the left-hand side is, as a function of z, a version of the Radon-Nikodym density on the right-hand side.

We remark for later use that the Corollary to Theorem 3 can, in view of Theorem 1, be strengthened to

$$\lim_{n\to\infty} \int_S \int_S |Q_n^m(x,B,z) - p^m(x,B)|\,\pi(dx)\, p^{m+n}(x,dz) = 0 \tag{13}$$

for every fixed $m \geq 1$. This can be proved by induction, since an obvious triangle inequality implies

$$|Q_n^{m+1}(x,B,z) - p^{m+1}(x, B)| \leq$$

$$\leq \int Q_{n+1}^m(x,dy,z)\, |Q_n^1(y,B,z) - p^1(y,B)| +$$

$$+ \left| \int Q_{n+1}^m(x,dy,z)\, p^1(y,B) - \int p^m(x,dy)\, p^1(y,B) \right|,$$

and hence

$$\int\int | Q_n^{m+1}(x,B,z) - p^{m+1}(x,B)|\pi(dx)\, p^{m+1+n}(x,dz)$$

$$\leq \int\int | Q_n^{1}(y,B,z) - p^{1}(y,B)|\pi(dy)\, p^{n+1}(y,dz) +$$

$$+ \int\int \left| \int Q_{n+1}^{m}(x,dy,z)\, p^{1}(y,B) - \int p^{m}(x,dy)\, p^{1}(y,B)\right|\pi(dx)\, p^{m+n+1}(x,dz)$$

where $p^1(y,B)$ in the second summand on the right can be approximated uniformly, as a function of y, by linear combinations of indicator functions.

We now turn to a result on the convergence of $p^n(x,\cdot)$ that will be needed in §4 below.

<u>3.3 Lemma</u>. For $\pi\otimes\pi$-almost every $(x_1,x_2)\in S^2$ there is an $n\geq 1$ such that the probability measures $p^n(x_1,\cdot)$ and $p^n(x_2,\cdot)$ are not mutually singular.

<u>Proof</u>. For each $N = 1,2,\ldots$ let $A_N^* = \{x\in A_N : p^{i_N+j_N}(x,A_N) > 0\}$. Then $\pi(A_N^*)\geq 2\pi(A_N) - 1$, since

$$\pi(A_N^*) \geq \int_{A_N^*} \pi(dx)\, p^{i_N+j_N}(x,A_N) = \int_{A_N}\int_{A_N} \pi(dx)\, p^{i_N+j_N}(x,dz)$$

$$= \int\int_{(A_N\times S)\cap(S\times A_N)} \pi(dx)\, p^{i_N+j_N}(x,dz)$$

$$\geq \left(\int_{A_N}\int_{S} + \int_{S}\int_{A_N}\right) \pi(dx)\, p^{i_N+j_N}(x,dz) - 1 = 2\pi(A_N) - 1\,. \tag{14}$$

It follows that $\pi(\bigcup_{N=1}^{\infty} A_N^*) = 1$. Now for π-almost all $x_1\in A_N^*$ and all $x_2\in A_N^*$, $B\in\mathcal{B}$

$$p^{i_N}(x_1,B) = \int_{z_1\in S}\int_{z_2\in S} Q_{j_N}^{i_N}(x_1,B,z_1)\, p^{i_N+j_N}(x_1,dz_1)\, p^{i_N+j_N}(x_2,dz_2)$$

$$\geq \int_{z_1\in A_N}\int_{z_2\in A_N} r_{j_N}^{i_N}(x_1,x_2,B,z_1,z_2)\, p^{i_N+j_N}(x_1,dz_1)\, p^{i_N+j_N}(x_2,dz_2)\ . \qquad (15)$$

Since the same is true of $p^{i_N}(x_2, B)$ for π-almost all $x_2 \in A_N^*$, the lemma will be proved if we show that the right-hand side of (15) is a non-trivial measure in B, i.e. that it is positive when $B = S$. However, this is obvious since $p^{i_N+j_N}(x_1,A_N) > 0$, $p^{i_N+j_N}(x_2,A_N) > 0$ and

$r_{j_N}^{i_N}(x_1,x_2,S,z_1,z_2) > 0$ for $x_1,x_2 \in A_N^*$ and $z_1,z_2 \in A_N$.

The assertion of Lemma 3.3 is exactly the hypothesis in a version of Orey's convergence theorem stated and proved as Theorem 2.1 in [6]. It follows, therefore, from that theorem that

3.4 Corollary For π-almost all $x \in S$

$$\lim_{n\to\infty} \|p^n(x,\cdot) - \pi(\cdot)\| = 0$$

4. Uniqueness

In the present section we shall prove Theorem 2.

Let $\ldots,X_{-1},X_o,X_1,\ldots$ and $\ldots,\tilde{X}_{-1},\tilde{X}_o,\tilde{X}_1,\ldots$ be two stationary Markov random fields associated with the irreducible system of kernels $\{Q_n^m, m \geq 1, n \geq 1\}$. Define $\pi(B)$ and $p^n(x,B)$ $(B \in \mathcal{B},\ x \in S,\ n \geq 1)$ as in §3 for the first process, and $\tilde{\pi}(B)$, $\tilde{p}^n(x,B)$ in a similar way for the second process. By Theorem 1 the Markov random fields $\{X_i, -\infty < i < \infty\}$ and $\{\tilde{X}_i, -\infty < i < \infty\}$ are in fact stationary Markov processes with transition functions $p^n(\cdot,\cdot)$ and $\tilde{p}^n(\cdot,\cdot)$ and stationary distributions π and $\tilde{\pi}$, respectively.

4.1 Lemma. The probability measures π and $\tilde{\pi}$ are not mutually singular.

Proof. With the notation of §2, for π-almost all $x_1 \in S$ and all $x_2 \in S$, $B \in \mathcal{B}$

$$p^{i_N}(x_1,B) = \int_{z_1\in S}\int_{z_2\in S} Q_{j_N}^{i_N}(x_1,B,z_1)\, p^{i_N+j_N}(x_1,dz_1)\, \tilde{p}^{i_N+j_N}(x_2,dz_2)$$

$$\geq \int_{z_1\in S}\int_{z_2\in S} r_{j_N}^{i_N}(x_1,x_2,B,z_1,z_2)\, p^{i_N+j_N}(x_1,dz_1)\, \tilde{p}^{i_N+j_N}(x_2, dz_2)\ .$$

Integrating both sides with respect to $\pi(dx_1)\ \tilde{\pi}(dx_2)$ over S^2 we obtain

$$\pi(B) \geq \int_S\int_S\int_S\int_S r_{j_N}^{i_N}(x_1,x_2,B,z_1,z_2)\ \pi(dx_1)p^{i_N+j_N}(x_1,dz_1)$$

$$\tilde{\pi}(dx_2)\tilde{p}^{i_N+j_N}(x_2,dz_2). \tag{16}$$

Since the same applies to $\tilde{\pi}(B)$, it follows that $(\pi\wedge\tilde{\pi})(B)$ is greater than or equal to the right-hand side of (16). If N is sufficiently large, (14) implies $\int_{A_N}\int_{A_N} \pi(dx_1)\ p^{i_N+j_N}(x_1,dz_1) > 0$ and similarly for $\tilde{\pi}$ and $\tilde{p}^{i_N+j_N}$.

Since $r_{j_N}^{i_N}(x_1,x_2,S,z_1,z_2) > 0$ for all $x_1,\ x_2,\ z_1,\ z_2 \in A_N$, the right-hand side of (16) defines a measure in B, which is positive for $B = S$. Hence $(\pi\wedge\tilde{\pi})(S) > 0$, proving the lemma.

Now, to establish Theorem 2 we show first that $\pi = \tilde{\pi}$. Suppose there is a $B \in \mathcal{B}$ such that $\pi(B) \neq \tilde{\pi}(B)$. For fixed $m \geq 1$, the measure $\pi(dx)\ p^{m+n}(x,dz)$ converges in total variation to $\pi(dx)\ \pi(dz)$ as $n \to \infty$, by Corollary 3.4. This and (13) imply

$$\lim_{n\to\infty} \int_S\int_S Q_n^m(x,B,z) - p^m(x,B)\ |\ \pi(dx)\ \pi(dz) = 0\ . \tag{17}$$

Now for each $i = 1,2,\ldots$ choose first an m_i such that

$$\int_S |p^{m_i}(x,B) - \pi(B)\ |\ \pi(dx) \leq \frac{1}{2^{i+1}}$$

(possible by Corollary 3.4) and then an n_i, such that

$$\int_S \int_S \left| Q_{n_i}^{m_i}(x,B,z) - p^{m_i}(x,B) \right| \pi(dx)\, \pi(dz) \le \frac{1}{2^{i+1}} .$$

From these inequalities it is clear that

$$\sum_{i=1}^{\infty} \int_S \int_S \left| Q_{n_i}^{m_i}(x,B,z) - \pi(B) \right| \pi(dx)\, \pi(dz) \le \sum_{i=1}^{\infty} \frac{1}{2^i} < \infty$$

and hence that $\lim_{i\to\infty} Q_{n_i}^{m_i}(x,B,z) = \pi(B)$ for $\pi \otimes \pi$ -almost all (x,z). It is easy to see that the sequence (m_i, n_i) can be chosen in such a way that we have at the same time $\lim_{i\to\infty} Q_{n_i}^{m_i}(x,B,z) = \tilde{\pi}(B)$ for $\tilde{\pi} \otimes \tilde{\pi}$ -almost all (x,z). Since we assumed $\pi(B) \neq \tilde{\pi}(B)$, it follows that the measures $\pi \otimes \pi$ and $\tilde{\pi} \otimes \tilde{\pi}$ are mutually singular. However, this contradicts Lemma 4.1. We therefore have $\pi(B) = \tilde{\pi}(B)$ for all $B \in \mathcal{B}$.

It is now clear that (17) and its counterpart for $\tilde{p}^m(x,B)$ imply that $p^m(x,B) = \tilde{p}^m(x,B)$ for π-almost all $x \in S$, and choosing a suitable countable class of sets $B \in \mathcal{B}$ we deduce that $p^m(x,\cdot) = \tilde{p}^m(x,\cdot)$ for π-almost all $x \in S$. This shows that the Markov processes $\{X_i, -\infty < i < \infty\}$ and $\{\tilde{X}_i, -\infty < i < \infty\}$ have the same distribution.

Before concluding this section we mention that Dang-Ngoc and Yor have given a uniqueness result in [3] for a special case, where the system of two-sided conditional probability kernels $\{\phi^k, k \ge 1\}$ admits a stationary Markov random field which is *assumed* to be a Markov process (in the absence of Theorem 1 there is no *a priori* reason to suppose that this is automatically the case), satisfying certain strong hypotheses. This result can be restated as follows. Let S be a locally compact, σ-compact space and $\{X_i, -\infty < i < \infty\}$ a stationary Markov process on S such that: (i) there are continuous and strictly positive functions $f^n(x,y)$ $(n = 1,2,\ldots)$ on S^2 such that $f^{m+n}(x,z) = \int_S f^m(x,y)\, f^n(y,z)\, \pi(dy)$ and $p^n(x,B) = \int_B f^n(x,y)\, \pi(dy)$ $(x,z \in S,\ B \in \mathcal{B},\ m \ge 1,\ n \ge 1)$, where π and $p^n(\cdot,\cdot)$ have the same meaning as in §3; (ii) for every compact subset K of S, $\sup_{x,y\in K} |f^n(x,y) - 1| \to 0$ as $n \to \infty$. Then $\{X_i, -\infty < i < \infty\}$ is the unique, up to equivalence in distribution, stationary Markov random field admitted by the system of kernels

$$Q_n^m (x,B,z) = \frac{\int_{y \in B} f^m(x,y)\ f^n(y,z)\ \pi(dy)}{f^{m+n}(x,z)} \tag{18}$$

$(x,z \in S,\ B \in \mathcal{B},\ m \geq 1,\ n \geq 1)$. The proof given in [3] follows that of Lemma 3 in [5]. If condition (ii) is replaced by the weaker assumption of Harris recurrence, then Corollary 4.5 in [3] implies that $\{X_i,\ -\infty < i < \infty\}$ is the unique, up to equivalence in distribution, stationary Markov process associated with the system of kernels (18). In view of Theorem 1 above, we now see that statements about the class of all stationary Markov processes associated with $\{Q_n^m,\ m \geq 1,\ n \geq 1\}$ can be formulated as statements about the class of all stationary Markov random fields associated with $\{Q_n^m,\ m \geq 1,\ n \geq 1\}$.

5. The countable case

Suppose that S is countable and that $Q_1^1\ (x,\{y\},z) > 0$ for all $x,\ y,\ z \in S$, an assumption implying irreducibility as pointed out in §2. It was shown by Spitzer ([7],[8]) that, in this case, there is a strictly positive matrix $M(x,y)$ $(x,\ y \in S)$ such that

$$Q_n^m\ (x,\{y\},z) = \frac{M^m(x,y)\ M^n(y,z)}{M^{m+n}(x,z)} \tag{19}$$

$(x,\ y,\ z \in S,\ m \geq 1,\ n \geq 1)$, where powers of M are understood in the sense of matrix multiplication. One can take for example

$$M(x,y) = \frac{Q_1^1(b,\{x\},y)}{Q_1^1(b,\{b\},y)} ,$$

where b is a fixed element of S. Such a representation is not available in the general case, unless one assumes for example that $Q_n^m(x,B,z)$ is already defined by a formula such as (18). Two matrices $\{M_1(x,y)\}$ and $\{M_2(x,y)\}$ define the same system $\{Q_n^m,\ m \geq 1,\ n \geq 1\}$ if and only if they are equivalent in the following sense: there exists a constant $\lambda > 0$ and a strictly positive function $v(x)$ $(x \in S)$ such that

$$M_2(x,y) = \frac{M_1(x,y)\, v(y)}{\lambda v(x)}$$

for all $x, y \in S$. Kesten ([5]), using among other things Vere-Jones's extension of Frobenius's theorem on positive matrices, proved that the system $\{Q_n^m, m \geq 1, n \geq 1\}$ admits a (unique) stationary Markov random field if and only if the matrix M in (19) is equivalent to a positive-recurrent stochastic matrix.

We explain here why this result follows from Theorem 1. The sufficiency of the condition is straightforward: if M is equivalent to a positive-recurrent stochastic matrix, then the stationary Markov chain having the latter as transition matrix is a stationary Markov random field admitted by $\{Q_n^m, m \geq 1, n \geq 1\}$. Uniqueness can be deduced either from Theorem 2 above or from Lemma 3 in [5]. Conversely, suppose that the system $\{Q_n^m, m \geq 1, n = 1\}$ admits a stationary Markov random field $\{X_i, -\infty < i < \infty\}$ By Theorem 1, the latter is a Markov chain and if, as in §3, we set $\pi(x) = P(X_i = x)$, $p(x,y) = P(X_{i+1} = y \mid X_i = x)$ $(x, y \in S)$, then

$$Q_n^m\,(x,\{y\},z) = \frac{p^m(x,y)\; p^n(y,z)}{p^{m+n}(x,z)}$$

and hence the matrices $\{M(x,y)\}$ and $\{p(x,y)\}$ are equivalent (cf. Theorem 2 in [8]). Since the strictly positive stochastic matrix $\{p(x,y)\}$ admits the invariant probability measure π, it is positive-recurrent and Kesten's theorem follows.

REFERENCES

1. G. Benfatto, E. Presutti and M. Pulvirenti, DLR measures for one-dimensional harmonic systems, *Z. Wahrscheinlichkeitstheorie und Verw. Gebiete 41* (1978), 305-312.

2. T. Cox, An example of phase transition in countable one-dimensional Markov random fields, *J. Appl. Probability 14* (1977), 205-211.

3. N. Dang-Ngoc et M. Yor, Champs Markoviens et mesures de Gibbs sur R, *Ann. Sci. École Norm. Sup.*, 4e série, *11* (1978), 29-69.

4. P.L.Dobrushin, The description of a random field by means of conditional probabilities and conditions of its regularity (in Russian), *Teor. Verojatnost. i Primenen. 13* (1968), 201-229. (Transl.: *Theor. Probability Appl. 13* (1968), 197-224.)

5. H. Kesten, Existence and uniqueness of countable one-dimensional Markov random fields, *Ann. Probability 4* (1976), 557-569.

6. F. Papangelou, A martingale approach to the convergence of the iterates of a transition function, *Z. Wahrscheinlichkeitstheorie und Verw. Gebiete 37* (1977), 211-226.

7. F. Spitzer, Random fields and interacting particle systems, 1971 Summer Seminar, Math. Assoc. of America.

8. F. Spitzer, Phase transition in one-dimensional nearest-neighbor systems, *J. Functional Analysis 20* (1975), 240-255.

A UNIFORM CENTRAL LIMIT THEOREM FOR PARTIAL-SUM PROCESSES INDEXED BY SETS

Ronald Pyke [1]

ABSTRACT

Let $\mathcal{A}$ be a family of subsets of the unit k-dimensional cube I^k and let $\{X_{\underset{\sim}{j}}\}$ be an array of independent random variables indexed by the lattice of non-negative integral k-tuples, $\underset{\sim}{j} = (j_1, \ldots, j_k)$. For each $n \geq 1$ and $A \in \mathcal{A}$ define the partial-sum $S_n(A)$ to be the sum (normalized by $n^{-k/2}$) of all $X_{\underset{\sim}{j}}$ for which $\underset{\sim}{j}/n \in A$. By classical results the finite dimensional distributions of the partial-sum process $\{S_n(A): A \in \mathcal{A}\}$ are asymptotically normal. It is proved in this paper that when appropriately smoothed, this process converges weakly to a Brownian process under weak restrictions on $\mathcal{A}$ and the moments of $\{X_{\underset{\sim}{j}}\}$. Applications of this result are made in later sections to derive weak convergence results for Poisson and Uniform empirical processes.

1. Introduction

The present paper considers the question of uniformity of convergence for the Central Limit theorem in the context of random measures on the plane or higher dimensional Euclidean spaces. Perhaps a point process type motivation is most appropriate. Consider a large forest in which some insect infestation is present. Suppose the forest is partitioned by a grid into relatively small equal area sections and then measurements are taken within each section of the amount of damage or number of insects or some other appropriate quantity. The resulting matrix of observations is of the type considered here. For any region A, not necessarily rectilinear, a measure of its damage could then be given by summing over the sections that approximate A. If the measures of disjoint sections are independent, the sum for large A would be approximately Normal by the classical Central Limit

---

[1] This work was supported in part by the National Science Foundation, Grant MCS-78-09858

theorem. The purpose of this paper is to derive a Uniform Central Limit theorem for the normalized sums when viewed as a process indexed by a large family of sets A.

Other situations in two or three dimensions are easily imagined with regard to census data, mineral deposits, flaws in steel plates, and so on. In Section 4 below we apply the results of Sections 1-3 to obtain a Uniform Central Limit theorem for Poisson processes. We then apply this result in Section 5 to illustrate a method for deriving a Central Limit theorem for Uniform empirical processes.

Following Pyke (1973), let $\{X_{\underset{\sim}{j}}: \underset{\sim}{j} \in J^k\}$ be a set of independent and identically distributed random variables where J^k denotes the set of k-tuples of positive integers $\underset{\sim}{j} = (j_1,\ldots,j_k)$. Let $I^k = [\underset{\sim}{0},\underset{\sim}{1}]$ denote the closed k-dimensional unit cube. If one views $X_{\underset{\sim}{j}}$ as a random mass at the point $\underset{\sim}{j}$, then for any subset $B \subset R^k$ one may define the *partial-sum* signed measure

$$S(B) = \Sigma_{\underset{\sim}{j} \in B} X_{\underset{\sim}{j}} . \tag{1.1}$$

The purpose of this paper is to derive a Uniform Central Limit theorem for the partial-sum processes $\{S(nA): A \in \mathcal{A}\}$ where $\mathcal{A}$ is a suitable family of closed subsets of I^k.

Assume throughout that $EX_{\underset{\sim}{j}} = 0$ and $\mathrm{var}(X_{\underset{\sim}{j}}) = 1$. For any Borel set B, let B^ε denote the ε-neighborhood of A with respect to Euclidean distance. Write B_ε for the "inner" neighborhood defined as the complement of the ε-neighborhood of the complement of B, namely $B_\varepsilon = ((B^c)^\varepsilon)^c$. Write $B(\varepsilon)$ for $B^\varepsilon \backslash B_\varepsilon$, the ε-annulus around the boundary ∂B of B. Lebesgue measure will be denoted by either λ or $|\cdot|$ as convenient.

The index families $\mathcal{A}$ considered in this paper are assumed to satisfy the two conditions:

A1. There is a constant $c > 0$ such that for all $\varepsilon > 0$ and $A \in \mathcal{A}$,

$$|A(\varepsilon)| = |A^\varepsilon \backslash A_\varepsilon| \le c\varepsilon .$$

A2. $\mathcal{A}$ is totally bounded with respect to the Hausdorff metric d_H defined by

$$d_H(A,B) = \inf\{\varepsilon > 0: A \subset B^\varepsilon \text{ and } B \subset A^\varepsilon\} .$$

Recall that the set of all closed subsets of I^k forms a complete and separable metric space with respect to d_H. (Cf. Debreu (1967) e.g.)

Let $\nu(\varepsilon)$ denote the smallest cardinality of an ε-net in A.

Write

$$S_n(A) = n^{-k/2}S(nA) \tag{1.2}$$

for the normalized partial sums. Then under the above assumptions, $S_n(A)$ has mean zero and variance $n^{-k}N(nA)$ where for any $B \subset R^k$ $N(B) := \text{card}\{\underset{\sim}{j}: \underset{\sim}{j} \in B\}$.

When A is the class of lower orthants, $[\underset{\sim}{0},\underset{\sim}{t}]$ for $\underset{\sim}{t} \in I^k$ one may more naturally write $S_n(\underset{\sim}{t})$ for $S_n([\underset{\sim}{0},\underset{\sim}{t}])$. Observe that then $\text{cov}(S_n(\underset{\sim}{t}), S_n(\underset{\sim}{s})) = |[\underset{\sim}{0},\underset{\sim}{t}] \cap [\underset{\sim}{0},\underset{\sim}{s}]|$ whenever $n\underset{\sim}{s}, n\underset{\sim}{t} \in J^k$. In Kuelbs (1968) it was shown for $k = 2$ and under certain additional moment assumptions, that $S_n \overset{L}{\to} Z$ where $Z = \{Z(\underset{\sim}{t}): \underset{\sim}{t} \in I^2\}$ is a 2-dimensional Brownian Sheet; that is, Z is a Gaussian process indexed by I^k with mean zero and

$$\text{cov}(Z(\underset{\sim}{t}), Z(\underset{\sim}{s})) = |[\underset{\sim}{0},\underset{\sim}{t}] \cap [\underset{\sim}{0},\underset{\sim}{s}]| .$$

In Wichura (1969), multidimensional versions of Kolmogorov and Skorokhod inequalities are obtained which enable Wichura to prove that $S_n \overset{L}{\to} Z$ under the finiteness of only the second moment. Here $\overset{L}{\to}$ denotes convergence in law (equivalent to the weak convergence of the image laws) with respect to the Skorokhod topology. The reader is referred to Pyke (1973, Sections 3.5 and 3.6) for more detailed discussion of these and related results.

The purpose of this paper is to show that analogous limit theorems are possible for more general index sets A. Problems of this nature were posed by the author about six years ago and preliminary results were obtained in 1977 and 1978. These results used the methods of Pyke (1977) and were unsatisfactory in that the restrictions on the tails of the distributions, namely generalized Gaussian, seemed to be unnecessarily strong. In the present paper the only conditions are in terms of finite moments of orders that increase as the "size" of the index families increase. It is still conceivable that a finite second moment is all that is required, but I believe that the orders assumed here are nearly optimal. In Erickson (1981), a detailed study of weak convergence in Lipschitz spaces is made and applied to partial-sum processes. This stronger convergence is obtained under generalized Gaussian assumptions for large index families and under a p-th moment condition $(p > 2)$ for smaller index families of similar size to the orthants.

The limiting process in the general case, $Z = \{Z(A): A \in \mathcal{A}\}$ will necessarily be the *Brownian process indexed by* $\mathcal{A}$ defined as a mean zero Gaussian process with

$$\mathrm{cov}(Z(A),\ Z(B)) = |A \cap B| \ ;\ A,B \in \mathcal{A}. \tag{1.3}$$

Let $C(\mathcal{A})$ be the set of all continuous real-valued functions defined on $(\mathcal{A},\ d_H)$. It is known (cf. Dudley (1973) that if $\mathcal{A}$ is not "too large", there exists a version of Z in $C(\mathcal{A})$. For example, if

$$\ln \nu(\varepsilon) \le K\varepsilon^{-r} \tag{1.4}$$

for some $r < 1$ and $K > 0$, and all sufficiently small $\varepsilon > 0$, then $Z \in C(\mathcal{A})$ is possible. In this context, the infimum of those r for which (1.4) holds is called the *exponent of metric entropy*.

A particular class of index families $\mathcal{A}$ for which (1.4) is true was introduced by Dudley (1974). To describe these, fix $\alpha > 0$ and $M > 0$. Then let $\mathcal{A}_\alpha \equiv \mathcal{A}(k,\alpha,M)$ denote the set of all closed subsets of I^k whose boundaries are represented parametrically as continuous mappings of the $k-1$ dimensional unit sphere, which have bounded (by M) derivatives of all orders up to but less than α and for which the derivatives of highest order satisfy a Lipschitz condition of order $\alpha - [\alpha] > 0$. (In this definition we use $[\alpha]$ for the greatest integer *less* than α.) Dudley (1974) shows that for $\mathcal{A}_\alpha$, (1.4) holds for any $r > (k-1)/\alpha$. For further discussion of $\mathcal{A}_\alpha$ see Sun and Pyke (1982). A closely related and more simply defined family, $\mathcal{R}(k,\alpha,M)$ say, of sets with smooth boundaries was introduced by Révész (1976). The exponent of metric entropy is also $(k-1)/\alpha$ for $\mathcal{R}(k,\alpha,M)$.

Other candidates for index families include $\mathcal{C}_k$, the class of closed convex subsets of I^k, and $\mathcal{P}_{k,m}$, the class of all closed polygons in K^k with no more than m vertices. The exponent of metric entropy of the former is $(k-1)/2$ (cf. Dudley (1974)) and of the latter is zero which may be deduced straightforwardly (cf. Erickson, 1981).

In order to derive the weak convergence of the partial-sum processes to Z it will be necessary to place conditions on $\mathcal{A}$ like those above, to insure that the limit $Z \in C(\mathcal{A})$ a.s. It will also be necessary, and at first more surprising, to restrict oneselves to continuous versions of the partial-sum processes themselves. It was not necessary to do this for the Wichura result for which $\mathcal{A}$ was the relatively small class of lower orthants. The reason that smoothing is needed in general is that with a larger $\mathcal{A}$ it may be possible to have very close sets whose boundaries weave in and around the lattice points $\underset{\sim}{j}/n$ in such a way that the S_n-measures of the symmetric differences are not uniformly small. This was discussed in Pyke (1977) and a specific example due to Dudley is included in Erickson (1981).

The natural continuous version of the partial-sum process is defined (cf. Pyke (1973), 4.3.5) as the random signed measure

$$Z_n(A) = n^{-k/2} X(nA) \tag{1.5}$$

where, for any Borel set $B \subset \mathbf{R}^k$,

$$X(B) = \Sigma_{\underset{\sim}{j} \in J^k} |B \cap C_{\underset{\sim}{j}}| X_{\underset{\sim}{j}}$$

where $C_{\underset{\sim}{j}}$ is the unit cube $(\underset{\sim}{j}-\underset{\sim}{1}, \underset{\sim}{j}]$. Thus X is a signed measure which is absolutely continuous with respect to Lebesgue measure λ and for which the density $dX/d\lambda$ is equal to $X_{\underset{\sim}{j}}$ on $C_{\underset{\sim}{j}}$. A similar description could also be given for Z_n.

<u>Lemma 1.1</u> The finite dimensional distributions of the Z_n-process converge weakly to those of Z.

<u>Proof</u>. Let both S_n and Z_n be defined in terms of the same array $\{X_{\underset{\sim}{j}}: j \in J\}$. It will suffice to show that $Z_n(A) - S_n(A) \overset{P}{\to} 0$ for each $A \in \mathcal{A}$, since the finite dimensional distributions of S_n are straightforwardly shown to converge to those of Z. To this end, write

$$Z_n(A) - S_n(A) = n^{-1/2} \Sigma_{\underset{\sim}{j}}: n^{-1}C_{\underset{\sim}{j}} \cap \partial A + \emptyset^{w_{\underset{\sim}{j}}} X_{\underset{\sim}{j}} \tag{1.6}$$

where $w_{\underset{\sim}{j}} = |nA \cap C_{\underset{\sim}{j}}| - I_A(\underset{\sim}{j}/n)$. The number of summands in (1.6), say m, can be bounded as follows. Each cube $C_{\underset{\sim}{j}}$ is of diameter $k^{1/2}$ and volume 1. Since the union of those cubes $n^{-1}C_{\underset{\sim}{j}}$ which intersect ∂A is therefore contained in the annulus $A(n^{-1}k^{1/2})$ it follows that

$$m\, n^{-k} \le |A(n^{-1}k^{1/2})| \le c\, n^{-1}\, k^{1/2}$$

where the last inequality uses Assumption A1. Thus since $|w_{\underset{\sim}{j}}| \le 1$,

$$\mathrm{var}[Z_n(A) - S_n(A)] \le n^{-k} \Sigma_{\underset{\sim}{j}:\, n^{-1} C_{\underset{\sim}{j}} \cap \partial A = \emptyset}\, w^2_{\underset{\sim}{j}}$$

$$\le m\, n^{-k} \le c\, n^{-1}\, k^{1/2} \to 0$$

as $n \to \infty$. This shows that $Z_n(A) - S_n(A) \overset{P}{\to} 0$ for each $A \in \mathcal{A}$, which in turn implies that the asymptotic finite dimensional distributions of the Z_n-processes are the same as for the S_n-processes which are those of Z. □

It is the fact that $Z_n(A) - S_n(A)$ need not converge to zero uniformly over $\mathcal{A}$ which necessitates our emphasis upon the continuized version of the partial-sum processes. The weak convergence of the Z_n-processes pertains to $(C(\mathcal{A}), d_H)$ and for this a tightness condition involving the modulus of uniform continuity is all that is needed in addition to Lemma 1.1. The study of tightness in the next section will then complete the proof of

Theorem 1. Let $\{X_{\underset{\sim}{j}}: \underset{\sim}{j} \in J^k\}$ be an array of independent, identically distributed random variables with zero mean and unit variance. If for $0<r<1$ and $s>2(1+r)/(1-r)$, $\ln \nu(\varepsilon)=O(\varepsilon^{-r})$ as $\varepsilon \searrow 0$ and $E(|X_1|^s) < \infty$, then under assumptions A1 and A2 the normalized continuous partial-sum process $\{Z_n(A): A \in \mathcal{A}\}$ converges weakly on $(C(\mathcal{A}), d_H)$ to $\{Z(A): A \in \mathcal{A}\}$, the Brownian process indexed by $\mathcal{A}$.

2. Tightness of the Z_n-processes

As in Pyke (1977) let $\mathcal{A}^F(\varepsilon)$ denote a finite ε-net in $\mathcal{A}$ of minimal cardinality $\nu(\varepsilon)$. For any A,B or C in $\mathcal{A}$ we will write A^F_ε, B^F_ε, or C^F_ε, respectively, to denote any member of $\mathcal{A}^F(\varepsilon)$ within ε of the given set. That is, $d_H(B, B^F_\varepsilon) \le \varepsilon$ and $B^F_\varepsilon \in \mathcal{A}^F(\varepsilon)$ for example.

To establish $Z_n \overset{L}{\to} Z$ it suffices in view of Lemma 1 to show that the image laws of the processes are tight. To this end, denote the modulus of continuity by

$$\omega(Z_n, \delta) = \sup\{|Z_n(B) - Z_n(C)| : B, C \in \mathcal{A},\ d_H(B,C) < \delta\} \tag{2.1}$$

To establish tightness it is sufficient in view of Lemma 1 (cf. Billingsley (1968), p.55) to show that for every $\varepsilon > 0$ there exists $\delta > 0$ and an integer n_o such that

$$P[\omega(Z_n, \delta) > \varepsilon] \le \varepsilon \quad \text{for all } n > n_o .$$

We will obtain this result by detailed approximations that require Z_n to be represented first as a telescoping series of values of Z_n evaluated at elements in the approximating nets. This is described as follows.

The supremum in (2.1) is over an infinite set of values. Nevertheless each $B \in \mathcal{A}$ can be approximated by a sequence of elements chosen from the ε-nets. To this end, choose $\delta_j \searrow 0$ and let $\delta_o = \delta$. For each B and C in (2.1) select $B^F_{\delta_j}, C^F_{\delta_j} \in \mathcal{A}^F(\delta_j)$ for each j. When $d_H(B,C) < \delta/2$ assume $B^F_{\delta_o} = C^F_{\delta_o}$. Then

$$\omega(Z_n, \delta/2) \le \Sigma_{j=1}^{k_n} \max\{|Z_n(B^F_{\delta_j}) - Z_n(B^F_{\delta_{j-1}})| + |Z_n(C^F_{\delta_j}) - Z_n(C^F_{\delta_{j-1}})| :$$

$$B, C \in \mathcal{A},\ d_H(B,C) < \delta/2\} + \sup\{|Z_n(A) - Z_n(A^F_{\delta_{k_n}})| : A \in \mathcal{A}\}. \quad (2.2)$$

Although card $\mathcal{A}^F(\delta_j)$ increases as δ_j decreases, it is nevertheless finite for every j and if the $\delta_{j-1} - \delta_j$ are sufficiently small we will be able to get sufficiently small uniform bounds on each of the terms in (2.2). This nesting approach was introduced by Strassen and Dudley (1969) and used subsequently by Dudley (1978) and Sun and Pyke (1982) in applications to empirical processes. For this approach to work here we will need to be able to choose $\{\delta_j\}$ and a sequence of bounds, $\{\eta_j\}$, so that for some k_n

$$\sum_{j=1}^{k_n} \nu(\delta_j)\nu(\delta_{j-1}) P[|Z_n(B^F_{\delta_j} \setminus B^F_{\delta_{j-1}})| > \eta_j] < \varepsilon \ , \quad (2.3)$$

$$P[\sup\{|Z_n(A) - Z_n(A^F_{\delta_{k_n}})| : A \in \mathcal{A}\} > \varepsilon] < \varepsilon \quad (2.4)$$

and

$$\sum_{j=1}^{k_n} \eta_j < \varepsilon \quad (2.5)$$

uniformly for $n \ge n_o$ for a preassigned $\varepsilon > 0$. Similar results will follow for the reversed differences $B^F_{\delta_{j-1}} \setminus B^F_{\delta_j}$ and for the analogous differences for the C-sets. In going from (2.2) to (2.3) we have used the fact that Z_n is finitely additive so that for any two sets,

$$Z_n(A_1) - Z_n(A_2) = Z_n(A_1\backslash A_2) - Z_n(A_2\backslash A_1) .$$

The summation in (2.3) can be replaced by one slightly simpler in form, namely,

$$\Sigma_j \nu(\delta_j)^2 P[|Z_n(B_j)|>\eta_j] , \tag{2.6}$$

since $\nu(\delta_j) \ge \nu(\delta_{j-1})$, where B_j represents any set which contains no sphere of diameter δ_{j-1}. Note that by construction, $d_H(B^F_{\delta_j}, B^F_{\delta_{j-1}}) \le 2\delta_{j-1}$.

Before approximating the summands in (2.6) we first truncate the array $\{X_{\underset{\sim}{j}}\}$. For given n, truncate each $X_{\underset{\sim}{j}}$ at τ_n and write $X_{\underset{\sim}{j}}(\tau_n)$ for the truncated version of $X_{\underset{\sim}{j}}$. Here, truncation means that the truncated values are replaced by *zero*. Choose τ_n so that

$$n^k P[|X_{\underset{\sim}{1}}|>\tau_n] > \varepsilon \tag{2.7}$$

in order to insure that

$$P[X_{\underset{\sim}{j}} = X_{\underset{\sim}{j}}(\tau_n): \quad \text{for all} \quad \underset{\sim}{j} \le n \underset{\sim}{1}] \ge 1 - \varepsilon.$$

Set $t_n = n^{-k/2}\tau_n$. For each B, introduce the coefficients $b_{\underset{\sim}{j}} = n^{-k/2}|nB\cap C_{\underset{\sim}{j}}|$ so that we may write

$$Z_n(B) = \Sigma_{\underset{\sim}{j}\in J^k} b_{\underset{\sim}{j}} X_{\underset{\sim}{j}} .$$

Let Z_n^{tr} denote the partial-sum process determined by the truncated array; thus

$$Z_n^{tr}(B) = \sum_{\underset{\sim}{j}\in J^k} b_{\underset{\sim}{j}} X_{\underset{\sim}{j}}(n^{k/2}t_n)$$

To minimize complications assume for now that the truncated variables have mean zero. At the end of the proof we will check that this assumption can be dropped.

Recall (cf. Bennett (1962)) that Bernstein's Inequality states that if S is a sum of independent mean zero r.v.'s each bounded by M, then for all $t > 0$,

$$P[S>t] < \exp\{-t^2/2(\sigma^2 + Mt/3)\}$$

where $\sigma^2 = \text{var}(S)$. For the sum $Z_n^{tr}(B)$ we have

$$b_j = n^{-k/2}|C_j \cap nB| \le n^{-k/2}; \quad |X_j(n^{k/2}t_n)| \le n^{k/2}t_n$$

and

$$\text{var } Z_n^{tr}(B) \le \text{var } Z_n(B) \le |B|. \tag{2.8}$$

Thus for the B_j of (2.6), $|B_j| \le 2c\delta_{j-1} := c'\delta_{j-1}$ by Al. Bernstein's Inequality may then be applied to show

$$P[Z_n^{tr}(B_j) > \eta_j] < \exp\{-\eta_j^2/2(c'\delta_{j-1} + t_n\,\eta_j/3\}$$

$$= \exp\{-\eta_j^2/2\delta_{j-1}(c' + t_n\,\eta_j/3\delta_{j-1})\}. \tag{2.9}$$

We choose geometric sequences

$$\delta_j = \delta\beta^j, \quad \eta_j = \eta\beta^{bj}, \quad \delta > 0, \ \eta > 0, \ b > 0 . \tag{2.10}$$

Substitution of (2.9) and (1.4) into (2.6) shows that (2.6) is bounded by

$$\sum_{j=1}^{k_n} \exp\{2K\delta_j^{-r} - \eta_j^2/2\delta_{j-1}(c' + t_n\,\eta_j/3\delta_{j-1})\}$$

$$= \sum_{j=1}^{k_n} \exp\{2K\delta^{-r}\,\beta^{-rj} - \eta^2\beta^{2bj-j+1}/2\delta(c'+t_n\beta^{\,bj-j+1}\eta/3\delta)\}. \tag{2.11}$$

Notice first of all that in order for this series to converge as $k_n \to \infty$, we must have $1 - 2b \ge r$ or $b \le (1 - r)/2$. Such a choice is possible since in order that $Z \in C(A)$ a.s. we have assumed $r < 1$. Secondly, if we write the exponent in (2.11) as

$$-2K\delta^{-r}\beta^{-rj}\left\{\eta^2\beta^{(r-1+2b)j+1}/4K\delta^{1-r}(c+t_n\beta^{bj-j+1}\,\eta/3\delta) - 1\right\} \tag{2.12}$$

it is seen that in order to insure convergence the term in parenthesis must remain positive. This will be satisfied in particular if we allow

$$t_n \beta^{(b-1)k_n} \to \infty \tag{2.13}$$

but postulate that

$$t_n \beta^{-(r+b)k_n} \to \gamma \tag{2.14}$$

for some suitably small constant $\gamma > 0$ to be specified later. Notice that if $1 - 2b > r$, (2.14) implies (2.13).

In Theorem 1 it is assumed that $E(|\underset{\sim}{X}_1|^s) < \infty$ for some $s > 2(1+r)/(1-r)$. This implies that

$$P[\,|\underset{\sim}{X}_1| > x\,] = o(x^{-s}) \ . \tag{2.15}$$

The truncation condition (2.7) is then satisfied if we choose $\{t_n\}$ so that

$$n^k \tau_n^{-s} \equiv n^k (n^{k/2} \ t_n)^{-s} \to 0$$

or equivalently

$$t_n = n^{-(1/2 \ - \ 1/s)k} \ \lambda_n \tag{2.16}$$

where $\{\lambda_n\}$ is any positive sequence diverging to $+\infty$. Together with (2.14) this implies that

$$\exp\{-(1/2 - 1/s)k \ln n + \ln \lambda_n + (-\ln \beta)(r+b)k_n\} \to \gamma \ .$$

It therefore suffices to choose k_n to satisfy

$$k_n = (-\ln \beta)^{-1}(r+b)^{-1} \{(1/2 - 1/s)k \ln n - \ln \lambda_n + \gamma_n\} \tag{2.17}$$

for some choice of $\gamma_n \to \ln \gamma$ that makes k_n a positive integer.

Another restriction upon k_n is implicit in (2.4). For the truncated process

$$\sup\{|Z_n^{tr}(A\backslash A^F_{\delta_{k_n}})|: A\in\mathcal{A}\} \le n^{-k/2}\tau_n \sup\{|n(A\backslash A_{\delta_{k_n}})|: A\in\mathcal{A}\}$$

$$\le t_n n^k c\, \delta_{k_n}$$

by A1. To satisfy (2.4) we will require that

$$t_n n^k \beta^{k_n} \to 0 . \tag{2.18}$$

Substitution of (2.16) and (2.17) into (2.18) shows that λ_n and γ_n must allow the divergence to $-\infty$ of

$$\begin{aligned}\ln(t_n n^k \beta^{k_n}) &= \ln \lambda_n + (1/2 + 1/s)k \ln n + (\ln \beta)k_n \\ &= \ln \lambda_n + (1/2 + 1/s)k \ln n - (r+b)^{-1}\{(1/2 - 1/s)k \ln n \\ &\qquad - \ln\lambda_n + \gamma_n\} \\ &= -(k \ln n)\{(1-r-b)/2 - (1+r+b)/s\}(r+b)^{-1} + \{1+(r+b)^{-1}\}\ln\lambda_n \\ &\qquad + (r+b)^{-1}\gamma_n .\end{aligned}$$

For this to happen it suffices to choose $\ln\lambda_n = o(\ln n)$ and to choose s so that the coefficient of $-k \ln n$ is positive; that is,

$$s > 2(1+r+b)/(1-r-b). \tag{2.19}$$

Since this bound is an increasing function of b we can always find $b > 0$ to satisfy (2.19) if

$$s > 2(1+r)(1-r) . \tag{2.20}$$

This latter inequality is the assumption made in Theorem 1.

It remains now to show that for the above choices of β, b, s, k_n, λ_n and γ_n, it is possible to choose values for the remaining parameters δ,η and γ so that (2.3), (2.4) and (2.5) hold for any preassigned $\varepsilon > 0$. Since

$$\sum_{j=1}^{k_n} \eta_j \le \eta\beta^b \sum_{j=1}^{\infty} \beta^{bj} = \eta\beta^b(1-\beta^b)$$

a choice of η sufficiently small will insure that (2.5) holds.

Expression (2.4) holds by construction since (2.18) insures that the event never occurs for large n for the truncated processes, while (2.16) in (2.7) insures that the probability of equal truncated and non-truncated processes can be made arbitrarily small. To check (2.3), observe that the exponent of the bound (2.11) when written in the form (2.12) is

$$-2K\delta^{-r}\beta^{-rj}\left\{\frac{\delta^r}{4K\eta+o(1)}\cdot\frac{\beta^{(r+b)j}}{t_n} - 1\right\} \tag{2.21}$$

where $o(1)$ indicates here that as j increases, the term decreases to its value for $j=k_n$, which value then converges to zero as $n\to\infty$ by (2.13). Also, by (2.14) the term $\beta^{(r+b)j}/t_n$ decreases to its value for $j=k_n$ which in turn converges to $1/\gamma$. Therefore, if we choose $\gamma=\delta^r/8K\eta$, the parenthetical term in (2.21) is positive and bounded away from zero uniformly in δ. Consequently the sum in (2.3) remains bounded as $n\to\infty$. The bound on these sums can then be made arbitrarily small by choosing δ sufficiently small, since the only effect of δ after our choice of γ is as the front multiplier in (2.21).

In the above derivations we assumed that $\mu_n := EX_{\underset{\sim}{1}}(\tau_n)=0$. If this is not the case use the bound

$$\omega(Z_n,\delta) \le \omega(Z_n - E(Z_n^{tr}),\delta) + \omega(E(Z_n^{tr}),\delta)\ . \tag{2.22}$$

Since for any B

$$E(Z_n^{tr}(B)) = \mu_n n^{k/2}|B|$$

the last term in (2.22) is bounded above by $2\mu_n n^{k/2}c\delta$ because of A1. But in view of (2.15)

$$|\mu_n| = |E(X_{\underset{\sim}{1}} I_{[|X_{\underset{\sim}{1}}|>\tau_n]})| \le C_1\tau_n^{-s+1}\ .$$

Thus by (2.16)

$$|\mu_n| n^{k/2} \le C_1 \, n^{k/2-(s-1)k/s} \, \lambda_n^{1-s} = o(1)$$

since $s > 2$ and $\lambda_n \to \infty$. This shows that the set of deterministic functions $\{E(Z_n^{tr}(\cdot)): n \ge 1\}$ is tight, demonstrating that our previous assumption was made without loss of generality.

3. Extensions to non-identically distributed arrays

In the preceding derivation of tightness for independent and identically distributed summands, the assumption of a common distribution was used only in the computation of variances such as in (2.8) and in the derivation of the truncation level τ_n as in (2.16). In what follows we describe briefly a natural generalization of these two steps that applies to non-identically distributed arrays. To this end let $\{X_{n\underset{\sim}{j}}: \underset{\sim}{j} \in J^k \cap nI^k\}$ be an array of independent random variables with $EX_{n\underset{\sim}{j}} = 0$ and $\mathrm{var}(X_{n\underset{\sim}{j}}) = \sigma^2_{n\underset{\sim}{j}} < \infty$ for each $\underset{\sim}{j}$. Define for each $n \ge 1$ and $B \in \mathcal{B}^k$

$$\Lambda_n(B) = \Sigma_{\underset{\sim}{j}} |n^{-1} C_{\underset{\sim}{j}} \cap B| \sigma^2_{n\underset{\sim}{j}} :$$

Then Λ_n is a measure. Let us assume that there exists a measure Λ for which

$$\lim_{n\to\infty} \Lambda_n(B) = \Lambda(B), \quad \text{uniformly for } B \in \mathcal{B}^k$$

and $\Lambda(B) \le C_o|B|$ for all $B \in \mathcal{B}^k$ and some constant C_o. (These conditions could have been stated in terms of $d\Lambda_n/d\lambda$, i.e. the function $\sigma_n^2(t) = \sigma^2_{n\underset{\sim}{j}}$ if $nt \in C_{\underset{\sim}{j}}$. The given statements are in a more natural form for application here.) In place of (2.9) one can now write

$$\mathrm{var}(Z_n^{tr}(B)) \le \mathrm{var}(Z_n(B)) = n^{-k} \Sigma_{\underset{\sim}{j}} |C_{\underset{\sim}{j}} \cap nB|^2 \sigma^2_{n\underset{\sim}{j}}$$

$$\le \Lambda_n(B) \le 2C_o|B|$$

for n sufficiently large, which is the inequality needed at that point in the proof.

For truncation, assume in place of (2.15) that

$$L_n(x) := n^{-k} \Sigma_{\underset{\sim}{j}} P[|X_{n\underset{\sim}{j}}| > x] = o(x^{-s}).$$

The proof of tightness then continues without further changes.

The analogue of Lemma 1 for the assumptions of this section is the following, derived straightforwardly as before by use of classical central limit theorems.

Lemma 2. The finite dimensional distributions of Z_n converge weakly to those of $W^\Lambda = \{W^\Lambda(A): A \in \mathcal{A}\}$, a mean zero Gaussian process with

$$\mathrm{cov}(W^\Lambda(A), W^\Lambda(B)) = \Lambda(A \cap B), \quad A,B \in \mathcal{A}.$$

Notice that as a consequence of the convergence $Z_n \xrightarrow{L} W^\Lambda$, one obtains the fact that $W^\Lambda \in C(\mathcal{A})$.

4. An Application to Poisson Processes

Let N denote a homogeneous Poisson-(1) process defined on $\mathcal{B}^k$, so that $N(B)$ is a Poisson random variable with $E\,N(B) = |B|$. For any $\lambda > 0$, write $N_\lambda(B) = N(\lambda^{1/k}B)$ so that N_λ is a homogeneous Poisson-(λ) process. Consider the normalized process $\{W_n(A): A \in \mathcal{A}\}$ defined by

$$W_n(A) = n^{-1/2}[N_n(A) - n|A|]. \tag{4.1}$$

Clearly the finite dimensional distributions of W_n converge weakly to those of the Brownian process Z. This holds even if $\mathcal{A} = \mathcal{B}^k$. To show that $W_n \xrightarrow{L} Z$ (where $\xrightarrow{L}$ must be clarified since $W_n \notin C(\mathcal{A})$) we will produce a matrix array $\{X_{n\underset{\sim}{j}}\}$ for which the corresponding Z_n-process is close to W_n. To this end set

$$X_{n\underset{\sim}{j}} = n^{(k-1)/2}\{N(n^{1/k-1}\,C_{\underset{\sim}{j}}) - n^{1-k}\},\ \underset{\sim}{j} \in J^k , \tag{4.2}$$

which is a Poisson - (n^{1-k}) random variable, normalized to have mean 0 and variance 1. Since these arrays $\{X_{n\underset{\sim}{j}}\}$ satisfy the conditions of Sections 2 and 3, the following result obtains.

Theorem 4.1 If $\{Z_n(A): A \in \mathcal{A}\}$ is the (continuous) partial-sum process formed from the array (4.2), then for any $\mathcal{A}$ satisfying A1, A2, and (1.4) for any $0<r<1$, $Z_n \xrightarrow{L} Z$ on $(C(\mathcal{A}), d_H)$.

We consider now the difference between W_n and Z_n. Although we stressed at the outset that one cannot in general obtain limit theorems without smoothing the partial-sum processes, it is possible in this Poisson case to show that W_n and Z_n are uniformly close because the expected number of discontinuities (atoms) of the Poisson process in I^k is only n, much less than the number n^k of summands. To verify this, write as in (1.6)

$$Z_n(A) - W_n(A) = n^{1/2}\Big[\Sigma_{\underset{\sim}{j}}|C_{\underset{\sim}{j}} \cap nA|\{N(n^{1/k-1}C_{\underset{\sim}{j}}) - n^{1-k}\}$$

$$- \Sigma_{\underset{\sim}{j}}\{N_n(n^{-1}C_{\underset{\sim}{j}} \cap A) - n|n^{-1}C_{\underset{\sim}{j}} \cap A|\}\Big]$$

$$= n^{-1/2}\,\Sigma_{\underset{\sim}{j}}\Big\{|C_{\underset{\sim}{j}} \cap nA|N_n(n^{-1}C_{\underset{\sim}{j}}) - N_n(n^{-1}C_{\underset{\sim}{j}} \cap A)\Big\}.$$

Thus, if A_n denotes the 'rectangular' approximation to A defined by $A_n = \cup\{n^{-1}C_{\underset{\sim}{j}}: n^{-1}C_{\underset{\sim}{j}} \cap A \neq \emptyset\}$

$$|Z_n(A) - W_n(A)| \le n^{-1/2}N_n(A_n \backslash A).$$

Now $A_n \backslash A \subset A^\delta \backslash A$ for any $\delta > k^{1/2}n^{-1}$, the diameter of each cube $n^{-1}C_{\underset{\sim}{j}}$. Thus for any $\delta > k^{1/2}n^{-1}$

$$P[\sup_{A \in \mathcal{A}}|Z_n(A) - W_n(A)| > \varepsilon] \le P[\sup_{A \in \mathcal{A}}N_n(A^\delta \backslash \dot{A}) > \varepsilon n^{1/2}]$$

$$\le P[\max_{A \in \mathcal{A}^F(\varepsilon)}N_n(A^{2\delta} \backslash A) > \varepsilon\, n^{1/2}]$$

$$\le \nu(\delta)\ P[Y_{2c\delta n} > \varepsilon\, n^{1/2}] \tag{4.2}$$

where $\nu(\delta)$ is the cardinality of $\mathcal{A}^F(\delta)$ and $Y_{2c\delta n}$ denotes a Poisson-$2c\delta n$ random variable. If we assume $\nu(\delta)$ is large and satisfies (1.4) for some $r \in (0,1)$ then a small exponential bound for the upper tail of the Poisson distribution is needed in order to make (4.2) small. To this end we prove

<u>Lemma 4.1</u> If Y_λ is a Poisson-λ random variable, then

$$P[Y_\lambda > b\lambda^\alpha] \le \exp\{b(1-\ln b)\lambda^\alpha - b(\alpha-1)\lambda^\alpha \ln\lambda - \lambda\}$$

for any $b > 0$ and $\alpha > 1$ satisfying $b\lambda^\alpha > \lambda$.

<u>Proof.</u> The proof is standard. By Chebichev's Inequality, for any $t > 0$.

$$P[Y_\lambda > b\lambda^\alpha] = P[e^{tY_\lambda} > e^{tb\lambda^\alpha}] \le e^{-tb\lambda^\alpha + \lambda(e^t-1)}\ .$$

This bound is minimized when $t = \ln(b\lambda^{\alpha-1})$, so that whenever $b\lambda^{\alpha-1} > 1$,

$$P[Y_\lambda > b^\alpha] \le \exp\{-b\lambda^\alpha \ln(b\lambda^{\alpha-1}) + b\lambda^\alpha - \lambda\}$$

which is the stated result. □

To apply this to (4.2), set $\lambda = 2c\delta n$. Write $\delta = n^{-1+q}$ so that $\lambda = 2cn^q$. Take $b = \varepsilon(2c)^{-\alpha}$ so that $\lambda^\alpha = (2c)^\alpha n^{q\alpha}$ must equal $(2c)^\alpha n^{1/2}$. Thus $\alpha = 1/2q$ and the bound of Lemma 4.1 becomes

$$\exp\{\varepsilon(1-\ln \varepsilon(2c)^{-1/2q})n^{1/2} - \varepsilon(1/2q-1)n^{1/2} \ln n - 2cn^q\} .$$

Substitution of this into (4.2) yields a bound of

$$\exp\{K\, n^{r(1-q)} - \varepsilon(1/2q-1)n^{1/2} \ln n + \varepsilon(1-\ln\varepsilon(2c)^{-1/2q})n^{1/2} - 2cn^q\}.$$

Clearly this bound can be made arbitrarily small for n sufficiently large if $r(1-q) \le 1/2$ and $2q < 1$. (Notice that the condition $b\lambda^\alpha > \lambda$ in this application becomes $\varepsilon n^{1/2} > 2cn^q$, which is satisfied for all sufficiently large n if $q < 1/2$.) Thus by choosing q to satisfy $1-1/2r < q < 1/2$, which is always possible since $r < 1$, one can make (4.2) arbitrarily small for all sufficiently large n. This establishes the fact that

$$\lim_{n\to\infty} P[\sup_{A\in\mathcal{A}} |Z_n(A) - W_n(A)| > \varepsilon] = 0 \tag{4.3}$$

for all $\varepsilon > 0$. This may be written as $\rho_{\mathcal{A}}(Z,W_n) \xrightarrow{P} 0$ where $\rho_{\mathcal{A}}(f,g) = \sup_{A\in\mathcal{A}} |f(A) - g(A)|$ is the uniform metric on $\mathcal{A}$.

Since all moments of a Poisson random variable are finite, it follows from Theorem 1.1 that for all $r \in (0,1)$, $Z_n \xrightarrow{L} Z$. Since $Z_n \in C(\mathcal{A})$, $n \ge 1$, and $Z \in C(\mathcal{A})$, a complete and separable metric space, the convergence in law, $\xrightarrow{L}$, of our previous theorems is defined by saying that $Eg(Z_n) \to Eg(Z)$ for all continuous bounded functions $g\colon C(\mathcal{A}) \to R^1$. For processes such as W_n that do not take values in a separable metric space, we use $\xrightarrow{L}$ to mean that $Eg(Z_n) \to Eg(Z)$ for all continuous bounded real valued functions g for which $g(Z_n)$, $n \ge 1$, and $g(Z)$ are random variables. (cf. Pyke and Shorack, 1968). For W_n we use the supremum metric $\rho_{\mathcal{A}}$ on any approved sample space that contains $C(\mathcal{A})$. Theorem 4.1 and (4.3) then verify the following result.

<u>Theorem 4.2</u> If W_n is defined as the normalized Poisson process of (4.1), then for any $\mathcal{A}$ satisfying A1, A2, and (1.4), $W_n \overset{L}{\to} Z$ with respect to $\rho_{\mathcal{A}}$.

5. An Application to Uniform Empirical Processes

Let $U_n = \{U_n(A): A \in \mathcal{A}\}$ be the Uniform empirical process indexed by $\mathcal{A}$. That is, if $V_1,\ldots,V_n$ are independent Uniform (I^k) random variables and $F_n = n^{-1}(\delta_{V_1}+\ldots+\delta_{V_n})$ is the empirical measure where $\delta_x(A) = 1$ or 0 according as $x \in A$ or $x \notin A$, then $U_n(A) = n^{1/2}(F_n(A) - |A|)$. If N_n denotes a Poisson -n process as in the previous section, define

$$T_n = \inf\{t: N_n(tI^k) > n\} .$$

Since $\{N_n(tI^k): t \geq 0\}$ is a non-homogeneous Poisson process with $E[N_n(tI^k)] = nt^k$, it is clear that $T_n^{1/k}$ is a $\Gamma(n+1: n)$ random variable so that $T_n^{1/k} \overset{P}{\to} 1$ as $n \to \infty$. Also, therefore, $T_n \overset{P}{\to} 1$. (Actually, by the strong law of Hsu and Robbins, we could claim $T_n \xrightarrow{a.s.} 1$ since the fourth moments of Exponential random variables are finite.)

Consider the process U_n^* defined on $\mathcal{A}$ by

$$U_n^*(A) = W_n(T_nA) - |A|W_n(T_nI^k) . \tag{5.1}$$

This process has the same finite dimensional distributions as does U_n, a fact we leave to the reader. To show weak convergence, we use Theorem 4.1 and (4.3) in conjunction with Skorokhod's construction (namely, if $Z_n \overset{L}{\to} Z$ on a complete separable metric space, (M,d) say, there exists equivalent processes, Z_n^*, Z^* say, on a common probability space for which $d(Z_n^*, Z^*) \xrightarrow{a.s.} 0$,) to show that without loss of generality we may assume that $\rho_{\mathcal{A}}(W_n, Z) \overset{P}{\to} 0$. Since $T_n \overset{P}{\to} 1$ it follows that for the strongly convergent versions

$$U_n^*(A) \overset{P}{\to} U(A) := Z(A) - |A|Z(I^k),$$

uniformly in $A \in \mathcal{A}$ where U is the 'tied-down' Brownian process associated with Z. Note that the mapping $A \to tA$ is continuous with respect to d_H. This is enough to establish

Theorem 5. If U_n is the Uniform empirical process and A satisfies A1, A2 and (1.4), then $U_n \overset{L}{\to} U$ with respect ρ_A.

For more direct approaches to this result see Dudley (1978) and Sun and Pyke (1982).

6. Concluding Remarks

I have focused in this paper on large index families satisfying (1.4). For many smaller classes such as the lower orthants or $P_{k,m}$ (see Section 1) the analogue of (1.4) would be

$$\nu(\varepsilon) \le K\varepsilon^{-r} \tag{6.1}$$

for constants $K > 0$ and $r > 0$. For example, for the orthants, $r = k$ and for $P_{k,m}$, $r = km$. In the case of (6.1) it should be possible to modify the arguments of Sections 1 and 2 to establish the uniform Central Limit theorem under the finiteness of second moments only. This problem will be considered separately.

The full Central Limit problem may also be considered, in which non-Gaussian limiting processes would be involved. Since these processes are not in $C(A)$, considerable care must be exercised concerning the discontinuiti near the boundaries of the sets in A, even if one still works with Z_n, the smoothed version of the partial-sum processes. The methods of this paper do not seem to be applicable to the general infinitely divisible situation. They may however be applicable in the possibly more applicable direction of dependent arrays which satisfy appropriate mixing conditions.

* * * * * * * * * *

It is a sincere privilege to dedicate this paper to Professor Kendall on the occasion of his sixty-fifth birthday. I have benefited greatly from his stimulating example in both the written and oral presentations of his research. The sabbatical leave of 1964-65 which I spent at the Statistical Laboratory of the University of Cambridge was an exceptional catalyst for my own research. In particular, my interest in processes with multi-dimensional index sets began about that time while I was establishing the weak convergence of empirical processes under random sample sizes, (cf. Pyke (1968)). The natural approach to this problem involved looking at two-dimensional empirical and Poisson processes, which in turn led to several problems including the one about higher-dimensional versions of Kolmogorov and Skorokhod inequalities resolved by Wichura (1969). (Further generalizations remain open as described in the next paragraph.) Another problem area regarding matrix arrays concerned fluctuation theory for their multidimensional partial-sums. Some of these problems were listed in Pyke (1970); see also Pyke (1973). In this latter paper the limiting Brownian Sheet was discussed and a sample-path investigation begun. A considerable literature now exists on this topic including a study of the contours of Brownian Sheet and Lévy's multidimensionally indexed Gaussian process by Professor Kendall's son, W.S. Kendall (1980). (See the recent monograph by Adler (1981) for other references and an excellent review of some of these topics.)

Open problems have always been an important emphasis of Professor Kendall, as indicated by the two notebooks of open problems which were regularl maintained in the library at Cambridge. Let me therefore conclude this paper by discussing a class of questions relating to generalizations of Kolmogorov and Skorokhod inequalities that pertain to the partial-sums of this paper. The original inequalities of this type pertain to maxima of the form

$$\max\{|S_j|: 1 \le j \le n\};\ S_j = X_1+\ldots+X_j \tag{6.2}$$

where $\{X_j\}$ are independent mean zero random variables. For an array $\{X_{\underset{\sim}{j}} \in \underset{\sim}{J}^k\}$ of independent mean zero random variables a natural extension of (6.2) would be

$$\max\{|S_{\underset{\sim}{j}}|: \underset{\sim}{1} \le \underset{\sim}{j} \le \underset{\sim}{n}\};\ S_{\underset{\sim}{j}} = \Sigma\{X_{\underset{\sim}{i}}: \underset{\sim}{i} \le \underset{\sim}{j}\}\ . \tag{6.3}$$

It was the question about maximal inequalities for these 'orthant' partial-sums that was resolved by Wichura (1969). The proofs of maximal inequalities for (6.2) rely on the linear ordering of the index set and focus on first-occurrence times. For (6.3), the one-dimensional methods are extended to the multidimensional case through induction, but this depends essentially on the rectangular shape of the orthants. There has been no success in obtaining inequalities for maxima over other regions; for example, in the 2-dimensional case it is not even known how to bound the maxima over lower triangles or lower half-planes.

In the notation of this paper, the general open question can be phrased in terms of finding bounds for probabilities of the form

$$P[\max\{|S(nA)|: A \in \mathcal{A}\} > \varepsilon]$$

for general families $\mathcal{A}$. Noting however that each partial sum is determined by an assignment of 0's and 1's to the given summands, one can more simply state the problem as follows: Let $X_1, X_2, \ldots, X_n$ be independent mean-zero random variables and let $\mathcal{J}$ be a class of subsets of $I_n = \{1,2,\ldots,n\}$. Find bounds for

$$P[\max_{J \in \mathcal{J}} |S(J)| > \varepsilon] \tag{6.4}$$

where $S(J) = \Sigma\{X_j : j \in J\}$. In particular, do constants $C_1(\mathcal{J})$ or $C_2(\mathcal{J})$ exist for which the probability in (6.4) is bounded above by

$$C_1(\mathcal{J}) \frac{\mathrm{var}(S(I_n))}{\varepsilon^2} \quad \text{or } C_2(\mathcal{J}) P[|S(I_n)| > \varepsilon] \ ?$$

This open question has been considered by several people over the past 15 years or so with little success. It has also arisen in contexts other than weak convergence. In particular, the problem was published for the first time in this form by Hanson, Pledger and Wright (1973, Section 5), it having arisen from their investigations into multidimensional regression.

There is a further extension of the problem that fits in with the context of this paper, and which was discussed in Pyke (1977). What was needed in the latter paper was a generalized Kolmogorov inequality for weighted sums with coefficients not restricted to 0's and 1's. To see what is intended, I will describe a special 3-dimensional case that will be more easily understood if the reader has a tiled floor (or ceiling) at which to gaze. Associate with each square one of a collection of independent random variables $\{X_{\underset{\sim}{j}}\}$. Consider the floor to be part of the boundary surface of one of the sets $A \in \mathcal{A}$. Consider the floor also to be one of the separating planes between rows of cubes $C_{\underset{\sim}{j}}$. Thus each tile serves simultaneously as the top and bottom of two adjacent cubes. Now, consider a perturbation of the floor by allowing some of the tiles to bulge upward smoothly. View the resulting bumpy floor as a part of the surface of another set $B \in \mathcal{A}$ that is close to A. Form a partial sum $\Sigma q_{\underset{\sim}{j}} X_{\underset{\sim}{j}}$ where the constants $q_{\underset{\sim}{j}}$ associated with the tiles are zero when their tiles are not bulged and elsewhere measure somewhat the amount of bulge exhibited. Even though the class $\mathcal{A}$ might impose bounds on the smoothness of the surfaces of its sets B, (as does $\mathcal{A}_\alpha$), it is clear that *every* subset of the tiles may still be realizable as a set of bulged tiles! However, the larger and more disconnected a subset is, the less pronounced must be the bulges, and hence the smaller must be the constants $|q_j|$. The open problem may be stated more precisely as follows: *Find upper bounds for*

$$P[\sup_{q \in Q}|\Sigma_{i=1}^{n} q_i X_i| > \varepsilon] \quad (6.5)$$

for independent mean zero $X_1,\ldots,X_n$ *and a subset* $Q \subset I^n$.

Each of the previous questions was a special case of (6.5) in which Q was a subset of the vertices of the unit cube I^n. In this general statement the coefficients q_i are not restricted to being 0's and 1's.

REFERENCES

Adler, R.J. (1981). *The Geometry of Random Fields*. John Wiley and Sons, New York.

Bennett, G. (1962). Probability inequalities for the sum of independent random variables. *J. Amer. Statist. Assoc. 57* 33-45.

Billingsley, P. (1968). *Convergence of Probability Measures*. Wiley, New York.

Debreu, G. (1967). Integration of correspondences. *Proc. Fifth Berkeley Symp. Math. Stat. Prob.*, Vol. II, Part 1, 351-372. University of California Press, Berkeley.

Dudley, R.M. (1973). Sample functions of the Gaussian process. *Ann. Prob. 1* 66-103.

Dudley, R.M. (1974). Metric entropy of some classes of sets with differentiable boundaries. *J. Approx. Th. 10* 227-236.

Dudley, R.M. (1978). Central limit theorems for empirical measures. *Ann. Prob. 6* 899-929.

Erickson, R.V. (1981). Lipschitz smoothness and convergence with applications to the central limit theorem for summation processes. *Ann. Prob. 9* 831-851.

Hanson, D.L., Pledger, G. and Wright, F.T. (1973). On consistency in monotonic regression. *Ann. Statist. 1* 401-421.

Kendall, W.S. (1980). Contours of Brownian processes with several dimensional time. *Z. Wahrschein. v. Geb. 52* 267-276.

Kuelbs, J. (1968). The invariance principle for a lattice of random variables. *Ann. Math. Statist. 39* 382-389.

Pyke, R. (1968). The weak convergence of the empirical process with random sample size. *Proc. Cam. Phil. Soc. 64* 155-160.

Pyke, R. and Shorack, G. (1968). Weak convergence of a two-sample empirical process and a new approach to Chernoff-Savage theorems. *Ann. Math. Statist. 39* 755-771.

Pyke, R. (Editor), (1970). *Proceedings of the Twelfth Biennial Seminar, Canadian Mathematical Congress, Montreal*.

Pyke, R. (1977). The Haar-function construction of Brownian Motion indexed by sets. Tech. Rpt. No. 35, University of Washington, NSF Grant MCS75-08557.

Révész, P. (1976). Three theorems of multivariate empirical process. *Lecture Notes in Math. Empirical Distributions and Processes. No. 566,* 106-126. Springer-Verlag, Berlin.

Strassen, V. and Dudley, R.M. (1969). The central limit theorem and ε-entropy. *Proc. Int'l Symp. Prob. and Inform. Th.* Lecture Notes in Mathematics, *89,* 224-231. Springer-Verlag, New York.

Sun, T.G. and Pyke, R. (1982). Weak convergence of empirical processes. To appear.

Wichura, M.J. (1969). Inequalities with applications to the weak convergence of random processes with multidimensional time parameters. *Ann. Math. Statist. 40* 681-687.

MULTIDIMENSIONAL RANDOMNESS

Brian D. Ripley

In my early days as a research student I was both impressed and influenced by David Kendall's enthusiasm for trying out ideas using computer-generated simulations. (His particular project at that time culminated in Kendall, 1974.) This can provide a bridge between what John Tukey calls "exploratory data analysis" and the more classical "confirmatory" aspects of statistics, model estimation and testing. Comparison of data with simulations is definitely "confirmatory" in that models are involved, yet it has much of the spirit of the "exploratory" phase, with human judgement replacing formal significance tests. Later I formalized this comparison to give Monte Carlo tests, discovered earlier by Barnard (1963) but apparently popularized by Ripley (1977).

Simulation has enabled progress to be made in the study of spatial patterns and processes which had previously seemed intractable. The starting point for all known simulation algorithms for spatial point patterns and random sets, such as those in Ripley (1981), is an algorithm to give independent uniformly distributed points in a unit square or cube. The purpose of this paper is to discuss the properties of these basic algorithms. Much of the material can be found scattered in the literature, but no one source gives a sufficiently complete picture.

CONGRUENTIAL RANDOM NUMBER GENERATORS

The usual way to produce approximately uniformly distributed random variables on (0,1) in a computer is to sample integers x_i uniformly from $\{0,1,\ldots,M-1\}$ or $\{1,2,\ldots,M-1\}$ and set $U_i = x_i/M$. Here M is a large integer, usually of the form 2^β. The approximation made is comparable with that made to represent real numbers by a finite set. *Congruential* generators generate the sequence (x_i) by

$$x_{i+1} = \{ax_i + c\} \bmod M$$

The choice of M as a power of 2 makes the modulus operation very easy on a binary computer. We will assume that this is the case. (Byte-based machines often choose M a power of 2^8.) For example, on our Cyber 174, $M = 2^{48}$, $c = 0$ and $a = 9,301,337,289,077$. Such sequences obviously have a period at most M; the period is M if and only if c is odd and $a \equiv 1 \bmod 4$. (This is proved by elementary number-theory arguments in Jansson, 1966, and Knuth, 1981.) Generators with c odd are said to be *mixed*. The other class commonly considered are *multiplicative* generators with $c = 0$ and $a \equiv 5 \bmod 8$. The study of these can be reduced to that of mixed generators. Let x_0 be the smallest value in the sequence. Then

$$(x_i - x_0) = a(x_{i-1} - x_0) + (a-1)x_0 \mod M$$

so by induction $x_i - x_0$ is a multiple of 4, say $4y_i$. Then

$$y_i = ay_{i-1} + [(a-1)/4]\, x_0 \mod M/4$$

which has period M/4 since $(a-1)/4$ is odd by hypothesis. Thus U_i's from this generator differ from those of the mixed generator with modulus M/4, the same a and $c = (a-1)x_0/4$, by a shift of x_0/M.

We now confine our attention to mixed genrators of full period M. These generate a sequence which is some permutation of $\{0,1,\dots,M-1\}$.

2 dimensions

The obvious way to generate points uniformly in the unit square is to take (U_0,U_1), (U_2,U_3),.... Figures 1 and 2 show examples from multiplicative congruential generators. Clearly there are M/2 pairs before the sequence repeats. The period of these vectors is that of the first component, for it alone determines the other component(s). Now

$$x_{2i} = a^2 x_{2i-2} + (a+1)c \mod M$$

and since $(a+1)c$ is a multiple of 2 but not of 4, x_{2i} has period M/2 by the argument above. Again take $x_0 = \min\{x_{2i}\}$ (0 or 1) and let $y_i = (x_{2i} - x_0)/2$. Then (y_i) takes all the values in the set $I = \{0,1,\dots,M/2-1\}$ and (x_{2i}, x_{2i+1}) takes all values in the set

$$\{(x_0 + 2s,\ 2as \oplus (c+ax_0)) \mid s \in I\}$$

Figure 1. (a) All 2048 pairs from the multiplicative generator with a = 125 and $M = 2^{14}$. (b) The first 200 pairs. Note the alignments.

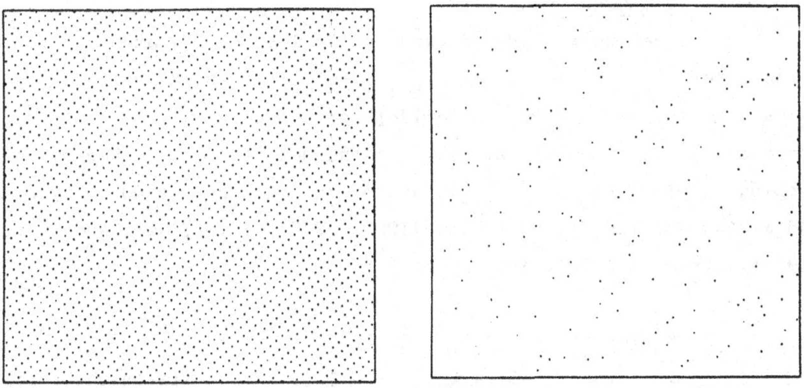

Figure 2. As Figure 1(a) with a = 29.

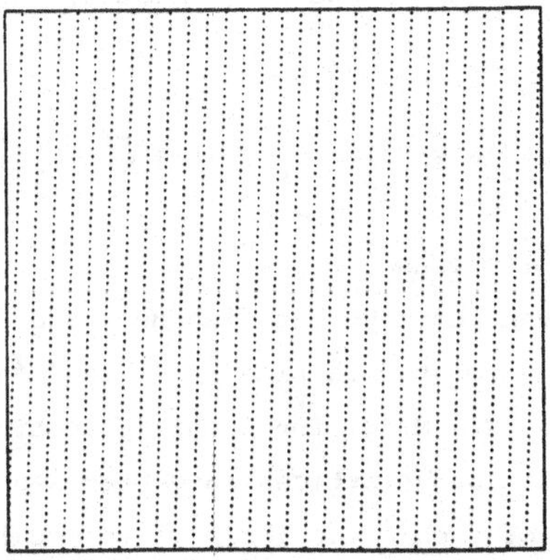

where $\oplus$ denote addition modulo M/2. This is the intersection with I^2 of of the shifted lattice

$$(x_0, x_1) + \{(2s,\ 2as + tM/2) \mid s,t \text{ integer}\}$$

If we now relate this to the points in the unit square, (U_{2i}, U_{2i+1}), we see that these take all values in the intersection of the unit square with

$$(U_0,U_1) + \{(2s/M,\ t + 2as/M) \mid s,t \text{ integer}\}$$

3 dimensions

The theory for 3 dimensions is less complicated since 3 is not a factor of M, and hence $(x_{3i},x_{3i+1},x_{3i+2})$ take M values, in fact all M values $(s,$ as $\oplus$ c, $a^2s \oplus \{a+1\}c)$ for $s = 0,1,\ldots,M-1$ where $\oplus$ denotes additional modulo M. This is the set of values taken by the overlapping triples (x_i,x_{i+1},x_{i+2}) and as such has been studied by Coveyou and MacPherson (1967), Marsaglia (1968, 1972), Beyer, Roof and Williamson (1971), Atkinson (1980) and Knuth (1981).

k dimensions

In general $(U_{ki}, U_{ki+1},\ldots,U_{k(i+1)-1})$ lie on the shifted lattice

$$(U_0,U_1,\ldots,U_{k-1}) + \{(\ell s/M, t_1+a\ell s/m,\ldots,t_{k-1}+a^{k-1}\ell s/m) \mid s,t_i \text{ integers}\}$$

where $\ell = \gcd(k,M)$. The proof generalises the arguments used for 2 and 3 dimensions. Thus if $N = M/\ell$ and $a^* = a \bmod N$, the k-tuples lie on a shift of the lattice

$$\Lambda = \{(s/N, t_1+a^*s/N,\ldots,t_{k-1}+(a^*)^{k-1}s/N) \mid s,t_i \text{ integer}\}$$

ANALYSING LATTICES

We have seen that the k-tuples generated by a congruential generator lie on a shift of the lattice Λ with $a^* = a \bmod N$, and $N = M/\ell$ for a mixed generator, $N = M/4\ell$ for a multiplicative generator, and $\ell = \gcd(k,M)$, and that such lattices have been studied for overlapping k-tuples. The 'cells' of such a lattice have volume $1/N$, so a side is of the order of $N^{-1/k}$. For a mixed generator with $\beta = 16$, this is 0.0055 (k=2) 0.025 (k=3) and 0.088 (k=4). To put these figures into perspective consider the minimum distance d between any pair from n independent uniformly distributed points in $[0,1]^k$. To a very good approximation, d^k has an exponential distribution with mean $\omega_k/[n(n-1)]$, where ω_k is the content of the unit ball in R^k (Ripley and Silverman, 1978, Silverman and Brown, 1978). This will be significantly larger than the order of a cell side if $N \geq 10\,n(n-1)$. For a multiplicative generator with $\beta = 16$, k=2, this implies $n \le 30$.

These figures will flatter the generator if the lattice cells have sides of rather disparate lengths or rather acute angles (as in Figure 2). Thus it is necessary to find the actual cells. We say vectors $(\underline{e}_1,\ldots,\underline{e}_k)$ form a *basis* of the lattice Λ if $\Lambda = \{t_1\underline{e}_1+\ldots+t_k\underline{e}_k \mid t_i \text{ integer}\}$. A basis is a *physical basis* if

(i) $\underline{e}_1$ is the shortest vector in Λ,

(ii) $\underline{e}_j$ is the shortest vector in Λ linearly independent of $\underline{e}_1,\ldots,\underline{e}_{j-1}$ for $2 \le j \le k$.

Thus $\underline{e}_1,\ldots\underline{e}_k$ are the sides of what one thinks of as the sides of a basic lattice cell in order of lengths, $|\underline{e}_i|$. We are particularly interested in $|\underline{e}_k|$ as a measure of the 'granularity' of the k-tuples. Physical bases are not unique; at least any $\underline{e}_i$ can be replaced by $-\underline{e}_i$. The problem of finding a good basis also occurs in number theory. Minkowski (1887) defined a basis as *reduced* if, for $1 \le j \le k$, $\underline{e}_j$ is the shortest of those vectors of the form $\sum_1^n t_i\underline{e}_i$ with $\gcd(t_j,\ldots,t_n) = 1$, and showed that, for $k \le 4$,

(1) a reduced basis is a physical basis,

(2) a basis is reduced if for every $\underline{e}_j$, $|\underline{e}_j| \le |\sum_1^n c_\ell \underline{e}_\ell|$ where $c_j = 1$ and the other $c_\ell \in \{-1,0,1\}$.

This provides a test for a basis to be a physical basis. (1) and (2) are thought to be false for $k \ge 5$.

One basis for Λ is obvious from its definition,

$$\underline{e}_1^T = (1,a,a^2,\ldots,a^{k-1})/N$$

$$\underline{e}_j = \text{jth unit vector} \qquad 2 \le j \le k$$

Obviously the powers of a in $\underline{e}_i$ can be reduced mod N. Two steps are available to 'improve' a basis:

Reduction step M. If $|\underline{e}_i| < |\underline{e}_j|$, consider replacing $\underline{e}_j$ by $\underline{e}_j - s\underline{e}_i$ for an integer s. The revised set of vectors will still be a basis. Choose s to make the new vector as short as possible by $s = \underline{\text{nint}}\,(\underline{e}_i^T\underline{e}_j/|\underline{e}_i|^2)$, where nint denotes 'the nearest integer to', halves being rounded towards zero. (This is a standard function in FORTRAN 77.)

Reduction step BRW. For $\underline{e}_i$, examine $|\sum c_\ell \underline{e}_\ell|$ with c_ℓ as in (2). If any choice gives a length less than $|\underline{e}_i|$, replace $\underline{e}_i$ by that linear combination. If there is more than one suitable combination, choose one as short as possible.

Marsaglia (1972) suggested (for overlapping k-tuples) applying step M to each pair (i,j) in some cyclic order until no further reductions are possible. This suggestion also occurs in Beyer, Roof and Williamson (1971), who give step BRW. Marsaglia asserts that his algorithm finds a "nearly optimal basis" although he does not state what he means by optimal. Atkinson (1980) appears to assume that this algorithm gives a physical basis. This is in general incorrect. Minkowski's test (2) shows that step M suffices for $k = 2$, but it can fail for $k \geq 3$. For example, step M applied to the pairs in the order $\{1,2\}, \{1,3\}, \ldots \{1,k\}, \{2,3\}, \{2,4\}, \ldots$ with $k = 3$, $a = 71365$, $N = 2^{32}$ finds the basis

$$\underline{e}_1 = (\ 662015,\quad 60219,\quad 2561639)/N$$
$$\underline{e}_2 = (\ 601869,\quad 2708225,\quad -1051195)/N$$
$$\underline{e}_3 = (1564727,\quad -2407341,\quad -1198625)/N$$

whereas step BRW shows that $\underline{e}_3$ can be replaced by the shorter $\underline{e}_1 + \underline{e}_2 + \underline{e}_3$.

Knuth (1981) also uses step M, but modifies the algorithm to provide a basis for $k = 2,3,4,\ldots$ in turn, applying the transformations used at dimension $k-1$ to the initial basis for dimension k before continuing with step M. This he suggests will make the algorithm more efficient, which is strongly contradicted by my experience. It does seem more likely to find a physical basis. It fails for $k = 4$, $a = 100485$ and $N = 2^{32}$ whereas the direct use of step M alone finds the physical basis. For $k \geq 5$ failures of both Marsaglia's method and Knuth's modification seem common (7 of the 12 values given by Atkinson (1980) for $k = 5$ in his Table 3 refer to bases which can be improved by step BRW). Unfortunately test (2) is invalid for $k \geq 5$, and the computation in step BRW is proportional to $k(3^{k-1}-1)$, which rapidly becomes excessive.

The *spectral test* of Coveyou and MacPherson (1967) finds the maximally spaced set of hyperplanes which contain Λ. This is another measure of 'granularity' closely related to $|\underline{e}_k|$. Knuth (1981) finds this by reducing a basis as far as possible with step M, taking the 'dual basis' of hyperplanes determined by the subsets of size (k-1) of basis vectors, then doing an exhaustive search which he states has never found a better basis in all the cases he has examined.

CONCLUSIONS

These considerations suggest that it is very dangerous to use any congruential generator with a period of less than 2^{20} to simulate spatial patterns in the obvious way. Those with a period of around this value should have their lattices checked carefully. For considerably larger moduli, say at least 2^{30}, all but the worst generators would seem satisfactory. One should be particularly suspect of random number generators in programmable calculators and microcomputer compilers whose details are often unspecified but seem to be based on at most 16 bits. The technique of applying a fixed permutation to blocks of *prime* length of the output of a generator with a short period to extend its period seems satisfactory, but its properties seem never to have been investigated thoroughly.

REFERENCES

Atkinson, A.C. (1980) Tests of pseudo-random numbers. *Applied Statistics* 29, 164-171.

Barnard, G. (1963) Contribution to the discussion of Bartlett's paper. *J. Roy. Statist. Soc.* B25, 294.

Beyer, W.A., Roof, R.B. and Williamson, D. (1971) The lattice structure of multiplicative congruential pseudo-random vectors. *Mathematics of Computation* 25, 345-363.

Coveyou, R.R. and MacPherson, R.D. (1967) Fourier analysis of uniform random number generators. *J. Assoc. Comput. Mach.* 14, 100-119.

Jansson, B. (1966) *Random Number Generators.* Almquist & Wicksell; Stockholm.

Kendall, D.G. (1974) Pole-seeking Brownian motion and bird navigation. *J. Roy. Statist. Soc.* B36, 365-402.

Knuth, D.E. (1981) *The Art of Computer Programming, volume 2, 2nd edition.* Addison-Wesley; Reading, Mass.

Marsaglia, G. (1968) Random numbers fall mainly in the planes. *Proc. Nat. Acad. Sci. U.S.A.* 61, 25-28.

Marsaglia, G. (1972) The structure of linear congruential sequences. In: *Applications of Number Theory to Numerical Analysis* ed. S.K. Zaremba, Academic Press; New York pp 249-285.

Minkowski, H. (1887) Zur Theorie der positiven quadratischen Formen. *J. Reine Angew. Math.* 101, 196-202.

Ripley, B.D. (1977) Modelling spatial patterns. *J. Roy. Statist. Soc.* B39, 172-212.

Ripley, B.D. (1981) *Spatial Statistics.* Wiley; New York.

Ripley, B.D. and Silverman, B.W. (1978) Quick tests for spatial regularity. *Biometrika* 65, 641-642.

Silverman, B.W. and Brown, T.C. (1978) Short distances, flat triangles and Poisson limits. *J. Appl. Probab.* 15, 815-825.

SOME PROPERTIES OF A TEST FOR MULTIMODALITY BASED ON KERNEL DENSITY ESTIMATES

B.W. Silverman

It is a great pleasure for me to have this opportunity to contribute a paper in honour of David Kendall. It was through David Kendall's lectures, writings and personal communications that I first became interested in density estimation and the other matters discussed in this paper. He is a great scientist and a great teacher and my debt to him is enormous. I wish him a very happy birthday!

1. Introduction

Silverman (1981) suggested and illustrated a way that kernel probability density estimates can be used to investigate the number of modes in the density underlying a given independent identically distributed real sample. Given an independent sample $X_1,\dots X_n$ from a univariate probability density f, define the *kernel density estimate* f_n with Gaussian kernel by

$$f_n(t,h) = \sum_{i=1}^{n} n^{-1}h^{-1}\phi\{(t-X_i)/h\} \quad ,$$

where the parameter h is the smoothing parameter or *window width* and ϕ is the standard normal density function. Kernel density estimates were introduced by Rosenblatt (1956) and Parzen (1962); the restriction to Gaussian kernels in this work is made for reasons given in Silverman (1981). Often the explicit dependence of f_n on h will be suppressed.

Consider the problem of testing the null hypothesis that f has k or fewer modes against the alternative that f has more than k modes. The statistic suggested for constructing such a test was the *k-critical window width* $h_{crit}(k)$, defined by

$$h_{crit}(k) = \inf\{h : f_n(\cdot,h) \ \text{ has at most } k \text{ modes}\} \quad .$$

In Silverman (1981) it was stated heuristically that large values of h_{crit} will tend to reject the null hypothesis. The results of this paper show that this procedure does indeed lead to a consistent test.

Subject to certain regularity conditions, it is shown that, under the null hypothesis, h_{crit} converges stochastically to zero, while this is not the case under the alternative hypothesis. The exact rate of convergence of h_{crit} to zero under the null hypothesis is found. It is perhaps interesting that this rate of convergence has precisely the same order as the rate of convergence for the optimum choice of window width for the uniform estimation of the density given, for example, by Silverman (1978b).

In Silverman (1981) a bootstrap procedure for assessing the significance of an observed value of h_{crit} was suggested. The idea of bootstrap methods in general is to construct a null hypothesis or model from the given data, rather than supplying it *a priori*. In our case the representative of the null hypothesis is constructed by smoothing the data up to the point where a density with k modes is just obtained; the resulting density is, of course, $f_n(.,h_{crit})$ and so has already been found! Simulating from $f_n(.,h_{crit})$ is straightforward; choose X uniformly (with replacement) from the original data and then add a perturbation ε to X, where ε is normally distributed with mean zero and variance h^2_{crit}.

Further details, together with an application, are given in Silverman (1981). It is interesting and apposite to draw some connections with David Kendall's work. Our technique is actually an example of a "smoothed bootstrap" technique as described by Efron (1979); there, however, the choice of smoothing parameter is entirely arbitrary. A somewhat related technique was used by Kendall and Kendall (1980) in their investigation of alignments in archaeological data. The 'over-unrounding' technique they used in their Section 5 was again a way of preserving the coarse structure of the data while erasing fine details which are only 'really' present if the null hypothesis is rejected. The over-unrounding technique differs from the smoothed bootstrap technique in several details, one of which is that the data points are no longer sampled with replacement and then perturbed, but instead *each* data point is perturbed by a random amount; thus the sampling is conducted without replacement. Another piece of data analysis worth mentioning in this context is D.G. Kendall's work on the megalithic yard (Kendall, 1974). Here again a density constructed from the data was used as a basis for simulation, with the aim of preserving the coarse features but not the fine structure of the data, though in the megalithic yard study a parametric model sufficed for this purpose.

A most interesting feature of the over-unrounding procedure used by Kendall and Kendall (1980) is that the parameter controlling the amount of random unrounding (or smoothing) is not chosen arbitrarily. Instead, a criterion *based on the phenomenon being studied* is used to choose how much to unround the data; in an informal way, the procedure seeks the smallest value of the unrounding parameter among those acceptable according to this criterion. This is of course also the case in the method studied in the present paper.

The remarks made above about the rate of convergence of h_{crit} to zero show that $f_n(.,h_{crit})$ is, in a certain sense, optimally uniformly consistent as an estimate of the true density f. This gives some theoretical justification for the bootstrap procedure, since, at least for large samples, the simulation density $f_n(.,h_{crit})$ is likely to be a good estimate (in the uniform norm) of the true underlying density. A possible drawback for small samples is the fact that the implied constant in the rate of convergence does not necessarily take its optimum value.

An interesting open question raised by this discussion is the possibility of using $h_{crit}(k)$ for some value of k in developing an automatic method for choosing the smoothing parameter in density estimation. Boneva, Kendall and Stefanov (1971) suggested choosing the window width where 'rabbits' or rapid fluctuations just started to appear. Such a window width would perhaps correspond to $h_{crit}(k)$ for some $k \geq j$; since $h_{crit}(k)$ converges to zero at the optimum rate for *all* $k \geq j$, a suitable formalization of the Boneva-Kendall-Stefanov procedure would give estimates which converged at the optimal rate, though not necessarily with the optimal constant multiplier. The fact that $h_{crit}(k)$ has the same rate of convergence for all $k \geq j$ provides some explanation for the observation made by Boneva, Kendall and Stefanov that the estimate seems suddenly to become noisy as the window width is reduced.

The use of kernel density estimates in mode estimation was originated by Parzen (1962). The 'gradient method' of cluster analysis is based on clustering towards modes in the estimated density; see, for example, Andrews (1972), Fukunaga and Hostetler (1975), and Bock (1977). Papers related to tests of multimodality are Cox (1966) and Good and Gaskins (1980).

2. The main result

In this section, the main result of this paper is stated and

and proved. It is convenient to use the convention throughout that all limits and implied limits are taken as n tends to infinity. Varying conventions will apply to unqualified suprema and infima in Propositions 1 and 2 below, and these will be introduced where necessary. The notations p lim inf and p lim sup will be used to signify the corresponding limits in probability as n tends to infinity, and O_p and o_p will denote probability orders of magnitude. Define, for $h > 0$,

$$\alpha(h) = h^{-5} \log(h^{-1}) \quad . \tag{1}$$

The main results are all contained in the following theorem.

Theorem

Suppose f *is a bounded density with bounded support* $[a,b]$, *and suppose that the following conditions are satisfied:*

(i) f *is twice continuously differentiable on* $[a,b]$

(ii) f *has exactly* j *local maxima on* (a,b)

(iii) $f'(a+) > 0,\ f'(b-) < 0$

(iv) $\min_{z:f'(z)=0} \dfrac{f''(z)^2}{f(z)} = c_0 > 0.$

Let $h_{crit}(k)$ *be the* k*-critical window width constructed from an i.i.d. sample of size* n *from* f. *Then, if* $k \geq j$, *defining* α *as in* (1) *above,*

$$\text{p lim inf } n^{-1}\alpha\{h_{crit}(k)\} \geq \frac{2}{3}\pi\sqrt{2}\, c_0 \tag{2}$$

and

$$\text{p lim sup } n^{-1}\alpha\{h_{crit}(k)\} < \infty \tag{3}$$

while if $k < j$ *then there exists a constant* $h_0(f,k)$ *such that*

$$P\{h_{crit}(k) > h_0\} \to 1 \quad . \tag{4}$$

Note that condition (iv) is equivalent, in the presence of the other conditions, to the condition that f is strictly positive on $[a,b]$ and f' has no multiple zeros on $[a,b]$.

It is convenient to prove the various assertions of the theorem separately. Except where otherwise stated, the conditions of the theorem on f will be assumed to be true throughout. The first proposition facilitates the proof of (2).

Proposition 1. *Given any* c_1 *with*

$$0 < c_1 < \frac{2}{3}\pi\sqrt{2}\, c_o \quad ,$$

suppose the sequence of window widths h_n *satisfies*

$$n^{-1}\alpha(h_n) \to c_1 \quad . \tag{5}$$

Then the number of maxima of f_n *tends in probability to* j.

It follows from Proposition 1 and Silverman (1981) that, for all $k \geq j$, provided (5) holds,

$$P\{h_{crit}(k) \leq h_n\} \to 1$$

and hence that (2) is satisfied.

The proof of Proposition 1 makes use of several lemmas, the first of which shows that, under certain conditions, maxima and minima of f_n can, eventually, only occur arbitrarily close to those of f.

Lemmà 1. *Let* I *be any closed interval contained in* $[a,b]$, *such that* I *contains none of the zeros of* f'. *Then, provided* $h_n \to 0$ *and* $n^{-1}h_n^2\alpha(h_n) \to 0$, *it will follow that*

$$P(f_n \text{ monotonic on } I \text{ in the same sense as } f) \to 1 \quad .$$

Proof. By slight adaptation of the results of Silverman (1978a), it can be seen that, provided f is bounded, we will have, if h_n satisfies the assumptions of Proposition 1,

$$\sup_{(-\infty,\infty)} |f_n' - Ef_n'| = \underline{o}_p\{n^{-\frac{1}{2}} h_n^{-1}\alpha(h_n)^{\frac{1}{2}}\}$$

$$= \underline{o}_p(1) \quad . \tag{6}$$

In Silverman (1978a) the uniform continuity of f was additionally assumed, but careful examination of the proofs of that paper shows that the derivation of the rate of stochastic convergence, though not of the exact constant implied in the $\underline{o}_p$, goes through under the assumption of bounded f.

Supposing without loss of generality that f is increasing on I, it follows from the continuity of f' on $[a,b]$ that f' is bounded away from zero on I and is non-negative on a neighbourhood of I, and hence by elementary analysis that

$$\liminf \inf_I Ef_n' > 0 \quad . \tag{7}$$

Combining (6) and (7) completes the proof of Lemma 1.

The next lemma shows that, under suitable conditions, f_n will eventually have exactly one maximum and no minima near each maximum of f, and exactly one minimum and no maxima near each minimum of f.

Lemma 2. *Suppose* $f'(z) = 0$ *and* f *has a local maximum (respectively minimum) at* z. *Suppose* $h_n \to 0$ *and*

$$n^{-1}\alpha(h_n) \to c_2 \in (0, \tfrac{2}{3}\pi\sqrt{2}\, f''(z)^2/f(z)) \quad . \tag{8}$$

Then, for all sufficiently small $\varepsilon > 0$, *the probability that* f_n' *has exactly one zero in* $(z-\varepsilon, z+\varepsilon)$, *and that this zero is a maximum (respectively minimum) of* f_n, *tends to one as* n *tends to infinity.*

Proof. Only the case of a local maximum will be considered. The proof for a minimum proceeds very similarly and is omitted. Throughout this proof unqualified infima and suprema will be taken to be over x in $[z-\varepsilon, z+\varepsilon]$. By the continuity of f and f", choose ε sufficiently small that

$$\frac{\inf f''(x)^2}{\sup f(x)} > \frac{3c_2}{2\pi\sqrt{2}} \tag{9}$$

and also $[z-\varepsilon, z+\varepsilon] \subseteq (a,b)$. It is then immediate that $f'(z-\varepsilon) > 0$ and $f'(z+\varepsilon) < 0$ since, by (9), f" cannot cross zero in $(z-\varepsilon, z+\varepsilon)$. Since f' is continuous at $z \pm \varepsilon$, by standard results on the consistency of f_n' (a combination of Parzen (1962) and Bhattacharya (1967))

$$P\{f_n'(z-\varepsilon) > 0 \text{ and } f_n'(z+\varepsilon) < 0\} \to 1 \quad . \tag{10}$$

Very slightly adapting the proofs of Silverman (1976 and 1978a) to cope with the fact that f" is only uniformly continuous on a neighbourhood of $[z-\varepsilon, z+\varepsilon]$ gives

$$n^{-\frac{1}{2}}\alpha(h)^{\frac{1}{2}} \sup|f_n''(x) - Ef_n''(x)| \overset{p}{\to} K_1$$

where

$$K_1^2 = 2 \sup f \int \phi''^2$$

$$= 3(2\pi\sqrt{2})^{-1} \sup f \quad .$$

Since, by elementary analysis, $\sup|Ef_n''(x) - f''(x)|$ converges to zero, it follows from (8) that $\text{p} \lim_n \sup \sup|f_n''(x) - f''(x)| \le K_1 c_2^{\frac{1}{2}}$

$$< \inf|f''(x)|$$

by (9). It is immediate that

$$P\{f_n''(x) < 0 \quad \text{for all} \quad x \quad \text{in} \quad [z-\varepsilon,\ z+\varepsilon]\} \to 1 \quad . \tag{11}$$

Combining (10) and (11) completes the proof of Lemma 2.

To complete the proof of Proposition 1, note first that no maxima of f_n can occur outside the interval (a,b). Let $z_1,\ldots,z_{2j-1}$ be the zeros of f' in (a,b) and choose ε sufficiently small to satisfy the conclusion of Lemma 2 for all z_i and to ensure that

$$a < z_1-\varepsilon < z_1+\varepsilon < z_2-\varepsilon < \cdots < z_{2j-1}+\varepsilon < b \quad . \tag{12}$$

Applying either Lemma 1 or Lemma 2 as appropriate to each of the intervals in the partition (12) of the interval (a,b) completes the proof of Proposition 1.

The next proposition leads to the proof of assertion (3), in a similar way to the derivation of (2) from Proposition 1.

Proposition 2. *Defining* α *as in* (1) *above, suppose that*

$$n^{-1}\alpha(h_n) \to \infty \quad \textit{and} \quad n^{-1}h_n^{-5} \to 0 \ . \tag{13}$$

Then the number of maxima in f_n *tends in probability to infinity.*

Given any k, it follows from this result and the corollary of Silverman (1981) that, provided (13) holds,

$$P\{h_{crit}(k) > h_n\} \to 1 \quad ;$$

assertion (2) follows at once.

To prove Proposition 2, suppose without loss of generality that f has a maximum at 0 in (a,b). Choose a sequence ℓ_n which satisfies

$$\ell_n \to 0, \quad h_n^{-1}\ell_n = \underline{o}\{n^{-1}\alpha(h_n)\} \quad ,$$

$$h_n^{-1}\ \ell_n \to \infty \quad \text{and} \quad |\log \ell_n| \ \ |\log h_n|^{-1} \to 1 \quad . \tag{14}$$

The explicit dependence of h and ℓ on n will often be suppressed. Let $I_{j,n}$ be the interval $[(j-1)\ell,\ j\ell]$ for integer $j \geq 0$.

Following Silverman (1978a) apply Theorem 3 of Komlos, Major and Tusnady (1975) to obtain

$$f_n'(x) = Ef_n'(x) + h^{-1}n^{-\frac{1}{2}}\ \rho_1(x) + \varepsilon_n'(x)$$

where ρ_1 is a Gaussian process with the same covariance structure as $n^{\frac{1}{2}}h(f_n' - Ef_n')$ and ε_n' is a secondary random error. The process ρ_1 is obtained by putting $\delta(u)$ equal to $\phi'(u)$ in Proposition 1 of Silverman (1978a). By elementary analysis and the arguments of Silverman (1978a) we have, in a neighbourhood of 0,

$$|Ef_n'(x) - f'(x)| = \underline{O}(h) \quad ;$$

$$|\varepsilon_n'(x)| = \underline{O}(n^{-1}h^{-2}\log n) \qquad \text{a.s.}$$

$$= \underline{o}(h^2) \quad \text{from (13) above} \quad ;$$

and $\quad |f'(x)| = \underline{O}(x) \quad ,$

since $f'(0) = 0$ and f'' exists. It follows that, a.s.,

$$\sup|Ef_n'(x) + \varepsilon_n'(x)| = \underline{O}(j\ell) + \underline{O}(h)$$

$$= \underline{o}\{n^{-1}h^{-5}\log(\ell/h)\}^{\frac{1}{2}} \tag{15}$$

by (13) and (14) above, where we adopt the convention, here and subsequently in this proof, that unqualified suprema are taken to be over the interval $I_{j,n}$, and that a fixed j is being considered.

We slightly adapt the argument of Silverman (1976) pp. 138-140 to investigate $\sup \rho_1$. Define

$$\sigma^2(x) = \text{var } \rho_1(x) = h^{-1} f(x) \int \phi'^2 (1 + \underline{o}(1))$$

$$= h^{-1} f(0) \int \phi'^2 (1 + \underline{o}(1)) \quad \text{for } x \text{ in } I_{j,n} \quad ,$$

since the end points of $I_{j,n}$ both converge to zero. Analogously to (12) of Silverman (1976), given any λ in $(0,2)$,

$$P[\sup \sigma^{-1} \rho_1 \le (1 - \tfrac{1}{2}\lambda) \{2 \log h^{-1} \ell)\}^{\frac{1}{2}}]$$

$$\le \underline{O}(\ell^{-2}) \log(h^{-1}\ell) \tag{16}$$

$$\times \iint_{I_{j,n}} |\chi| \exp\{2 \log(h^{-1}\ell)(1 - \tfrac{1}{2}\lambda)^2 |\chi|/(1 + |\chi|)\}$$

where $\chi(x,y) = \text{corr}\{\rho(x), \rho(y)\}$. Using a similar argument to that following (12) of Silverman (1976), but allowing the interval I to vary, shows that the expression in (16) is dominated by

$$\underline{O}(\ell^{-2}) \log(h^{-1}\ell) \{\sigma^2(0) + \underline{o}(1)\}^{-1} \{h^{-1}\ell\}^{(1 - \frac{1}{2}\lambda)^2} \underline{O}(\ell)$$

$$= (h^{-1}\ell)^{-\lambda + \frac{1}{4}\lambda^2} \log(h^{-1}\ell) \to 0$$

by (14) above.

It follows that, setting $K = \{2f(0) \int \phi'^2\}^{\frac{1}{2}}$,

$$p \lim \inf \sup\{h^{-1} \log(h^{-1}\ell)\}^{\frac{1}{2}} \rho_1 \ge K \tag{17}$$

and that the same result holds if ρ_1 is replaced by $-\rho_1$, giving a corresponding result for $\inf \rho_1$. It follows from (15), (17) and the corresponding result for $\inf \rho_1$ that

$$P\{\rho_1 \text{ crosses } -n^{\frac{1}{2}}h(Ef_n' + \varepsilon_n') \text{ in } I_{j,n}\} \to 1 ,$$

and hence that

$$P\{f_n' \text{ crosses zero in } I_{j,n}\} \to 1 . \qquad (18)$$

Since (18) holds for all j, the number of maxima in f_n tends in probability to infinity, completing the proof of Proposition 2.

The final proposition of this section deals with the case where the alternative hypothesis is true, and shows that h_{crit} will remain bounded away from zero.

Proposition 3. *If* $k < j$ *then there exists a constant* $h_o > 0$, *depending on* f *and* k, *such that*

$$P\{h_{crit}(k) > h_o\} \to 1 .$$

Proof. By arguments analogous to those of the proof of the theorem of Silverman (1981), making use of the variation diminishing properties of the Gaussian kernel and the continuity properties of Ef_n, the number of maxima in $Ef_n(\cdot,h)$ is a right continuous decreasing function of h, for $h \geq 0$. By choosing h_0 sufficiently small, we can ensure that $Ef_n(\cdot,h_0)$ has, independently of n, exactly j maxima. Because of the conditions imposed on f in the statement of the Theorem above, we can also ensure that $Ef_n''(\cdot,h_0)$ is non-zero at all stationary points of $Ef_n(\cdot,h_0)$.

The argument of Lemma 2.2 of Schuster (1969), which does not in fact require the convergence to zero of the sequence of window widths, then implies that, with probability one,

$$f_n'(x,h_0) - Ef_n'(x,h_0) \quad \text{and} \quad f_n''(x,h_0) - Ef_n''(x,h_0)$$

both converge to zero uniformly over x. By an argument similar to that used in Proposition 1 above, it follows that the number of maxima of $f_n(\cdot,h_0)$ on $[a,b]$ tends almost surely to j, the number of maxima of $Ef_n(\cdot,h_0)$. Applying the corollary of Silverman (1981) completes the proof of Proposition 3.

Discussion

It is natural to enquire to what extent the conditions of the theorem above can be relaxed without affecting the conclusions. In particular it seems intuitively clear that the condition of bounded support for the density f should be able to be replaced by some condition on the tails of f, though the present method of proof cannot deal with this case. Condition (iv) appears to be more fundamental to the result; if, for example, $f'(0) = f''(0) = 0 \neq f'''(0)$, then an examination of f_n and Ef_n near zero seems to indicate that, under suitable regularity conditions, there will be no maximum of f_n near zero provided $|f_n''' - Ef_n'''|$ remains small. A heuristic argument suggests that a result corresponding to the theorem of Section 2 can be proved, but with $\alpha(h)$ replaced by $h^{-7}\log(h^{-1})$, so that h_{crit} converges to zero more slowly. Even slower convergence will occur for higher order zeros in f'.

The interest in this discussion lies in the fact that the bootstrap density constructed using the critical window width will not only have infinite tails of similar weight to those of the corresponding normal kernels but will also have a stationary point which is a point of inflexion. The slower convergence of zero of h_{crit} provides support for the remark of Silverman (1981) that the bootstrap test may be conservative; it also bears out the intuition of P.Huber (private communication) that the bootstrap procedure may be excessively conservative, though the difference between $n^{-\frac{1}{5}}$ and $n^{-\frac{1}{7}}$ convergence is very slight in practice.

The methods of this paper can also be used to study the asymptotic properties of a corresponding test for the number of points of inflexion in the density. Both Cox (1966) and Good and Gaskins (1980) prefer to use points of inflexion as an indication that the density is a mixture. The critical window width will now be the smallest window width for which the density has k maxima. Under suitable conditions a result corresponding to the theorem of Section 2 can be proved, but again, among other changes, $\alpha(h)$ will be replaced by $h^{-7}\log(1/h)$ since f_n'' will be replaced by f_n''' in much of the argument of the proofs of Propositions 1 and 2.

Acknowledgements

Part of this work was completed while the author was visiting the Mathematics Research Center at the University of Wisconsin-Madison, supported by the United States Army under Contract No. DAAAG29-80-C-0041. The author is most grateful to H.H. Bock, P. Deheuvels and P.J. Huber for helpful discussions and suggestions.

REFERENCES

Andrews, H.C. (1972), *Introduction to mathematical techniques in pattern recognition*. Wiley, New York.

Bhattacharya, P.K. (1967), Estimation of a probability density function and its derivatives. *Sankhyā*, Series A, *29*, 373-382.

Bock, H.H. (1977), On tests concerning the existence of a classification. *Proc. First International Symposium on Data Analysis and Informatics*, Versailles, 1977, Institut de Recherche d'Informatique et d'Automatique, Domaine de Voulceau, Le Chesnay, France, 449-464.

Boneva, L.I., Kendall, D.G. and Stefanov, I. (1971), Spline Transformations. *J. Roy. Statist. Soc. B*, *33*, 1-70.

Cox, D.R. (1966), Notes on the analysis of mixed frequency distributions. *Brit. J. Math. Statist. Psych.*, *19*, 39-47.

Efron, B. (1979), Boostrap methods - another look at the jack-knife. *Ann. Statist.*, *7*, 1-26.

Fukunaga, K. and Hostetler, L.D. (1975), The estimation of the gradient of a density function with applications in pattern recognition. *IEEE Trans. Inform. Theory*, IT-21, 32-40.

Good, I.J. and Gaskins, R.A. (1980), Density estimation and bump-hunting by the penalized likelihood method exemplified by scattering and meteorite data. *J. Amer. Stat. Assoc.*, *75*, 42-56.

Kendall, D.G. (1974) Hunting quanta. *Phil. Trans. R. Soc. London A276*, 231-266.

Kendall, D.G. and Kendall, W.S. (1980), Alignments in two-dimensional random sets of points. *Adv. Appl. Prob.*, *12*, 380-424.

Komlos, J., Major, P. and Tusnady, G. (1975), An approximation of partial sums of independent random variables and the sample distribution function. *Z. Wahrscheinlichkeitsth. und Verw. Gebiete*. *32*, 111-131.

Parzen, E. (1962), On estimation of a probability density function and mode. *Ann. Math. Statist.*, *33*, 1065-1076.

Rosenblatt, M. (1956), Remarks on some non-parametric estimates of a density function. *Ann. Math. Statist.* *27*, 832-837.

Schuster, E.F. (1969), Estimation of a probability density and its derivatives. *Ann. Math. Statist.*, *40*, 1187-1196.

Silverman, B.W. (1976), On a Gaussian process related to multivariate probability density estimation. *Math. Proc. Cambridge Philos. Soc.*, *80*, 135-144.

Silverman, B.W. (1978a), Weak and strong uniform consistency of the kernel estimate of a density and its derivatives. *Ann. Statist.*, *6*, 177-184.

Silverman, B.W. (1978b), Choosing the window width when estimating a density. *Biometrika*, *65*, 1-11.

Silverman, B.W. (1981), Using kernel density estimates to investigate multimodality. *J. Roy. Statist. Soc. B*, *43*, 97-99.

CRITERIA FOR RATES OF CONVERGENCE OF MARKOV CHAINS, WITH APPLICATION TO QUEUEING AND STORAGE THEORY

R.L. Tweedie

ABSTRACT

This paper first presents a review of the connections between various rate of convergence results for Markov chains (including normal Harris ergodicity, geometric ergodicity and sub-geometric rates for a variety of rate functions ψ), and finiteness of appropriate moments of hitting times on small sets. We then present a series of criteria, analogous to Foster's criterion for ergodicity, which imply the finiteness of these moments and hence the rate of convergence of the Markov chain.

These results are applied to random walk on $[0,\infty)$, and we deduce for example that this chain converges at rate $\psi(n) = n^{\alpha} (\log n)^{\beta}$ if the increment variable has finite moment of order $n^{\alpha+1} (\log n)^{\beta}$. Similar results hold for more general storage models.

1. Introduction

It is not always possible to pinpoint the beginnings of a major research area, and yet there can be little doubt that for the bulk of the vast quantity of mathematics known as queueing theory one can trace both the types of problems and the methods of solution to the single fundamental paper of David Kendall [8]. In that paper he introduced the idea of analysing queueing systems using embedded Markov chains, and the interplay between Markov chain theory and queueing theory has since been of vital importance in the development of both areas.

One of the most immediate, simplest and most elegant products of this interplay was the discovery by F.G. Foster (then a student of Kendall) of a criterion for positive recurrence of Markov chains, ensuring that these chains would have stationary distributions. The work of Foster ([4], [5]) has been extended in various directions (see, for example, Pakes [20], Mauldon [14], Marlin [13], Tweedie [26],[27]); all of these extensions attempt to quantify the idea that if a Markov chain moves, on average, towards the centre of its state space, then it is positive recurrent.

A second major development of Kendall in Markov chain theory was the introduction of the idea of geometric ergodicity in [9]. Kendall, and later in [28] Vere-Jones (again, at the time, a student of Kendall) investigated the way in which the distribution of a Markov chain might converge geometrically quickly to its limit. These results have been strengthened and extended recently by Nummelin and Tweedie [18] and Nummelin and Tuominen [16]; and in [16] and also for countable space chains in Popov [23], analogues of Foster's criterion are given which guarantee that a chain will be geometrically ergodic. These quantify the idea that the chain must move towards the centre of the space geometrically quickly if the convergence to stationarity is to be geometric.

Foster's original conditions, and the geometric ergodicity criteria, are particularly suited to the analysis of the models in queueing theory. In [16] they are applied to deduce simply the known conditions for random walk on $[0,\infty)$ to be geometrically ergodic, and the methods of [25] can be used to extend them to a very general range of storage models.

Recently, much more general rates of convergence have been studied for Markov chains on countable spaces (Pitman [21] and Lindvall [12]) and on general spaces (Nummelin and Tuominen [17]). In this paper, we provide a unified account of various aspects of all these rates of convergence results. We then prove that for the general rates of [17], an analogue of Foster's criterion exists. This rounds out a cycle of results for rates of convergence within which those of Kendall and Foster can be seen as primary and archetypical.

In the second part of the paper we apply these results to random walk and queueing and storage models, and give simple criteria for convergence at such general subexponential rates as

$$n^{\alpha} (\log n)^{\beta} .$$

We see that, as has been clear since [5], it is for these examples that our type of criteria function best.

2. Convergence of Markov chains and hitting time moments

We will consider a temporally homogeneous Markov chain $\{X_n\}$ on a general state space (S,F); the σ-field F of subsets of S will be assumed to be countably generated, but no other restrictions are placed on

S. We write

$$P^n(x,A) = \mathbb{P}(X_n \in A \mid X_o = x), \quad x \in S, A \in F,$$

for the n-step transition probabilities of $\{X_n\}$. We assume that $\{X_n\}$ is ϕ-*irreducible*; that is, for some measure ϕ on F, we have $\sum_n P^n(x,A) > 0$ for every $x \in S$ and every $A \in F$ with $\phi(A) > 0$.

The fundamental papers of Nummelin [15] and Athreya and Ney [1] show that such a chain can be expected to exhibit asymptotic properties completely analogous to those known for chains on a countable space. This has been observed, although proofs of the properties were much harder in the seemingly very much more general context; however, in [15] and [1] it is shown that for ϕ-irreducible chains it is possible to construct an "artificial atom" for $\{X_n\}$ which enables the use of discrete-time renewal theory to carry over properties of countable chains.

In particular, ϕ-irreducible chains can be classified as transient, null and positive recurrent in a natural way [26]. We shall be mainly concerned with the last named, and for convenience shall also assume that $\{X_n\}$ is *aperiodic* (see [19], p. 10), although all our results essentially hold without this restriction.

For our purposes it will be most appropriate to define $\{X_n\}$ to be *Harris ergodic* (often called aperiodic and positive recurrent) if, for some probability measure π on F, and every $x \in S$,

$$||P^n(x,\cdot) - \pi|| \to 0, \quad n \to \infty, \tag{2.1}$$

where $||\cdot||$ denotes total variation of signed measures on F.

In this paper, our main interest will lie in the rate of convergence in (2.1). We will define $\{X_n\}$ to be *geometrically ergodic* if there exists a $\rho < 1$ such that

$$\rho^{-n} ||P^n(x,\cdot) - \pi|| \to 0, \ n \to \infty, \tag{2.2}$$

for every $x \in S$.

For more general rates we shall consider the class of sequences Λ studied in [12], [17]. Let Λ_o denote those sequences ψ which are such that

(i) ψ is non-decreasing, with $\psi(j) \geq 2$ for all $j \geq 0$,

(ii) $\log \psi(j)/j \downarrow 0$ as $j \to \infty$;

then Λ is the class of sequences for which there exists $\psi_0 \in \Lambda_0$ with

$$\liminf_{n\to\infty} \psi(n)/\psi_0(n) > 0, \quad \limsup_{n\to\infty} \psi(n)/\psi_0(n) < \infty .$$

Examples of sequences in Λ are

$$\psi(n) = n^{\alpha}(\log n)^{\beta} \exp(\gamma n^{\delta})$$

for $0 < \delta < 1$ and either $\gamma > 0$ or $\gamma = 0$ and $\alpha > 0$.

We shall call $\{X_n\}$ *ergodic of order* ψ if

$$\psi(n) \, ||P^n(x,\cdot) - \pi|| \to 0, \; n \to \infty \quad , \tag{2.3}$$

for every $x \in S$.

The definitions of (2.2) and (2.3) are slightly stronger than those in [18], [16] and [17], where the limits are only required to hold outside a π-null set. We shall see that the criteria in Section 3 in fact imply the stronger, and more satisfactory, forms above.

In some contexts it is natural to require (2.1)-(2.3) to hold only π-almost everywhere. For example, suppose that S contains an *atom* Δ; that is, a point with $\phi(\Delta) > 0$. (When S is countable and $\{X_n\}$ is irreducible in the usual sense, then every point is an atom.) Then (2.1) holds for π-almost all x [25] if for some $\pi_\Delta > 0$, $P^n(\Delta,\Delta) \to \pi_\Delta$; and (2.2) holds for π-almost all x [18] when this local convergence occurs at a geometric rate. A similar result, extending pointwise convergence to a global "solidarity" property, is still lacking for ergodicity of order ψ. The proofs in the Harris and geometric ergodicity cases depend on generating function arguments (such as Kendall's Lemma [18]) and these are not available in the general ψ case where coupling proofs are used.

We will be concerned with deriving criteria for (2.1)-(2.3) based on relationships between these rates of convergence and moments of return times. Call a set $B \in F$ *small* if $\phi(B) > 0$ and for every $A \in F$ with $\phi(A) > 0$ there exists j such that

$$\inf_{x \in B} \sum_{1}^{j} P^n(x,A) > 0.$$

Note that every finite collection of atoms is small, as a consequence of ϕ-irreducibility; whilst in [16] it is pointed out that under fairly general continuity conditions on the transition probabilities $P(x,\cdot)$, every relatively compact set of positive ϕ-measure is also small.

Let us denote the hitting time on a set B by $\tau_B = \inf(n \geq 1: X_n \in B)$, and write $\mathbb{E}_x$ and $\mathbb{P}_x$ respectively for expectation and probability conditional on $X_0 = x$. Our first result summarises known relations between rates of convergence and moments of τ_B.

Theorem 1: (i) *If* B *is small and* $\sup_{x\in B} \mathbb{E}_x(\tau_B) < \infty$, *and* $\mathbb{P}_y(\tau_B < \infty) = 1$ *for all* y, *then* $\{X_n\}$ *is Harris ergodic.*

(ii) *If* B *is small and for some* $r > 1$, $\sup_{x\in B} \mathbb{E}_x(r^{\tau_B}) < \infty$, *and* $\mathbb{E}_y(r^{\tau_B}) < \infty$ *for all* y, *then* $\{X_n\}$ *is geometrically ergodic.*

(iii) *If* B *is small and for some* $\psi \in \Lambda$, $\sup_{x\in B} \mathbb{E}_x(\psi^0(\tau_B)) < \infty$, *where* $\psi^0(n) = \sum_1^n \psi(j)$, *and* $\mathbb{E}_y(\psi(\tau_B)) < \infty$ *for all* y, *then* $\{X_n\}$ *is ergodic of order* ψ

Proof: (i) is well known if B is an atom, and follows in general from Nummelin's splitting technique and Proposition 3 of [25], as in the proof of Theorem 3(ii) of [16]; (ii) is Theorem 3(ii) and 3(iii) of [16]; and (iii) is Theorem 4(i) and 4(ii) of [17]. □

Converses to all these results hold, and for completeness we give

Theorem 2: (i) *If* $\{X_n\}$ *is Harris ergodic, then for any* $A \in F$, $\int_A \pi(dy)\, \mathbb{E}_y(\tau_A) < \infty$.

(ii) If $\{X_n\}$ *is geometrically ergodic, then for any set* $A \in F$, *there exists* $r > 1$ *such that* $\int_A \pi(dy)\, \mathbb{E}_y(r^{\tau_A}) < \infty$.

(iii) If $\{X_n\}$ *is ergodic of order* ψ, *with* $\psi \in \Lambda$, *then for any set* $A \in F$, $\int_A \pi(dy)\, \mathbb{E}_y(\psi(\tau_A)) < \infty$, *and for* π*-almost all* $x \in S$, $\mathbb{E}_x(\psi^0(\tau_A)) < \infty$.

Proof: (i) is classical (see for example [3]); (ii) is Theorem 3(i) of [16]; and (iii) is Theorem 3 of [17]. □

In [16] and [17] more detailed results are also given concerning, in particular, the range of measures λ and μ for which the quantities

$$\int \int \lambda(dx)\, \mu(dy)\, ||P^n(x,\cdot) - P^n(y,\cdot)||$$

have the appropriate limiting behaviour.

We shall not give these details here. Rather, having summarised what we believe to be the most important connections between rates of convergence and hitting time moments, we shall in the next section describe the analogues of Foster's criterion for these various moments to be finite.

3. Criteria for finiteness of hitting time moments

Suppose that $g(x)$ is a non-negative measurable function on S; g will be a test function for the various types of ergodicity above.

Theorem 3: (i) *If, for some* $\varepsilon > 0$, *and some* $A \in F$,

$$\int_S P(x,dy)\, g(y) \le g(x) - \varepsilon, \qquad x \in A^c, \tag{3.1}$$

then

$$\mathbb{E}_x(\tau_A) \le g(x)/\varepsilon, \qquad x \in A^c. \tag{3.2}$$

(ii) *If* $g(x) \ge 1$ *for* $x \in A$, *and for some* $\varepsilon > 0$

$$\int_S P(x,dy)\, g(y) \le [1 - \varepsilon] g(x), \quad x \in A^c, \tag{3.3}$$

then

$$\mathbb{E}_x(r^{\tau_A}) \le g(x)/[1 - r(1-\varepsilon)], \qquad x \in A^c, \tag{3.4}$$

for any $r < (1-\varepsilon)^{-1}$.

(iii) *If* $g_1(x) \ge \mathbb{E}_x(\psi(\tau_A))$ *for* $x \in A^c$, *and*

$$\int_S P(x,dy)\, g(y) \le g(x) - g_1(x), \ x \in A^c, \tag{3.5}$$

then

$$\mathbb{E}_x(\psi^0(\tau_A)) \le g(x). \tag{3.6}$$

Proof: (i) is Theorem 6 of [27]. To prove (ii), define the usual taboo probabilities ${}_AP^n(x,B) = \mathbb{P}_x(X_n \in B,\ \tau_A \ge n)$. Assume inductively

$$\int_{A^c} {}_A P^m(x,dw)\; g(w) \le (1-\varepsilon)^m\; g(x); \tag{3.7}$$

(3.3) shows that this holds for $m = 1$, whilst iterating (3.3) gives

$$\int_{A^c} {}_A P^n(x,dy)\; g(y) = \int_{A^c} \int_{A^c} {}_A P^{n-1}(x,dw)\; P(w,dy)\; g(y)$$

$$\le \int_{A^c} {}_A P^{n-1}(x,dw)\; g(w)\; (1-\varepsilon)$$

so that if (3.7) holds for $m = n-1$ it also holds for $m = n$. Similarly

$$\mathbb{P}_x(\tau_A = n) \le \int_S {}_A P^n(x,dy)\; g(y) \tag{3.8}$$

$$\le \int_{A^c} {}_A P^{n-1}(x,dw)\; g(w)\; (1-\varepsilon)$$

$$\le (1-\varepsilon)^n\; g(x)$$

from $g(x) \ge 1$ on A and (3.7). Multiplying both sides of (3.8) by r^n for $r < (1-\varepsilon)^{-1}$ and summing gives the result.

To prove (iii), note that by iterating (3.5) we have, using Fubini's Theorem,

$$g(x) \ge \sum_{j=0}^{\infty} \int_{A^c} {}_A P^j(x,dy)\; g_1(y)$$

$$\ge \sum_{0}^{\infty} \psi(j) \sum_{n=j}^{\infty} \mathbb{P}_x(\tau_A = n)$$

$$= \sum_{0}^{\infty} \mathbb{P}_x(\tau_A = n)\; \psi^0(n).$$

□

From Theorem 3 and the results of the previous section we have

Theorem 4: *Suppose that* A *is a small set, and that*

$$\sup_{x \in A} \int_S P(x,dy)\; g(y) < \infty. \tag{3.10}$$

Then

(i) *if* g *and* A *satisfy (3.1) then* $\{X_n\}$ *is Harris ergodic.*

(ii) *if* g *and* A *satisfy (3.3) then* $\{X_n\}$ *is geometrically ergodic.*

(iii) *if* g *and* A *satisfy (3.5) then* $\{X_n\}$ *is ergodic of order* ψ.

Proof: (i) follows from Theorem 1(i) and

$$\mathbb{E}_x(\tau_A) \le 1 + \int_{A^c} P(x,dy)\, \mathbb{E}_y(\tau_A) \ ;$$

(ii) follows from Theorem 1(ii) and

$$\mathbb{E}_x(r^{\tau_A}) \le r \int_{A^c} P(x,dy)\, \mathbb{E}_y(r^{\tau_A});$$

and for (iii), use Theorem 1(iii) and the fact that for $\psi \in \Lambda_0$,

$$\mathbb{E}_x(\psi^0(\tau_A)) = \psi(1) + \int_{A^c} P(x,dy)\, \mathbb{E}_y(\psi^0(\tau_A) + \psi(\tau_A + 1))$$

$$\le \psi(1) + \int_{A^c} P(x,dy)\, \mathbb{E}_y(\psi^0(\tau_A)[1+\psi(1)])$$

since $\psi(m+n) \le \psi(m)\,\psi(n)$ from [17], (1.4). □

Theorem 4(i) is Foster's criterion in the more general setting; however, the realisation that $\mathbb{E}_x(\tau_A)$ is essentially the minimal solution to the inequalities (3.1) (see Theorem 6.2 of [27]) renders the criterion much more transparent, as Foster's original proof [5] showed only that the chain was not null, so that ergodicity followed by a classification argument. Theorem 4(ii) is a general state space version of Theorem 1 of Popov [23]; a slight variation, with (3.3) replaced by

$$\int_S P(x,dy)\, g(y) \le (1 - \varepsilon)\, g(x) - \varepsilon,$$

is given in [16]; the present result is slightly stronger. Theorem 4(iii) is new, and as we shall see, provides an iterative form of Foster's criterion for convergence rates such as $\psi(n) = n^r$ for $r > 1$. Pitman has pointed out that Theorem 3(iii) holds, as does Foster's criterion, because the probabilistic solution $\mathbb{E}_x(\psi^0(\tau_A))$ is the minimal solution of an identity, which is in this case the occupation time identity studied in [22].

4. Convergence rates for continuous time processes

Let $\{X_t\}$ be a Markov process with continuous time parameter on $(S,\mathcal{F})$, with transition probability semigroup $P^t(x,A) = \mathbb{P}_x(X_t \in A)$.

Following [25], we will call $\{X_t\}$ Harris ergodic if for some probability measure π and all x,

$$||P^t(x,\cdot) - \pi|| \to 0, \qquad t \to \infty \quad ;$$

whilst $\{X_t\}$ is called exponentially ergodic if for all x,

$$||P^t(x,\cdot) - \pi|| = O(e^{-\alpha t}), \qquad t \to \infty,$$

for some $\alpha > 0$.

Let $\bar{\Lambda}$ denote the class of functions ψ from $[0,\infty)$ to $(0,\infty)$ which are monotone increasing and for which ψ, considered as a function on $\{1,\ldots,\}$, lies in Λ. We shall say that $\{X_t\}$ is ergodic of order ψ if

$$\psi(t)\ ||P^t(x,\cdot) - \pi|| \to 0, \qquad t \to \infty,$$

for all x.

For the analysis of queueing models in Section 6 we shall investigate properties of continuous time processes using the discrete skeletons $\{X_{nh}, h > 0\}$. We summarise the results of Theorem 1 of [25], together with the order ψ analogue which follows similarly using the non-increasing character of $||\nu P^t||$ as a function of t for bounded signed measures ν, in

Theorem 5: *If any skeleton chain* $\{X_{nh}\}$ *is Harris, geometrically or order* ψ *ergodic, then* $\{X_t\}$ *is Harris, exponentially or order* ψ *ergodic.* □

It is worth remarking that the ease with which continuous time process results follow from those for skeletons is very much a product of the total variation norm convergence in (2.1)-(2.3). On a countable space, using pointwise convergence results that seem more "natural", the extension from skeletons to processes is non-trivial (see [10]). It must be stressed that in general pointwise and total variation convergence are equivalent.

5. Criteria for convergence rates for random walk on a half-line.

We shall illustrate our results by applying them to the simple random walk on $[0,\infty)$ defined by

$$X_n = (X_{n-1} + Y_n)^+ \tag{5.1}$$

where $\{Y_i\}$ is a sequence of independent and identically distributed random variables. We shall denote by Y the generic variable with the distribution of any of the Y_i.

Theorem 6: (i) *Suppose that* $\mathbb{E}(Y) < 0$. *Then* $\{X_n\}$ *is Harris ergodic.*

(ii) *If also, for some* $s > 0$, $\mathbb{E}(e^{sY}) < \infty$, *then* $\{X_n\}$ *is also geometrically ergodic.*

Proof: (i) dates back using the methods here to Foster [5] when S is countable, and follows from Theorem 4(i) using $g(x) = x$; see [11] for details of this and related models. Using the test function $g(x) = e^{tx}$ for $t < s$ yields (i), as in Section 3 of [16]. □

The use of Theorem 4(iii) for general $\psi \in \Lambda$ is not obvious, but it can be used to develop criteria for a large class of convergence rate functions. We shall build them up in several steps.

Let us assume from now on that $\{X_n\}$ is Harris ergodic, and that $\mathbb{E}(Y) < 0$ exists. From the proof of Theorem 6(i) we have

$$\mathbb{E}_x(\tau_0) \leq x/|\mathbb{E}(Y)|, \tag{5.2}$$

and for simplicity we will take $\mathbb{E}(Y) = -1$.

We shall exploit the following stochastic monotonicity property of $\{X_n\}$, which holds for any non-decreasing function f:

$$\mathbb{E}_y(f(\tau_{[0,b]})) \leq \mathbb{E}_x(f(\tau_{[0,b]})), \qquad y \leq x, \tag{5.3}$$

and

$$\mathbb{E}_x(f(\tau_{[0,b]})) = \mathbb{E}_{x-b}(f(\tau_0)), \qquad x \geq b, \tag{5.4}$$

which holds because of the translation invariant nature of random walk. From Theorem 1(iii) and the fact that $\{0\}$ is an atom, we have that $\{X_n\}$ will be ergodic of order ψ if for some $x \in [0,\infty)$

$$\infty > \mathbb{E}_x(\psi^0(\tau_0)) \geq \mathbb{E}_0(\psi^0(\tau_0)) \tag{5.5}$$

from (5.3).

Let $\bar{\Lambda}$ be the continuous analogue of Λ described in Section 4, and assume from here on that ψ' is defined everywhere; clearly this is always possible when we are extending $\psi \in \Lambda$. Write $\psi^0(x) = \int_0^x \psi(y)\,dy$.

Proposition 1: *If* ψ *satisfies*

$$\mathbb{E}_x(\psi(\tau_0)) \le c_1\,\psi(x) \tag{5.6}$$

for some c_1 *and all large enough* x, *then*

$$\mathbb{E}_x(\psi^0(\tau_0)) \le c_0\,\psi^0(x) \tag{5.7}$$

for some c_0 *and all large enough* x, *provided both* ψ *and* ψ' *are in* Λ *and*

$$\mathbb{E}(Y^2\,\psi'(Y^+)) < \infty. \tag{5.8}$$

Proof. Note that $\psi' \in \Lambda$ implies $\psi'(x) > 0$ for all $x \ge 0$, and hence (5.8) implies

$$\mathbb{E}(Y^2) < \infty. \tag{5.9}$$

We will verify that (3.5) holds with $g_1(x) = c_1\,\psi(x)$ and $g(x) = c_0\,\psi^0(x)$ where c_0 is yet to be chosen.

By a Taylor series expansion

$$\int_0^\infty P(x,dy)\;g(y) \quad = \int_{-x}^\infty \mathbb{P}(Y\in dy)\;c_0\,\psi^0(x+y) \tag{5.10}$$

$$\le c_0\,\psi^0(x) + c_0\int_{-x}^\infty \mathbb{P}(Y\in dy)\;y\;\psi(x)$$

$$+\;c_0/2\int_{-x}^\infty \mathbb{P}(Y\in dy)\;y^2\;\psi'(x+\xi)$$

where $\xi \in (0,y)$. Now by (5.2), we can ensure that for x, c_0 large enough

$$c_0\int_{-x}^\infty \mathbb{P}(Y\in dy)\;y\;\psi(x) \le -2\,c_1\,\psi(x).$$

Moreover, since $\psi' \in \overline{\Lambda}$, from (5.8) and [17], (1.4), for some c,c',

$$\int_{-x}^{\infty} \mathbb{P}(Y\in dy)\; y^2\; \psi'(x+\xi) \le c \int_{-x}^{\infty} \mathbb{P}(Y\in dy)\; y^2\; \psi'(x) \tag{5.11}$$

$$+ c \int_{o}^{\infty} \mathbb{P}(Y\in dy)\; y^2\; \psi'(x)\; \psi'(y)$$

$$\le c' \quad \psi'(x).$$

From (2.15) of [17], $\psi'(x) = o(\psi(x))$, and so this term can be bounded by $c_1\ \psi(x)$ for large enough x; and hence (3.5) holds as required, and for all $x \ge b$ for some b

$$\mathbb{E}_x(\psi^o(\tau_{[0,b]})) \le c_o\ \psi^o(x).$$

From (5.4),

$$\mathbb{E}_x(\psi^o(\tau_o)) \le c_o\ \psi^o(x+b)$$

$$\le c_o\ [c'' + (1+c'')\ \psi^o(x)]$$

for some c'' ([17], (1.5)); and hence the proposition is proved. □

From Proposition 1 we will be able to use an iterative argument and we now set up the first steps in this iteration.

Proposition 2: *If* ψ *is such that* ψ^o *is concave on* $[b,\infty)$ *for some* b *then* $\{X_n\}$ *is ergodic of order* ψ.

Proof: Let θ be a concave function on $[0,\infty)$ agreeing with ψ^o on $[b,\infty)$. By Jensen's Inequality and (5.2), there exist constants c' and c'' such that

$$\mathbb{E}_x(\psi^o(\tau_o)) \le c' + \mathbb{E}_x(\theta(\tau_o)) \tag{5.12}$$

$$\le c' + \theta(x)$$

$$\le c'' + \psi^o(x)$$

and the result follows from (5.5). □

Proposition 3: *Suppose that* $\psi \in \overline{\Lambda}$ *is concave on* $[b,\infty)$ *for some* b. *Then for large enough* x, c

$$\mathbb{E}_x(\psi^0(\tau_0)) \le c\, \psi^0(x)$$

provided $\mathbb{E}(Y^2) < \infty$, *and hence* $\{X_n\}$ *is ergodic of order* ψ.

Proof: From (5.12), the concavity of ψ implies that (5.6) holds. The proof of Proposition 1 can be carried through, except that the bound (5.11) fails; notice that ψ concave means that ψ' is ultimately decreasing and hence cannot be a member of $\overline{\Lambda}$. However, ψ' is bounded, and so for some c'

$$\int_{-x}^{\infty} \mathbb{P}(Y\epsilon dy)\; y^2\; \psi'(x+\xi) \le c'\, \mathbb{E}(Y^2);$$

since $\psi(x) \to \infty$ as $x \to \infty$, the result follows. □

Define now $\overline{\Lambda}_k \subseteq \overline{\Lambda}$, $k \ge 1$, as the set of functions that are $(k+1)$ times differentiable and for which

$$\psi^{(1)}, \psi^{(2)}, \ldots, \psi^{(k)} \in \overline{\Lambda},$$

$$\psi^{(k)} \text{ is concave on } [b,\infty) \text{ for some } b.$$

Theorem 7: *If* $\psi \in \overline{\Lambda}$ *and* $\mathbb{E}(Y^2\, \psi'(Y^+)) < \infty$, *then* $\{X_n\}$ *is ergodic of order* ψ.

Proof: From Proposition 3, since $\psi^{(k)}$ is concave $[\psi^{(k)}]^0$ satisfies (5.6). Now for any j, $[\psi^{(j)}]^0 = \psi^{(j-1)}$; so $\psi^{(k-1)}$ satisfies (5.6). But now $\psi^{(k)} \in \overline{\Lambda}$ and we can use Proposition 1 inductively to deduce that (5.7) holds; note that $\mathbb{E}(Y^2\, \psi^{(j)}(Y^+))$ is a monotone increasing function as j decreases and hence is finite for all relevant j from (5.8). From (5.5) the theorem is proved. □

It is easy to check that $\overline{\Lambda}_k$ contains functions of magnitude

$$\psi(x) \sim x^{\alpha}\, (\log x)^{\beta} \tag{5.13}$$

where $k = [\alpha] \ge 1$, and $\beta \ge 0$ are arbitrary. For the rate functions (5.13) with $0 \le \alpha \le 1$ we can use Proposition 2 immediately since $\psi(x)$ is ultimately concave.

The conditions of Theorem 7 have the particularly simple form, since $\psi'(x) \sim x^{\alpha-1}(\log x)^{\beta}$, given in

Theorem 8: *When* ψ *is given by (5.13), then* $\{X_n\}$ *is ergodic of order* ψ *provided* $\mathbb{E}(Y^2)$ *is finite and*

$$\mathbb{E}((Y^+)^{\alpha+1}(\log^+ Y^+)^\beta) < \infty. \qquad \square$$

We conjecture that the appropriate condition for ergodicity of order ψ is that $\mathbb{E}(\psi^o(Y^+)) < \infty$, as in Theorem 8. To prove this, one would seem to need a more delicate expansion than the simple Taylor series which we have used at (5.10).

6. Convergence rates for queueing and storage models

We conclude by applying the results of the previous section to some queueing and storage models. Let $\{X_t\}$ be a continuous time Markov process satisfying the storage equation

$$X_t = X_o + A_t - \int_o^t r(X_s)\, ds$$

where $\{A_t\}$ is an input process which we shall assume to be a non-negative additive process and r is a state dependent release rate with $r(0) = 0$. Such processes are discussed in [25], where conditions for Harris ergodicity and exponential ergodicity are derived. Here we use similar methods to investigate ergodicity of order ψ.

(i) Suppose that $r(x) = 1$, $r > 0$, and that $\{A_t\}$ is a compound Poisson process with jump rate λ and jump size distribution F. Then $\{X_t\}$ is the virtual waiting time process in an M/G/1 queue in the notation introduced by Kendall; the arrival rate is λ and the service time distribution is F.

Theorem 9: *Suppose that* $\psi \in \bar{\Lambda}_k$, *and that* $\mathbb{E}(Y^2\, \psi'(Y^+)) < \infty$, *where* Y *has the distribution* F. *If* $\{X_t\}$ *is Harris ergodic then it is ergodic of order* ψ.

Proof: As in Theorem 2 of [25], we will use Theorem 5 and also compare the hitting times of the skeletons $\{X_n\}$ of unit time step with those of the random walk $\{\hat{X}_n\}$ given by

$$\hat{X}_n = \hat{X}_{n-1} + Z_n$$

where $Z_n = A_n - A_{n-1} - 1$, $n = 1,2,\ldots$.

If $Z = A_1 - 1$, then $Z + 1$ is a random sum of independent and identical variables with distribution F, and the number of terms in the sum is Poisson with parameter λ. Hence $\mathbb{E}(Z^2 \psi'(Z^+)) < \infty$; to see this simply, note that Lemma 2.2 of [17] shows a geometric sum of such variables to retain the appropriate moment, and use a stochastic comparison argument.

From Theorem 7, we thus have that $\{X_n\}$ is ergodic of order ψ, and in particular

$$\mathbb{E}_0(\psi^0(\hat{\tau}_0)) < \infty.$$

The proof of Theorem 2 of [25] shows that $\mathbb{P}_x(\tau_{[0,1]} \geq n) \leq \mathbb{P}_0(\hat{\tau}_0 \geq n)$, $x \leq 1$, and so

$$\sup_{x\leq 1} \mathbb{E}_x(\psi^0(\tau_{[0,1]})) < \infty.$$

Conclude with a geometric trials argument: for each n with $X_n \in [0,1]$, there is probability at least $e^{-\lambda}$ that $X_{n+1} = 0$, and so $\mathbb{E}_0(\psi^0(\tau_0)) < \infty$. The details are as in Theorem 2 of [25], with Lemma 2.2 of [17] providing the appropriate analogue of Proposition 3 of [25]. □

(ii) Let $\{A_t\}$ remain a compound Poisson process, but now let r be any function satisfying the conditions that r is strictly positive, left continuous and has strictly positive right limits everywhere on $(0,\infty)$. We assume that

$$\int_0^x [r(y)]^{-1}\, dy < \infty ,$$

$x > 0$, and that $r(x) > \mathbb{E}(A_1)$ for all large enough x, so that from Theorem 3 of [25] $\{X_t\}$ is Harris ergodic.

This is a common model in storage theory, and has been studied in detail in [6], [7] and [2].

<u>Theorem 10</u>: *Under the conditions of Theorem 9,* $\{X_t\}$ *is ergodic of order* ψ.

<u>Proof</u>: This is identical with that of Theorem 3 of [25], with Theorem 9 above taking the part of Theorem 2 of [25]. □

Acknowledgements

This work was begun at Colorado State University and continued at the CSIRO Division of Mathematics and Statistics. I am grateful to Pekka Tuominen and Geoff Eagleson for useful conversations.

REFERENCES

[1] Athreya, K.B. and P.E. Ney, A new approach to the limit theory of recurrent Markov chains, *Trans. Amer. Math. Soc. 245* (1978), 493-501.

[2] Brockwell, P.J., S.I. Resnick and R.L. Tweedie, Storage processes with general release rule and additive input, *Adv. Appl. Prob.* (to appear).

[3] Cogburn, R., A uniform theory for sums of Markov chain transition probabilities, *Ann. Probab. 3* (1975), 191-214.

[4] Foster, F.G., Discussion to [8].

[5] Foster, F.G., On the stochastic matrices associated with certain queueing processes, *Ann. Math. Statist. 24* (1953), 355-360.

[6] Harrison, J.M. and S.I. Resnick, The stationary distribution and first exit probabilities of a storage process with general release rule, *Math. Operat. Res. 1* (1976), 347-358.

[7] Harrison, J.M. and S.I. Resnick, The recurrence classification of risk and storage processes, *Math. Operat. Res. 3* (1978), 57-66.

[8] Kendall, D.G., Some problems in the theory of queues, *J. Roy. Statist. Soc.* (B) *13* (1951), 151-185.

[9] Kendall, D.G., Unitary dilations of Markov transition operators and the corresponding integral representation for transition-probability matrices, In *Probability and Statistics* (U. Grenander, ed) (1959), Almqvist and Wiksell, Stockholm.

[10] Kingman, J.F.C., Ergodic properties of continuous-time Markov processes and their discrete skeletons, *Proc. London Math. Soc.* (3) *13* (1963), 593-604.

[11] Laslett, G.M., D.B. Pollard and R.L. Tweedie, Techniques for establishing ergodic and recurrence properties of continuous valued Markov chains, *Nav. Res. Log. Quart. 25* (1978), 455-472.

[12] Lindvall, T., On coupling of discrete renewal processes, *Z. Wahrscheinlichkeitstheorie verw. Geb. 48* (1979), 57-70.

[13] Marlin, P.G., On the ergodic theory of Markov chains, *Operat. Res. 21* (1973), 617-622.

[14] Mauldon, J.G., On non-dissipative Markov chains, *Proc. Camb. Phil. Soc. 53* (1958), 825-835.

[15] Nummelin, E., A splitting technique for Harris recurrent Markov chains, *Z. Wahrscheinlichkeitstheorie verw. Geb. 43* (1978), 309-318.

[16] Nummelin, E. and P. Tuominen, Geometric ergodicity of Harris recurrent Markov chains with application to renewal theory, *Stoch. Proc. Applns.* (to appear).

[17] Nummelin, E. and P. Tuominen, The rate of convergence in Orey's theorem for Harris recurrent Markov chains with applications to renewal theory (submitted for publication).

[18] Nummelin, E. and R.L. Tweedie, Geometric ergodicity and R-positivity for general Markov chains, *Ann.Probab.* *6* (1978), 404-420.

[19] Orey, S., *Lecture Notes on Limit Theorems for Markov Chain Transition Probabilities,* (1971), van Nostrand Reinhold, London.

[20] Pakes, A.G., Some conditions for ergodicity and recurrence of Markov chains, *Operat. Res.* *17* (1969), 1058-1061.

[21] Pitman, J.W., Uniform rates of convergence for Markov chain transition probabilities, *Z. Wahrscheinlichkeitstheorie verw. Geb.* *29* (1974), 194-227.

[22] Pitman, J.W., Occupation measures for Markov chains, *Adv. Appl. Prob.* *9* (1977), 69-86.

[23] Popov, N., Conditions for geometric ergodicity of countable Markov chains, *Soviet Math. Dokl.* *18* (1977), 676-679.

[24] Rosberg, Z., A note on the ergodicity of Markov chains, *J. Appl. Prob.* 18 (1981), 112-121.

[25] Tuominen, P. and R.L. Tweedie, Exponential ergodicity in Markovian queueing and dam models, *J. Appl. Prob.* *16* (1979), 867-880.

[26] Tweedie, R.L., A relation between positive recurrence and mean drift for Markov chains, *Austral. J. Statist.* *17* (1975), 96-102.

[27] Tweedie, R.L., Criteria for classifying general Markov chains, *Adv. Appl. Prob.* 8 (1976), 737-771.

[28] Vere-Jones, D., Geometric ergodicity in denumerable Markov chains, *Quart. J. Math. Oxford 2nd Ser.* 13 (1962), 7-28.

COMPETITION AND BOTTLE-NECKS

P. Whittle

One of the most tattered volumes on my bookshelf is the *Symposium on Stochastic Processes*, reprinted from the 1949 volume of the Journal of the Royal Statistical Society, Series B. This consists of the three substantial and highly personal survey articles contributed to the symposium by Joseph Moyal, Maurice Bartlett and David Kendall. Together they provided magnificent reading and reference at a time when the stochastic process literature was a meagre scattered one, and continued to do so for many readers over many years. In doing so, they confirmed, if they did not indeed found, the recognisable British tradition in stochastic processes.

David Kendall's article in this volume, "Stochastic processes and population growth", is only one of many he has written in the course of a distinguished career, and may well not be the one he prizes highest. I single it out, however, both for the seminal effect it had upon me personally, and because it gives an early example of his fascination with the capturing of an applied problem, having both intrinsic interest and an aspect of intangibility, in the meshes of a theory which is light but strong. David's trophies in this benign hunt are legion and living.

1. Introduction

There have been many models of competition between species, deterministic and stochastic. The most satisfactory of these are those in which the mechanism of competition is made explicit, as competition for limited resources. Deterministic versions of such models lead to the conclusion which O.M. Phillips (1973, 1974) terms the *principle of competitive exclusion*. This principle is that, if s species are competing for r critical resources, then in equilibrium at most r species will survive, if one disallows circumstances which are so special as to be structurally unstable.

Which of the s species will survive must be determined by examination of the local stability of the possible equilibrium points. In the case $r = 1$ there is a simple extremal characterisation of the surviving species: it is the species which can survive at the lowest ambient concentration of the unconsumed resource. It would be attractive if one could produce such an extremal characterisation of the relevant equilibrium in general; to do so is one of my aims. The other is to construct a stochastic version of the model, and so derive a stochastic version of the principle of competitive exclusion.

Let resource j be denoted R_j $(j=1,2,\dots,r)$ and species k be denoted S_k $(k=1,2,\dots,s)$. Let the total abundance of R_j be denoted b_j, and the number of S_k by n_k. I consider a closed system, in which resources are recycled rather than consumed. That is, it is supposed that an individual of S_k has locked up within him an amount a_{jk} of R_j $(j=1,2,\dots,r)$, which is returned to general availability upon his death. Such an individual cannot be constituted unless at least these amounts of resources are free in the system.

In a stochastic treatment the n_k are necessarily integer-valued; I shall assume the same true of the b_j and the a_{jk}. This is no effective constraint if one is prepared to choose the "quantum" of resource small enough. Define column vectors $b = (b_j)$, $n = (n_k)$ and the $r\times s$ matrix $A = (a_{jk})$. The element m_j of the vector

$$m = b - An \tag{1}$$

is then the amount of free R_j.

2. The simplest stochastic model, and its primal characterisation

I shall begin by suggesting an equilibrium distribution of n for prescribed b, and then demonstrate that this corresponds to a plausible model which can be powerfully generalised. The distribution manifests the principle of competitive exclusion and in itself implies an extremal characterisation of the equilibrium configuration.

Suppose, initially, that the system is an open one, in that resources can enter and leave it, and that b is then a random variable jointly with n. We shall obtain the distribution of n for the closed system by calculating the conditional distribution $P(n|b)$ for the open system. Let

$$\Pi(z,w) = E\left(\prod_{jk} z_j^{b_j} w_k^{n_k}\right)$$

be the joint probability generating function (p.g.f.) of these two sets of variables in equilibrium.

We shall assume, to begin with, the prescription

$$\Pi(z,w) = \prod_j (1-e^{d_j} z_j)^{-1} \prod_k (1-e^{c_k} w_k \prod_j z_j^{a_{jk}})^{-1} \tag{2}$$

That is, that the variables n_k are independent geometric variables, as are the amounts m_j of free resource. The z-arguments in the second square bracket of (2) account for the resource which is locked up in living individuals.

The same approach was taken in an earlier calculation (Whittle 1971, p.188) of the distribution of types of polymer for prescribed abundances of the elements from which the polymers are formed. The kinetics of polymer formation were such that the analogue of (2) was based upon independent Poisson variables rather than independent geometric variables. The difference in "kinetics" that make geometric distributions appropriate in the present case are that (a) species increase by reproduction, not by association as in the polymer case, and (b) resources can only be absorbed by individuals of the species at a limited rate, even if these resources are present in gross excess. That distributions should be exactly geometric is not crucial, the concept of *geometric type* behaviour is, as will emerge in §4.

It follows from (2) that for a fixed value of b and variable n

$$P(n|b) \propto P(n,b) = \begin{cases} \exp\left(\sum_k c_k n_k + \sum_j d_j m_j\right) & (m \geq 0, n \geq 0) \\ 0 & (\text{otherwise}) \end{cases} \tag{3}$$

where m is the amount of free resource, determined by (1).

Let $\lambda_k(n)$ be the total birth intensity and $\mu_k(n)$ the total death intensity for species k, as functions of n. There will be a dependence upon b as well, but b is regarded as fixed for the moment. Then distribution (3) would be consistent with the prescription

$$\mu_k(n) = \gamma_k n_k$$

$$\lambda_k(n) = \begin{cases} \delta_k(n_{k+1}) & (m_j \geq a_{jk};\ j=1,2,\ldots,r) \\ 0 & \text{Otherwise} \end{cases} \tag{5}$$

where

$$\delta_k = \gamma_k e^{c_k - \sum_j d_j a_{jk}}$$

and γ_k has an arbitrary positive value. That is, of a fixed individual death rate γ_k, and of an individual birth rate which is δ_k or zero according as to whether the constituents of another individual of S_k are or are not available. To this extent does resource-limitation manifest itself. A more graduated response can be achieved in any degree; see §4.

Once one has calculated a distribution one automatically has an extremal principle. We see from (3) that the most probable value of n is determined by the extremal problem:

$$\left.\begin{array}{ll} \text{Maximise} & (c'-d'A)n \\ \text{subject to} & An \leq b \\ & n \geq 0 \end{array}\right\} \tag{6}$$

Of course, there is also the constraint of integer values, and the actual value of n will vary stochastically about the value determined by (6). However, the relative magnitude of these effects will tend to zero if we increase the system by increasing b by a scale factor, when the value of n determined by (6) will increase in proportion.

In (6) we recognise a linear programme. The relation (1) represents r equality constraints on the non-negative variables m, n. One can then say that in the extremal solution just r components of (m,n) will be non-zero (save in exceptional, structurally unstable cases). One can express this as: the number of surviving species equals the number of resources which are fully utilised. This is the statement of the principle of competitive exclusion for this particular stochastic model.

3. Interpretation of the dual

The two classical statistical mechanical techniques for determining the prevailing configuration n in the "thermodynamic limit" $b \to \infty$ are to calculate the most probable value and to calculate the conditional expectation $E(n|b)$. The p.g.f. of n conditional on prescribed abundances b is

$$E(\prod_k w_k^{n_k} | b) = \frac{\oint \Pi(z,w) \prod_j z_j^{-b_j-1} dz_j}{\oint \Pi(z,1) \prod_j z_j^{-b_j-1} dz_j} \tag{7}$$

where paths of integration can be taken as $|z_j| = \varepsilon$ $(j=1,2,\ldots,r)$ in complex z-space for some small positive ε. In calculating conditional moments of n one is then calculating integrals of the type occurring in the denominator of (7).

In the case of large b one will look for a saddle-point evaluation. If the saddle-point value $\bar{z}$ may be supposed real and positive (as usually seems to be the case) then it will be the real, positive value minimising $\prod_j z_j^{-b_j}$ subject to convergence of $\Pi(z,1)$. If we set $z_j = e^{-y_j}$, and specify Π by (2) then the saddle-point is characterised by the extremal problem:

$$\left.\begin{array}{ll} \text{Minimise} & y'b \\ \text{subject to} & y'A \geq c' \\ & y \geq d \end{array}\right] \tag{8}$$

This is just the dual of problem (6). The constraints which are active in (8) correspond respectively to the species which survive and the resources which are in surplus. These represent the points at which resources accumulate in the equilibrium configuration, and so represent what one would regard as "bottlenecks" for resource if circulation of resources were regarded as the ideal. Just as in Whittle (1968) a bottleneck manifests itself by the divergence of a p.g.f. (in that the active constraints in (8) mark how the saddle-point $\bar{z}$ lies against the boundary of the region of convergence of $\Pi(z,1)$).

4. Generalisations

The essence of assumption (2) is not that distributions are geometric, but that they have finite radius of convergence, so that the bottleneck phenomenon does indeed emerge. Consider the generalisation of (2) to

$$\Pi(z,w) = \Phi(z) \prod_k \Psi_k(w_k \prod_j z_j^{a_{jk}}) \tag{9}$$

where

$$\Phi(z) = \sum_b e^{\phi(b)} \prod_j z_j^{b_j}$$

$$\Psi_k(\omega) = \sum_{i=0}^{\infty} e^{\psi_k(i)} \omega^i$$

The equilibrium distribution corresponding to (3) is then

$$\log P(n|b) = \text{const.} + \sum_k \psi_k(n_k) + \phi(b-An) \tag{10}$$

and maximisation of this with respect to n constitutes the "primal" problem.

The effect of the ϕ-generalisation is to graduate the response of birth-rate to abundance of free resources. This response will be increasing, but ultimately limited if Φ has a finite region of convergence. The effect of the Ψ-generalisation is to allow the individual birth-rate for S_k to respond to species abundance n_k, but again having an essentially limited value for large n_k if Ψ_k has finite radius of convergence. The fact that the w_k occur in separate factors of (9) corresponds to the assumption that different species interact only via the mediation of resource competition.

The dual approach of §3 remains meaningful. Consider for simplicity the case where only Φ is generalised, and the Ψ_k have the geometric form of (2). The "primal" problem is then

$$\left.\begin{array}{ll} \text{Maximise} & c'n + \Phi(b-An) \\ \text{subject to} & n \geq 0 \end{array}\right] \tag{11}$$

The generalisation of the "dual" characterisation (8) is

$$\left.\begin{array}{ll} \text{Minimise} & y'b \\ \text{subject to} & y'A \geq c' \\ & y \in D \end{array}\right] \qquad (12)$$

where D is the set of real y for which

$$\Phi(e^{-y_1}, e^{-y_2}, \ldots, e^{-y_r}) < \infty$$

The formal dual of the "dual" problem (12) is

$$\left.\begin{array}{ll} \text{Maximise} & c'n + \Delta(b-An) \\ \text{subject to} & n \geq 0 \end{array}\right] \qquad (13)$$

where

$$\Delta(m) = \inf_{y \in D} (y'm) \qquad (14)$$

Problems (11) and (13) agree in the case of linear ϕ, as we have seen, but not in general otherwise. The function $\Delta(m)$ is to be regarded as an "essential" version of $\phi(m)$, conveying only the radius-of-convergence information implied by ϕ. That is, if Φ and Φ' have the same D then they have the same Δ. For example, if

$$e^{\phi'(m)} = \alpha(m)e^{\phi(m)}$$

where α is of less than exponential growth and exceeds a fixed positive lower bound for a sufficiently dense set of m then Δ will have the same evaluation in the two cases.

References

Kendall, D.G. (1949) Stochastic processes and population growth. *J. Roy. Statist. Soc., B, 11*, 230-264.

Phillips, O.M. (1973) The equilibrium and stability of simple marine biological systems. I. *The American Naturalist, 107*, 73-93.

Phillips, O.M. (1974) The equilibrium and stability of simple marine biological systems. II. *Arch. Hydrobiol., 73*, 310-333.

Whittle, P. (1968) Equilibrium distributions for an open migration process. *J. Appl. Prob.*, *5*, 567-571.

Whittle, P. (1971) *Optimization under constraints*. Wiley Interscience.

CONTRIBUTORS

Aldous, Dr D.J.	Department of Statistics, University of California, Berkeley, Ca. 94720, U.S.A.
Bartlett, Prof. M.S., F.R.S.	Priory Orchard, Priory Avenue, Totnes, Devon TQ9 5HR.
Bingham, Dr N.H.	Department of Mathematics, Westfield College, Kidderpore Avenue, Hampstead, London NW3 7ST.
Cranston, Prof. M.	School of Mathematics, University of Rochester, Rochester, NY 14627, U.S.A.
Dynkin, Prof. E.B.	Department of Mathematics, White Hall, Cornell University, Ithaca, N.Y. 14853, U.S.A.
Eagleson, Dr G.K.	Division of Mathematics and Statistics, CSIRO, P.O. Box 218, Lindfield, N.S.W., Australia 2070.
Gundy, Prof. R.F.	Department of Statistics, Hill Center for the Mathematical Sciences, Rutgers University, Busch Campus, New Brunswick, New Jersey 08903, U.S.A.
Hammersley, Dr J.M., F.R.S.	Trinity College, Oxford.
Hawkes, Dr J.	Department of Statistics, University College of Swansea, Singleton Park, Swansea, Wales SA2 8PP.
Kelley, Dr F.P.	Statistical Laboratory, University of Cambridge, 16 Mill Lane, Cambridge CB2 1SB.

Kent, Dr J.T.	Department of Statistics, University of Leeds, Leeds LS2 9JT.
Kingman, Prof. J.F.C., F.R.S	Mathematical Institute, 24-29 St. Giles, Oxford OX1 3LB.
Moran, Prof. P.A.P., F.R.S.	Department of Statistics, Australian National University, Canberra, Australia.
Orey, Prof. S.	School of Mathematics, University of Minnesota, Minneapolis, Minnesota 55455, U.S.A.
Papangelou, Prof. F.	Statistical Laboratory, Department of Mathematics, The University, Manchester M13 9PL.
Pitman, Prof. J.W.	Department of Statistics, University of California, Berkeley, Ca. 94720, U.S.A.
Pyke, Prof. R.	Department of Mathematics, University of Washington, Seattle, Washington 98195, U.S.A.
Ripley, Dr B.D.	Department of Mathematics, Imperial College of Science and Technology, Queen's Gate, London SW7 2BZ.
Rösler, Prof. U.	Elmweg 14, 3400 Göttingen, West Germany.
Silverman, Dr B.W.	School of Mathematics, University of Bath, Claverton Down, Bath BA2 7AY.
Tweedie, Prof. R.W.	SIROMATH, 1 York Street, Sydney, NSW 2000, Australia.
Whittle, Prof. P., F.R.S.	Statistical Laboratory, 16 Mill Lane, Cambridge CB2 1SB.